Experimental Organic Chemistry

Laboratory Manual

Experimental Organic Chemistry

Laboratory Manual

Joaquín Isac-García
Universidad de Granada, Granada, Spain

José A. Dobado
Universidad de Granada, Granada, Spain

Francisco G. Calvo-Flores
Universidad de Granada, Granada, Spain

Henar Martínez-García
Universidad de Valladolid, Valladolid, Spain

ELSEVIER

AMSTERDAM • BOSTON • HEIDELBERG • LONDON
NEW YORK • OXFORD • PARIS • SAN DIEGO
SAN FRANCISCO • SINGAPORE • SYDNEY • TOKYO
Academic Press is an imprint of Elsevier

Academic Press is an imprint of Elsevier
125 London Wall, London, EC2Y 5AS, UK
525 B Street, Suite 1800, San Diego, CA 92101-4495, USA
225 Wyman Street, Waltham, MA 02451, USA
The Boulevard, Langford Lane, Kidlington, Oxford OX5 1GB, UK

British Library Cataloguing in Publication Data
A catalogue record for this book is available from the British Library

Library of Congress Cataloging-in-Publication Data
A catalog record for this book is available from the Library of Congress

ISBN: 978-0-12-803893-2

For information on all Academic Press publications
visit our website at http://store.elsevier.com/

Working together
to grow libraries in
developing countries

www.elsevier.com • www.bookaid.org

Preface

This textbook is designed as a manual to start training students in Organic Chemistry lab work. It also aims to help teachers to find a wide enough variety and complexity of experiments so that they can properly plan the experiences associated with the development of an experimental course and where they can adapt these practices to the required level, time, and resources available at each center. The book seeks to cover the theoretical foundations as well as practical aspects of the discipline. This text includes novel materials such as Green Chemistry and microscale synthesis, without overlooking traditional educational topics such as qualitative organic analysis and some basic principles of structural elucidation by spectroscopic methods.

Organic Chemistry is a highly experimental science based on a body of theory that is consolidated in very basic aspects but at the same time is in constant development. The reactions reflected in any regular publication on Organic Chemistry or theories, principles, postulates, or rules found in the numerous texts dedicated to teaching the discipline are the result of work done by Organic Chemists in the laboratories. Therefore, it is essential for future professionals to acquire a solid background in the laboratory as soon as possible, developing from the outset good habits and skills needed to address the challenges of the experimental work.

From this standpoint, the book is divided into several parts. In the first, safety issues are addressed in the laboratory, both in terms of attitude and responsible behavior and in handling of the chemicals and equipment most frequently used. It also includes some tips for reporting practices and keeping laboratory notebooks, both being critical instruments to accurately reflect the results of the teaching laboratories and/or research.

In the second part, the most common laboratory equipment and the main basic operations that can be undertaken in an Organic Chemistry laboratory are described. Special care has been taken in each case to include tips and warnings that may be useful to students, which we have gleaned as the result of our teaching and research experience. Some of these are too obvious to practitioners of Organic Chemistry but may be of interest to a student starting experimental work in a laboratory. Next, some basic concepts of spectroscopic techniques and how to prepare the corresponding samples for each branch of spectroscopic methods are discussed. This part of the book ends with qualitative organic analysis, which has been included, despite disappearing from some recent texts, because it offers a valuable didactic component, and therefore, we have dedicated a chapter to this item.

One hundred laboratory experiments form the main body of the book. In every experiment a goal, background, and a detailed procedure description are presented. In each case, two types of tables are included: the first one with the basic lab operations necessary for following the procedure, the time needed, and a difficulty code using a flask icon (from one for the simplest to three for the more difficult) and the second one with relevant physical-chemical data and safety information related to the reagents, products, and solvents of every experiment. The experiments are divided into groups, according to the following criteria:

In the first group the experiments are designed to provide the basic knowledge of the main techniques of an Organic Chemistry laboratory, the so-called basic laboratory operations. In this case, the reactions or the products involved are secondary concepts. The next group of experiments are constituted by synthetic procedures split on two levels, basic and advanced, according to their complexity, both regarding the reaction type and/or the equipment or procedure necessary.

A second group of experiments is devoted to a full set of microscale and Green Chemistry laboratory experiments. For microscale experiments an introductory chapter includes the description of specific material for this purpose and the basic laboratory operations adapted to it. For Green Chemistry experiments, some of the principles and the basic ideas related to this subject are described. It is important to point out that, to our knowledge, this is the first time that in a general experimental Organic Chemistry text, these types of experiments are presented. In an up-to-date scientific context, we consider it valuable to offer to the potential undergraduate students a solid foundation on Green Chemistry, not only from a theoretical perspective but also from a practical orientation.

We would like to express our appreciation to Garceta Editorial and its staff (our special thanks to Mrs. Isabel Capella) for their facilities that have given us permission to use the copyright of the former Spanish version of the book.

Last, but not least, we would like to give our special thanks to Dr. Ángel Sánchez-González for the photograph of the front cover and to Mr. David Nesbitt for his invaluable work on the revision of the English manuscript.

This book is dedicated to Dr. Jorge Fidel López Aparicio (b. Osuna, Spain, 1918–2005) graduated in Chemistry at the University of Seville, received the PhD in Oxford, England (1952), and earned a second doctorate at the University of Seville. In 1958 he became full professor of Organic Chemistry at the University of Valladolid, Spain, where he became Dean of the Faculty of Science (1960–1963) and Rector (1963–1965). In 1968 he became full professor of Organic Chemistry at the University of Granada, Spain, a position he held until his retirement in 1986. At the University of Granada he was Dean of the Faculty of Sciences (1971–1974). Among other merits earned were the Grand Cross of Alfonso X the Wise and in 1980 the Gold Medal of the Royal Spanish Society of Chemistry for his research. He was the director of numerous doctoral theses, including those of two of the authors of this book (JIG and FGC-F). He taught that cleanliness, order, observation, rationality, and precision in the preparation of a report, as well as in laboratory work, are vital qualities of a professional in Organic Chemistry. This is our humble contribution in his memory.

THE AUTHORS
In Granada, July 2015

Table of Contents

Chapter 1

Laboratory Safety

Work in the Organic Chemistry laboratory involves a number of inherent risks due, first, to chemicals and solvents that are handled and, second, to the techniques used in conducting experiments. Although laboratory practices are intended to minimize these risks, there is always the possibility that an accident of varying severity can occur (e.g. cuts, burns, splashes, spills, or, even worse, fires and explosions).

Therefore, it is necessary from the beginning for students to acquire safe work habits and learn the steps of basic safety regulations. They should be able to identify potential sources of accidents, and, if an accident occurs, to act appropriately, minimizing the consequences of such incidents.

This chapter describes in a general way the most relevant aspects of safety in the laboratory, although specialized monographs are available dealing with all aspects of safety in the laboratory [1,2].

Moreover, due to the continuous changes in the international regulations regarding risk and safety (Globally Harmonized System or GHS), we have also included a few comments dedicated to the latest version of the labeling of chemicals. Likewise, a section on the dangers inherent in potentially explosive and flammable substances is included.

We end with a review of the issue of classifying, treating, and disposing of waste in chemistry labs.

1.1. The lab as a safe place

Before starting to work in a laboratory, students should have a thorough knowledge of the location of the different emergency equipment needed in situations where speed of response is critical. All students should know where to go in case of an accident, being completely sure of evacuation areas of both the laboratory and the building. In addition, they should know whom to call in case of an accident, and the emergency phone numbers of fire and medical departments, should be posted with good visibility.

1.1.1. Safe behavior and habits

Students should learn the best practices to conduct lab work safely with chemicals and materials that can be dangerous in case of carelessness or negligence. This involves taking appropriate precautions at all times and asking the instructor any questions regarding handling and how to proceed. It is especially important at all time for students to maintain their beach clean and tidy, to avoid accidental spills, broken glassware, etc.

The risk of accidents can be minimized as follows:

- Consult the Material Safety Data Sheets (MSDS) and labeling for all reagents that will be handled in a laboratory experiment or practice, knowing the potential dangers associated with handling, as well as how to neutralize or reduce the hazards in case of an accident.

- Prior to the experiment, study and document techniques and procedures (know in detail all basic operations and manipulations that take place) in the laboratory.

- Follow all safety instructions specific to each practice or experiment to be conducted in the laboratory.

- Know the location and proper operation of all general emergency equipment in the laboratory.

- Before using a particular compound, be sure that it is the desired one. Read the label twice, if necessary.

- Never return used waste products to their original containers without consulting the instructor.

- Any newly prepared solution should be stored in a clean container and properly labeled or tagged.

- Do not touch chemicals with your hands, and never place any in your mouth. Do not engage in mouth pipetting; use the bulb syringe, pipette, or dispenser.

- Do not directly smell any chemicals, as it can be irritating, harmful, and can provoke tears, etc.

- To dilute an acid, such as sulfuric acid, always add the sulfuric acid to the water. Never add the water to the acid because violent splashing or heat releases can occur.

- Keep flammable solvents away from heating elements such as hot plates, stoves, radiators, etc.

- When handling glassware, be extremely careful with sharp edges and points. Always keep glassware away from the eyes and mouth.

- Protect your hands with gloves or rags when a plug is inserted into a glass tube.

- Hot glassware is no different at first glance from cold. If glassware is inside an oven or in contact with a heat source, prevent burns by using tongs, tweezers, or other protective devices to manipulate the item and allow it to cool before touching it directly or using it.

- Never heat closed containers, as it may cause an explosion.

- Every time glassware is assembled, check the clamping and assembly before starting the experiment

- Do not eat, drink, or chew gum in the laboratory.

- Always pay attention to the work being done and maintain a responsible attitude.

1.1.2. Causes of accidents in the laboratory

Most accidents in the Organic Chemistry laboratory are caused by improper handling of hazardous chemicals, glassware, and laboratory equipment. Therefore, prior to any practice session, the practice session should be thoroughly prepared and the compounds and laboratory equipment should be correctly used. This will greatly diminish the likelihood of such accidents.

The most common causes of accidents in the laboratory involve the following:

- Lack of orderliness in the workplace.

- Inappropriate use of personal protective equipment.

- Negligent use of glassware.

- Improper transfer of liquids in general.

- Heating plates at high temperatures and/or other sources of heat.

- Distillations with an open collector next to a flame or heat source.

- Purification of solvents that can contain peroxides (e.g. ethers).

- Spontaneous ignition of catalyst residues or Zn wastes from reductions.

- Oil bath heated to temperatures above 160 °C.

- Negligent destruction of Na, K, $NaNH_2$, $LiAlH_4$, and CaH_2 wastes.

- Applying a vacuum in an Erlenmeyer flask or other container not prepared for vacuum.

- Failure to follow instructions.

1.1.3. What to do in case of an accident

Although not exhaustive, some recommendations in order to minimize the consequences of an accident are as follows:

- Accident (in general). Immediately notify the instructor or head of the laboratory. In case of a serious accident call emergency phone number. Warn anyone nearby about the nature of the emergency. Do not move any injured persons, except in case of fire or chemical exposure.

- Accidental ingestion of chemicals. Immediately contact a local poison-control center and go to the emergency hospital with the label and the product. If the victim is unconscious, turn the person's head sideways. Do not give liquids to ingest or induce vomiting.

- Inhalation of chemicals. Immediately evacuate the affected area and go to a place with fresh air; then go immediately to the nearest hospital emergency room.

- Fire in the laboratory. Evacuate following instructions from the person in charge. If the fire is small and localized, efforts can be made to put it out with a fire blanket or fire extinguisher. Flammable chemicals that are near the fire should be removed. Never use water to extinguish a fire caused by organic solvents.

- Fire on the body. If clothes catch fire, ask for help immediately, lie down, and roll over and over to try to put out the flames. Never run or try to reach a safety shower unless this is very close. Remember that it is your responsibility to help someone who is burning, using a fire blanket.

- Burns. Small burns can be treated by washing with cold water. In the case of severe burns, go immediately to the nearest hospital emergency room.

- Cuts. Wash well with running water. If they are small and stop bleeding quickly, apply an antiseptic and cover with appropriate dressing. If they are large and will not stop bleeding, immediately go to the nearest hospital emergency room.

- Chemical splashes on the skin. Wash immediately with plenty of water. Take a shower and flush eyes with water when a large area of the body is affected and washing in a sink would be insufficient. Quick and thorough washing is very important to reduce damage. In the case of corrosive products in contact with the eyes, the reaction time for action is critical (less than 10 s). Seek medical assistance, even though the injury may look minor.

- Chemical spills. For discharges of liquid products, act quickly for neutralization, absorption, and elimination. Evacuate the laboratory, if necessary, using the corresponding protection material. Absorb the spill with an inert

Figure 1.1: Personal protective equipment.

material (vermiculite, sand, etc.). If the liquid is flammable, cut off all possible sources of ignition in the area and absorb with charcoal or another specific absorbent. Never use sawdust because of its flammability. Discharges of strong acids must be absorbed quickly by using a neutralizing absorbent. Alternatively, sodium bicarbonate can be used as a neutralizer. For strong base discharges, specific commercial products can be used. Or, plenty of water should be used in order to change the pH to slightly acidic.

1.2. Personal protective equipment

The personal protective items in the Organic Chemistry laboratory should consist of the following: goggles, gowns, gloves, and mask in certain cases (see Figure 1.1). Concerning the type of clothing that should be worn during the experiments or work in the laboratory, the student is especially urged not to wear shorts, sandals, or any other type of uncovered footwear. It is advisable to tie up long hair. The student should avoid wearing pendants, necklaces, bracelets, scarves, or shirts with wide sleeves which can be caught in machines or devices or that can make contact with flames or heat sources.

1.2.1. Safety glasses

Safety glasses protect the eyes from splashes and thus should be worn at all times in the Organic Chemistry laboratory. Do not use contact lenses, since the effect of chemicals on the eyes is exacerbated if the chemicals get between the lens and cornea. Furthermore, prescription glasses do not guarantee adequate protection, so prescription safety glasses may be needed. Organic lenses exposed to solvent vapors or splashes can be significantly damaged. Safety glasses are specially made for this purpose. Additionally, safety glasses should be designed to provide good front and side protection. These should be as comfortable as possible, should be aligned with the nose and face, and should not interfere with the user's movements.

1.2.2. Lab coat

The lab coat protects the body and clothes from chemical splashes. These garments are usually made of cotton or natural fibers and synthetic fibers that resist corrosive substances. Cotton or natural fibers, when burned, do not stick to the body, so that a corrosive substance can be removed easily. The lab coat should always be buttoned and should cover the body to below the knee.

1.2.3. Gloves

Protective gloves should be used, especially when corrosive and hazardous chemicals are handled. Depending on the type of glove, specific precautions are needed because no material can protect against all chemicals. Before use (especially latex), make sure that the gloves are in good condition, i.e. without holes, punctures, or tears. Gloves that have been pierced by reagents should be disposed of in the appropriate waste container.

Depending on the use, they are available in different market types. The most common ones in the laboratory cover the wrist, while other models protect the forearm or the entire arm. The most commonly used materials to make gloves are:

- Plastic. Protects against mild corrosives and irritants.

- Latex. Provides light protection against irritants (some people may be allergic to latex).

- Natural rubber. Protects against mild corrosives and electric shock.

- Neoprene. Protects against solvents, oils, or slightly corrosive substances.

- Cotton. Absorbs perspiration and keeps objects clean as they are handled; often with a fire retardant.

- Zetex. Protects when manipulating small hot objects.

When working with extremely corrosive materials (e.g. HF), wear thick gloves.

Gloves should be removed before leaving the laboratory to avoid contamination and also before touching everyday objects such as cell phones, notebooks, pens, computers, etc. In addition, care must be taken when remove gloves. The correct way is to pull from the wrist to the fingertips, making sure that the outside of the gloves does not touch the skin. Disposable gloves should be discarded in the containers designated for that purpose.

1.2.4. Masks

Protective masks guard the air passages against particles, gases, and vapors. Depending on the level of safety needed, a wide range of models are commercially available. The most common are:

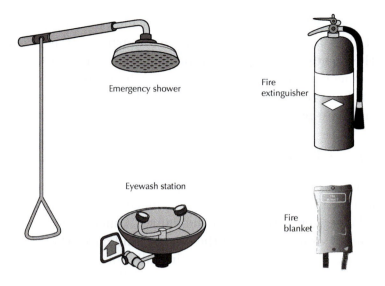

Figure 1.2: Laboratory emergency equipment.

- Masks. Act as a barrier against particles and dust. They are easy to use and the most widespread. For example, alumina or silica gel transfer requires a mask or should be performed in a fume hood.

- Filtering devices. Consist of a filter and facepiece. The filter is designed to act as a barrier to particles (mechanical filters) such as vapors and/or gases (chemical filters), or a combination of the two (mixed filters), while the adapter can be a face mask or mouthpiece.

1.3. Emergency equipment for general purpose

These safety elements, commonly used, must be present in a laboratory because they minimize the accident risks. The student needs to know both the location of the equipment as well as its operation, so that emergency action can be taken efficiently (see Figure 1.2).

1.3.1. Emergency shower

This is commonly used in cases of bodily contamination by chemicals with risk of chemical burns, and even ignition of clothing. The water flow must be sufficient to wet a person quickly with an easily opened tap.

1.3.2. Emergency eyewash

An emergency eyewash is designed to clean the eyes of a person following an accident with harmful substances, enabling quick and effective cleaning of the eyes and face. Eyewash stations are composed of two separate, opposing taps projecting water jets at low pressure to avoid causing pain or injury to the eye. Water is collected in a sink designated for that purpose. It is essential to be fully aware of the location and operation of this device in emergency so that it can be used as quickly as possible in situations where vision may be partially or fully impeded. For proper use, the eyes should be kept open with the help of the fingers to facilitate proper washing under the eyelids.

1.3.3. Fireproof blankets

Fireproof blankets or fire blankets are designed to extinguish small fires. They consist of a sheet of fireproof material to be placed over the fire to suffocate it, by blocking the oxygen supply from the flames. Quick action is necessary at the outbreak of fire. Completely open the blanket and cover the whole object or area on fire, taking care not to burn hands (initially fold the blanket around the hands). If used correctly and quickly, the blanket will snuff out fires without the need of fire extinguishers, since the latter can cause extensive incidental damage to material and electrical equipment by spreading pressure jets of foam in the laboratory. Fireproof blankets are usually folded in a particular way for quick release, and they come manufactured with fireproof fibers such as Nomex® or fiberglass and are sometimes impregnated with flame retardants.

1.3.4. Fire extinguishers

These are bottle-shaped metal containers that emit a fire-extinguishing agent under pressure. When a valve is opened, the agent exits through a nozzle which is directed to the base of the fire (not to the top of the flame). They usually have a seal to prevent accidental activation. Before use, it is important to tap the base of the extinguisher on the floor, so that the contents are homogeneous.

When a small fire in the laboratory cannot be controlled by fireproof blankets, quickly use a fire extinguisher. However, be aware that there are different types of extinguishing agents appropriate for different kinds of fires. Using an extinguisher against the wrong kind of fire can be counterproductive. Therefore, the user needs to know in each case the appropriate type of fire extinguisher to use.

1.3.4.1. Extinguishing agents

These are substances of special composition used in fire extinguishers to put out fires.

- Fire class H, halogen. Allowed only in certain military applications because its composition destroys the ozone layer. It contains a choking agent that destroys the oxygen, so it is recommended to use indoors without the presence of life.

- Class N fire extinguishers. This neutralizes fire by the formation of gases by chemical or WMD (weapons of mass destruction). They are composed of the corresponding neutralizing agent powder.

- Water spray. They are designed to protect all areas containing Class A fire risks (solid fuels) efficiently and safely.

- Demineralized water. This is suitable for Class C fires (flammable gases), connected devices, and chemical or bacteriological fire risks.

- Water and foam (AFFF). These are designed to protect areas containing Class A (solid fuel) and Class B (liquid and gaseous fuels) fire hazards.

- Carbon dioxide (CO_2). This is used to protect areas containing Class B (combustible liquids) and Class C (flammable gases) fire hazards.

- Versatile chemical powder (ABC). This dry chemical (ammonium monophosphate 75%) is used to fight Class A (solid fuels), energy (liquid fuels), Class C (flammable gases) fire hazards.

- Dry chemical powder (BC). This is designed to protect areas containing Class B (combustible liquids) and Class C (flammable gases) fire hazards.

- Chemical powder (D). Formulated to protect areas containing Class D (combustible metal) fire risks, it contains lithium, sodium, sodium-potassium alloys, magnesium, and metal compounds. It is loaded with compound sodium borate powder.

Most extinguishers used in the laboratory contain the versatile chemical powder. In addition, for extinguishing combustible metal fires, either a special powder type or dry sand should be used.

1.3.5. Chemical adsorbent

In the case of spills of hazardous liquids, adsorbent solids made with a special formulation (highly active surface adsorption) that controls spills quickly are commonly used, preventing the spill from covering a large area. The use of sawdust as adsorbent material is strongly discouraged because of its easy combustion. The most common adsorbents used, granules of different sizes, can be classified into:

- Compounds with oxygen and polar hydrophilic nature (silica gel or zeolites). Silica gel is usually used to dry liquids or gases, and also for adsorption of high-molecular-weight hydrocarbons in the natural gas. Zeolites are also used in drying gases and CO_2 removal from natural gas.

- Carbonaceous compounds and non-polar hydrophobic nature (activated carbon and graphite). Activated carbon is used for the adsorption of organic molecules and non-polar compounds.

- Polymeric materials having polar or non-polar functional groups in a porous polymer matrix.

1.3.6. Security screens

These are portable panels that are generally made with polycarbonate and offer protection against splashes and projections of reagents as well as against small explosions and/or implosions of glassware, etc. Screens are typically used, for example, in rotary evaporators, flash chromatography, etc.

1.3.7. Fume hoods

These provide respiratory safety during experiments as they have a suction system that removes harmful fumes and hazardous or malodorous gases that may arise from chemical reactions or the handling of toxic substances. They have a safety glass screen to guard against splashing liquids and glassware that can break, small explosions, etc. (see Figure 1.3). All basic operations that require a screen should be held within a fume cupboard, since it has the services of a laboratory table, as the water or vacuum circuit takes gas and electricity supplies or devices for fastening and assembly laboratory material.

Fume hoods allow safe work in the laboratory because they:

- Collect emissions from hazardous chemicals.

- Protect against splashes and projections.

- Enable work in an area of the laboratory safe against ignition sources.

- Depending on the design, protect against small explosions.

- Facilitate the renovation of laboratory air.

- Create a depression in the laboratory that prevents the flow of pollutants into adjoining areas.

- Allow the display of the experiment under way.

Recommendations for proper use of fume hoods:

- Do not use them to store hazardous reagents.

- The interior of the fume hood must at all time remain as clear and clean as possible.

- Avoid the generation of greenhouse gases at high speeds, so that they can rise properly. In this sense, always manipulate small amounts of reagents.

- Fume hoods must be open to a minimum to allow work to be properly undertaken. Do not leave harmful gases unattended. Never work while placing your head in such enclosures.

- At the end of the experiment, leave the fume hoods running with the windows closed to completely remove the remains of gases and vapors generated.

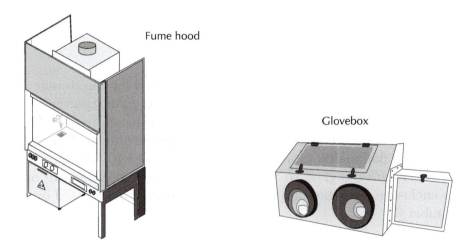

Fume hood

Glovebox

Figure 1.3: Safety equipment in the laboratory.

1.3.8. Glovebox (dry box)

These sealed boxes are designed to enable handling of reagents and objects so that the contents are sealed off from the operator at all times. Gloves are arranged so that the operator can place hands and arms inside of them and manipulate the contents of the box without breaking the insulation. The boxes are usually transparent to allow an adequate view of what is being handled (see Figure 1.3).

Depending on the use, the boxes are classified into two types:

- Handling of hazardous substances (radioactive or infectious-disease agents, etc.).

- Handling of substances and reactions that must remain in an inert atmosphere, sterile, dry, dust-free, or with very high purity.

1.3.9. Fireproof cabinet

These cabinets are specifically designed with insulation and non-combustible materials that protect the products stored inside from the fire. The protection level is labeled RF (resistance to fire) followed by a number indicating the minutes that can elapse after the fire ignites without the inside temperature exceeding 180 °C.[1] Examples of resistance levels are RF15, RF30, RF60, RF90,

---

[1]The UL Class 350 1-hour fire rating means that when exposed to external temperatures of over 1,700 °C, the internal temperature of the safe will not exceed 350 °C for at least one hour.

or RF120. They usually have other protection systems such as expansion joints which when heated to over 50 °C expand to seal cracks in doors and vents, protecting stored products from an external fire or smothering a fire inside. These cabinets are used to store materials that are flammable, explosive, toxic, corrosive, solvent, etc.

1.3.10. Cold lab chamber

The use of domestic refrigerators is not recommended for the storage of volatile and flammable organic solvents, especially in large quantities. For these, cold lab chambers have sufficient capacity for storing solvents or chemicals at low temperatures. They are specially designed to protect against explosions and fires caused by flammable vapors and come equipped with air-renewal systems. The chambers are designed so that the electrical components are outside the storage enclosure, and the compressor must be sealed and located on the top of the chamber.

- Safety containers: When it is necessary to transport or store volatile and flammable organic solvents, containers specially designed for this purpose are used. These are made of metal (steel with a reinforced raised bottom for better impact resistance) and have an automatic closing system with a recoil spring plug that prevents leakage. Moreover, this system functions as a closing overpressure valve and has a double safety pilot filter, so that even burning vapors do not reach the stored products.

- Dewar vessel: This is a container designed to provide thermal insulation, and is used to store hot or cold liquids. This is constructed with double-wall silver glass (reflected radiation) and inside has a vacuum vessel (prevents heat loss by convection and conduction). This type can also be manufactured in stainless steel, which has the advantage of being more robust and better able to withstand temperature changes. It is often used to store, for example, liquid nitrogen (b.p. -196 °C) or liquid oxygen (b.p. -183 °C).

1.4. GHS for classification and labeling of chemicals

This section provides a brief introduction to the proposed United Nations system of classifying and labeling of chemicals, called the globally harmonized system (GHS), and the definition of a set of harmonized criteria on the danger of chemicals.

In an attempt to standardize and harmonize the global classification and labeling of chemicals the GHS system was established. The system defines and classifies the potential hazards of chemicals for use in labeling and in so-called safety data sheets (material safety data sheets, MSDS), which report on the hazards and risks of chemicals in the handling, storage, and transport.

The main objective of the GHS system is to improve the protection of human health and the environment through adequate information on the dangers of chemicals to users, suppliers, and carriers.

One of the key elements within this harmonized system are the so-called hazard pictograms that are formed by:

- An icon or symbol. Here a total of nine symbols of normalized hazards are used. There are three groups that represent physical hazards, hazards to human health, and environmental hazards. These icons are part of the set of symbols used in the United Nations Recommendations on the Transport of Dangerous Goods (Model Regulations) except the symbol representing the danger to health, the exclamation mark and the fish and tree, which relate to the environment (see Table 1.1).

- A geometric figure formed by a square rotated to stand on one corner with red border.

- A legend consists of a series of signs and words of danger or warning, expressed in the so-called H phrases, using labeling.

The harmonized criteria allow:

- Sorting of chemicals by the danger involved.

- Labeling of chemicals using statements and world-standardized hazard pictograms.

GHS implementation has begun around the world, and many countries have already adopted this system.

Table 1.1: Globally harmonized system (GHS) icons

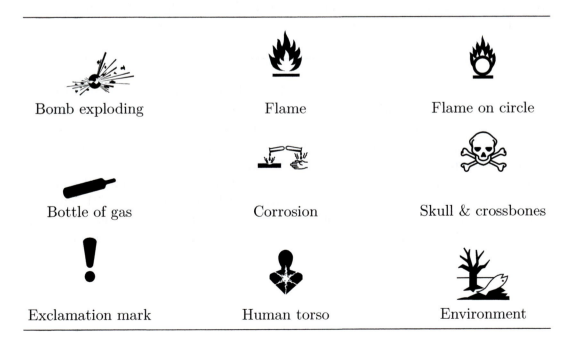

Bomb exploding   Flame   Flame on circle

Bottle of gas   Corrosion   Skull & crossbones

Exclamation mark   Human torso   Environment

In Table 1.2, the pictograms are shown along with their legends that form the GHS system. The pictograms are composed of nine icons inside a square set on one apex, with a red border on a white background.

Table 1.2: Pictograms of the globally harmonized system (GHS)

GHS01
Danger, Unstable, Explosive

GHS02
Danger, or Warning, Flammable

GHS03
Danger, or Warning, Oxidizing

GHS04
Warning, Compressed gas

GHS05
Danger, or Warning, Corrosive cat. 1

GHS06
Danger, Toxic cat. 1–3

GHS07
Warning, Toxic cat. 4, Irritant cat. 2 or 3, Lower systemic health hazards

GHS08
Danger, or Warning, Systemic health hazards

GHS09
Warning, (for cat. 1), (for cat. 2 no signal word), Environmental hazard

Additionally, the format of the labels of chemicals has been standardized. Each label contains a product identifier, a pictogram, an identifying word, phrase hazard, and supplier information. Manufacturers or suppliers of chemicals can give additional information on the label. The category designation is a letter or a number, shown in Tables 1.4, 1.5, and 1.6 and included in the MSDS. There are nine pictograms and two words: "Danger" or "Warning." Examples of warning phrases are "May cause respiratory irritation," "Toxic in contact with skin" or "Heating may cause a fire or explosion." The nine GHS pictograms are shown in Table 1.2. In addition, another set of pictograms is used for transport (see Table 1.3).

Table 1.3: Transport pictograms of the globally harmonized system (GHS)

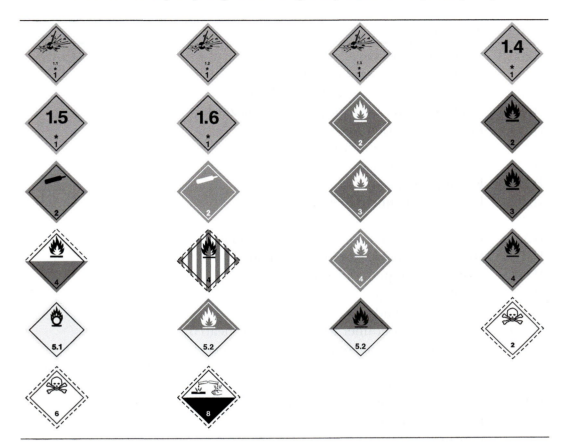

1.4.1. Hazard statements

Hazard statements are phrases assigned to a hazard class and category that describes the nature of the hazards of a substance and, where appropriate, the degree of danger. They are part of the GHS system for standardizing a series of statements relating to dangerous chemicals that can be translated into different languages. The United Nations has published the list of H statements in all official languages.

Each hazard statement is associated with a code starting with the letter H and three digits that follow. Sentences corresponding to related hazards are grouped together with a code number, so that the numbering is not consecutive. The code is used as a reference, and the phrase itself must appear on the labels and MSDSs.

Tables 1.4, 1.5, and 1.6 list different hazard statements grouped in three categories: physical hazards, health hazards, and environmental hazards.

Table 1.4: Hazard statements: physical hazards

Code	Statement
H200	Unstable explosive
H201	Explosive; mass explosion hazard
H202	Explosive; severe projection hazard
H203	Explosive; fire, blast or projection hazard
H204	Fire or projection hazard
H205	May mass explode in fire
H220	Extremely flammable gas
H221	Flammable gas
H222	Extremely flammable aerosol
H223	Flammable aerosol
H224	Extremely flammable liquid and vapour
H225	Highly flammable liquid and vapour
H226	Flammable liquid and vapour
H227	Combustible liquid
H228	Flammable solid
H229	Pressurized container & may burst if heated
H230	May react explosively even in the absence of air
H231	May react explosively even in the absence of air at elevated pressure and/or temperature
H240	Heating may cause an explosion
H241	Heating may cause a fire or explosion
H242	Heating may cause a fire
H250	Catches fire spontaneously if exposed to air
H251	Self-heating; may catch fire
H252	Self-heating in large quantities; may catch fire
H260	In contact with water releases flammable gases which may ignite spontaneously
H261	In contact with water releases flammable gas
H270	May cause or intensify fire; oxidizer
H271	May cause fire or explosion; strong oxidizer
H272	May intensify fire; oxidizer
H280	Contains gas under pressure; may explode if heated
H281	Contains refrigerated gas; may cause cryogenic burns or injury
H290	May be corrosive to metals

Table 1.5: Hazard statements: health hazards

Code	Statement
H300	Fatal if swallowed
H301	Toxic if swallowed
H302	Harmful if swallowed
H303	May be harmful if swallowed
H304	May be fatal if swallowed and enters airways
H305	May be harmful if swallowed and enters airways
H310	Fatal in contact with skin
H311	Toxic in contact with skin
H312	Harmful in contact with skin
H313	May be harmful in contact with skin
H314	Causes severe skin burns and eye damage
H315	Causes skin irritation
H316	Causes mild skin irritation
H317	May cause an allergic skin reaction
H318	Causes serious eye damage
H319	Causes serious eye irritation
H320	Causes eye irritation
H330	Fatal if inhaled
H331	Toxic if inhaled
H332	Harmful if inhaled
H333	May be harmful if inhaled
H334	May cause allergy or asthma symptoms or breathing difficulties if inhaled
H335	May cause respiratory irritation
H336	May cause drowsiness or dizziness
H340	May cause genetic defects
H341	Suspected of causing genetic defects
H350	May cause cancer
H351	Suspected of causing cancer
H360	May damage fertility or the unborn child
H361	Suspected of damaging fertility or the unborn child
H361d	Suspected of damaging the unborn child
H362	May cause harm to breast-fed children
H370	Causes damage to organs
H371	May cause damage to organs
H372	Causes damage to organs through prolonged or repeated exposure
H373	May cause damage to organs through prolonged or repeated exposure

Table 1.6: Hazard statements: environmental hazards

Code	Statement
H400	Very toxic to aquatic life
H401	Toxic to aquatic life
H402	Harmful to aquatic life
H410	Very toxic to aquatic life with long lasting effects
H411	Toxic to aquatic life with long lasting effects
H412	Harmful to aquatic life with long lasting effects
H413	May cause long lasting harmful effects to aquatic life
H420	Harms public health and the environment by destroying ozone in the upper atmosphere

1.4.2. Material safety data sheets (MSDS)

A material safety data sheet is a document that provides detailed information about the chemicals, including descriptions of the precautions to be taken for handling, the immediate emergency measures for these products, and their potential risk to humans and the environment. They contain vital data to transport these dangerous goods and to laboratory personnel. These sheets offer information on prevention and control measures. Therefore, the aim of this information is to reduce the risk of accidents to both people and the environment.

In a laboratory, it is mandatory to have the appropriate tabs. These should be developed by the manufacturer of chemicals and kept up to date. Moreover, the format of these records varies depending on the manufacturer or the laws of the different countries. In addition, many products necessarily include the product safety record on the label.

1.5. Chemical incompatibility and storage

Many chemicals, besides presenting risks alone, can lead to dangerous reactions in contact with other products, with uncontrolled chemical reactions that can result in:

- Emission of toxic gases.

- Emission of corrosive or flammable gases.

- Formation of corrosive liquid.

- Explosive reaction.

- Formation of products sensitive to friction or shock.

- Exothermic reaction.

- Explosion and/or fire.

- Generation of gases that can break the containment vessel.

- Heating substances or an uncontrolled start of a decomposition reaction.

- Reduction of the thermal stability of a substance.

- Quality degradation of stored products.

- Damage to containers (holes, effaced labels, etc.).

Especially in storage areas, either of chemicals used as raw material or any final chemical products, the risk exists of chemical incompatibility.

In Table 1.7, chemicals are classified into 23 groups, including specific examples of each and groups that are incompatible. This is not an exhaustive description, but rather a general guide for handling and storing these products [2].

The most common chemical incompatibilities are summarized in the following list:

- Strong acids with strong bases. For example, NaOH with H_2SO_4.

- Strong acids with weak acids that give off toxic gases. For example, HCl with cyanides and sulfides.

- Oxidizers with reducing agents. For example, HNO_3 with organic compounds.

- Water with various compounds. For example, boranes, anhydrides, carbides, trichlorosilanes, hydrides, and alkali metals.

- Oxidants with nitrates, oxyhalogenic, oxides, peroxides, and fluorine.

- Reducing agents with flammable materials, carbides, nitrides, hydrides, sulfides, metal-alkyls, aluminum, magnesium, and zinc powder.

- H_2SO_4 with sugar, cellulose, perchloric acid, potassium permanganate, chlorates, and sulfocyanides.

Table 1.7: Chemical incompatibility

Group (Name)	Incompatible with the following groups
1 (Inorganic acids)	2, 3, 4, 5, 6, 7, 8, 10, 11, 13, 14, 16, 17, 18, 19, 21, 22, 23
2 (Organic acids)	1, 3, 4, 7, 14, 16, 17, 18, 19, 22
3 (Alkali)	1, 2, 5, 6, 7, 8, 13, 14, 15, 16, 17, 18, 20, 22, 23
4 (Amines and hydroxylamines)	1, 2, 5, 7, 8, 13, 14, 15, 16, 17, 18, 23
5 (Halogen compounds)	1, 3, 4, 11, 14, 17
6 (Alcohols, glycols, & glycol ethers)	1, 3, 7, 14, 16, 20, 23
7 (Aldehydes)	1, 2, 3, 4, 6, 8, 15, 16, 17, 19, 20, 23
8 (Ketones)	1, 3, 4, 7, 19, 20
9 (Saturated hydrocarbons)	20
10 (Aromatic hydrocarbons)	1, 20
11 (Olefins)	1, 5, 20
12 (Oil derivatives)	20
13 (Esters)	1, 3, 4, 19, 20
14 (Polymerizable monomers)	1, 2, 3, 4, 5, 6, 15, 16, 19, 20, 21, 23
15 (Phenols)	3, 4, 7, 14, 16, 19, 20
16 (Alkylene oxides)	1, 2, 3, 4, 6, 7, 14, 15, 17, 18, 19, 23
17 (Cyanohydrins)	1, 2, 3, 4, 5, 7, 16, 19, 23
18 (Nitriles)	1, 2, 3, 4, 16, 23
19 (Ammonia and derivatives)	1, 2, 7, 8, 13, 14, 15, 16, 17, 20, 23
20 (Halogens)	3, 6, 7, 8, 9, 10, 11, 12, 13, 14, 15, 19, 21, 22
21 (Ethers)	1, 14, 20
22 (Phosphorous)	1, 2, 3, 20
23 (Acid anhydrides)	1, 3, 4, 6, 7, 14, 16, 17, 18, 19

1.6. Substances with risk of fire and explosion

1.6.1. General aspects

The use of substances or mixtures of highly flammable or explosive substances is relatively common in Organic Chemistry laboratories. If proper precautions are taken, danger is minimal, but carelessness can lead to hazards. As general guidelines, the following considerations should be taken into account:

- If a substance is explosive, find an alternative when possible.

- When the use of an explosive or flammable substance is strictly necessary, the smallest possible amounts should be used, taking appropriate precautions:

  - Work in a hub case without flames or heat sources.

- Use proper cooling.
- Employ an inert atmosphere.
- Use shields.
- Add reagents slowly.

1.6.2. Explosive substances

The following compounds or compound groups can be explosive, by themselves or with heat, impact, or friction:

- Acetylene and their metal salts such as silver or copper.

- Polyacetylenes.

- Hydrazoic acids.

- Azides and their organic and inorganic salts (only sodium azide is safe).

- Diazonium salts and diazo compounds.

- Inorganic nitrates, especially ammonium nitrate.

- Polynitrocompounds.

- Metal salts of nitrophenols.

- Peroxides, concentrated hydrogen peroxide solution.

1.6.3. Mixtures of dangerous compounds

Strong oxidizers are dangerous when mixed, such as:

- Perchloric acid and perchlorates.

- Chromium trioxide, chromates, and dichromates.

- Nitric acid and nitrates.

- Permanganates.

- Concentrated hydrogen peroxide.

- Liquid air and liquid oxygen.

The following are especially dangerous when mixed with easily oxidizable organic substances such as:

- Alcohols and polyols.

- Carbohydrates.

- Materials with cellulose (e.g. paper).

- Clothes containing wool or cotton.

These oxidants are also dangerous when mixed with elements such as sulfur, phosphorus, or finely divided metals such as magnesium.

1.6.4. Special danger of explosion

The following product may pose some specific danger of explosion.

1.6.4.1. Peroxides

This is one of the most common risks of explosion in the Organic Chemistry laboratory. Common solvents such as diethyl ether, dioxane or THF, and diiso-propyl ether form peroxides which are exposed to air or light, although these are not the only ones (see Table 1.8).

Due to extensive use in laboratories, ethers deserve special mention. When purified by distillation, the residual peroxide progressively increases, with the consequent risk of explosion. To minimize this risk the following precautions should be taken into account:

- Use peroxide inhibitors (most ethers are marketed with such inhibitors).

- In case of prolonged storage, add new inhibitors.

- Use safety containers.

- Never store these solvents for long periods of time.

- In the case of ethyl ether, there is doubt about the presence of peroxides, so proceed with extraction using sodium sulfite.

1.6.4.2. Sodium amide and metal potassium

Oxidation products are formed on the surface of both substances. If they are removed with a knife or spatula, explosions may occur. Do not crush in a mortar or other type of bowl. Use them directly. The excess sodium amide can be destroyed with propan-2-ol. Furthermore, the potassium metal is destroyed by mixing with solid ammonium chloride.

1.6.4.3. Alkali metals with chlorinated solvents

Alkali metals (lithium, sodium, and potassium) and others such as aluminum or magnesium, especially when finely divided, react violently with halogen compounds such as chlorinated solvents (e.g. CCl_4). Therefore, the shavings or residues of these compounds should never be washed with chlorinated solvents.

Table 1.8: Chemical incompatibility

Type A	Substances that form explosive peroxide levels without concentrating the sample	Butadiene[a], chloroprene[a], vinylidene chloride, divinylacetylene, isopropyl ether, tetrafluoroethylene[a]
Type B	Substances that form explosive peroxide levels when concentrating the sample	Acetal, acetaldehyde, benzyl alcohol, butan-2-ol, cyclohexanol, cumene, 2-cyclohexen-1-ol, cyclohexene, decahydronaphthalene, diacetylene, dicyclopentadiene, diethyl ether, diethylene glycol dimethyl ether, dioxane, ethylene glycol dimethyl ether, 1-phenylethanol, 2-phenylethanol, heptan-4-ol, hexan-2-ol, isopropanol, methylacetylene, 3-methylbutan-1-ol, methylcyclopentane, methyl isobutyl ketone, 4-methylpentan-2-ol, pentan-2-ol, 4-penten-1-ol, THF, tetrahydronaphthalene, vinyl ethers
Type C	Substances that can self-polymerize as a result of a process of accumulation of peroxides	Vinyl acetate, acrylic acid[a], acrylonitrile[b], butadiene[b], chloroprene[c], chlorotrifluoroethylene, styrene, vinyl chloride, methylmethacrylate[b], tetrafluoroethylene[c], vinyl acetylene, vinyl chloride, vinyl pyridine

[a] When stored as a liquid monomer.
[b] Although these form peroxides, no explosions have been reported from these monomers.
[c] When stored in liquid form explosive peroxide levels without concentrating. Store in a gaseous state in steel cylinders.

1.6.4.4. Perchloric acid

Perchloric acid can react violently with materials such as cork, rubber, wool, or natural fibers (cotton, linen, etc.). If it makes contact with such materials, it can be absorbed quickly by them, causing a high risk of explosion.

1.6.4.5. Chromic acid/nitric acid

Mixtures of chromic and nitric acids are often used as agents for cleaning glassware in the lab. Therefore, it may be dangerous heating these mixtures to wash residues in flasks or in dirty glass material.

1.6.4.6. Azides

Azides are explosive, except for the sodium azide. The inadvertent appearance in some reactions, such as the Sandmeyer one, can cause explosions.

1.6.4.7. Liquid nitrogen

Liquid nitrogen (b.p. $-196\,°C$) contains some liquid oxygen (b.p. $-186\,°C$). Excessive evaporation can increase the oxygen content. It is extremely dangerous when this oxygen makes contact with flammable or combustible materials.

1.6.4.8. Glass joints under high vacuum

If the glass material which is used to work under vacuum is not in perfect condition, or its thickness and strength are not adequate, implosions may occur once the vacuum is established. Pay special attention to rotary evaporators.

1.6.4.9. Opening glass ampoules

Some reagents are supplied in sealed glass ampoules. If it is suspected that such reagents are volatile, the ampoules should be cooled before opening.

1.6.4.10. Compressed gas cylinders

Under certain circumstances, metal cylinders containing compressed gases (N_2, NH_3, O_2, CO_2, etc.) may constitute a risk of explosion, despite their robust construction. A sudden rupture releases an immense volume of gas that can also, according to their nature, fan flames, etc. Therefore, it is advisable to anchor gas cylinders with a chain.

1.6.4.11. Experimental work under pressure

In the case of flash chromatography, plastic jackets are commonly used to protect glass columns or in order to minimize risks of broken glass columns.

Liquid waste
container

Solid chemical
waste container

Biohazard waste
container

Figure 1.4: Different types of containers for waste storage.

1.7. Waste treatment and disposal

Wastes generated in the Organic Chemistry laboratory must be adequately managed, according to their potential risk. These facilities must have a plan for waste collection. Under no circumstances should they be discharged into drains or bins. The laboratory should be equipped with suitable containers for storage and further processing of these wastes (see Figure 1.4). Their location and use should be known by students.

1.7.1. Waste classification

Since not all wastes have the same treatment, and in order to reduce management costs of hazardous wastes, these are classified according to their nature into 10 groups:

1.7.1.1. Group I (halogenated solvents)

These are organic liquids, highly toxic, irritating, and in some cases carcinogenic, with a content of more than 2% halogen. They may also include mixtures of halogenated and non-halogenated solvents (provided that the halogen content of the mixture is greater than 2%). For example, CH_2Cl_2, $CHCl_3$, $CHCl_3$, $PhBr$, etc. Do not use any aluminium container for storage.

1.7.1.2. Group II (non-halogenated solvents)

These are flammable and toxic organic liquids containing less than 2% of halogen. For example, $R-OH$, $R-CHO$, $R-CO-R'$, $R-COO-R'$, $R-H$, $R-CN$, etc.

1.7.1.3. Group III (aqueous solutions)

Because this is a very large group, and since aqueous solutions include both organic and inorganic compounds, they are subdivided into:

- Inorganic. Basic aqueous solutions. Aqueous solutions of heavy metals. Aqueous solutions of hexavalent chromium (VI).

- Organic. Aqueous solutions of dyes. Solutions of organic fixatives. Mixtures of water/solvent.

1.7.1.4. Group IV (acids)

Inorganic acids and their concentrated aqueous solutions > 10% by volume (v/v) are included. Extreme caution must be taken, since the mixture of some of these acids can produce some chemical exothermic reactions or release of toxic gas.

1.7.1.5. Group V (oils)

Group V consists of mineral oils derived from the maintenance of heating baths.

1.7.1.6. Group VI (solids)

This group is composed of chemical products in the solid state (organic and inorganic), which should not be mixed together. These are subdivided into:

- Organic solids. Organic chemicals or contaminated with organic chemicals, such as activated carbon or silica gel impregnated with organic solvents.

- Inorganic solids. Chemicals of inorganic origin such as salts of heavy metals.

- Contaminated disposable material: contaminated with various chemical materials.

1.7.1.7. Group VII (special products)

All chemicals (solid or liquid) that for their high toxicity or hazard cannot be included in any of the remaining groups, as well as outdated or expired pure reagents, are included in this group. Do not mix with each other. Whenever possible, they should be kept in their original container if the amounts are less than one liter.

- Oxidizing agent: peroxides.

- Pyrophoric compounds: magnesium metal powder.

- Highly reactive compounds: fuming acids, alkali metals, hydrides, peroxidizable compounds, unlabeled reaction byproducts, active halogen compounds, polymerizable compounds.

- Highly toxic compounds: osmium tetroxide, chromic mixture, cyanides, sulfides, etc.

- Unknown compounds.

1.7.1.8. Group VIII (contaminated glassware)

The group includes glassware contaminated with traces of chemicals (including closed empty glass bottles with traces of chemicals). The following materials are not included in this group: glass pipettes or other sharp implements; clean or uncontaminated glass.

1.7.1.9. Group IX (biohazardous)

The following sanitary wastes and related substances belong in this group:

- Microbiological cultures.

- Waste of infectious animal anatomical waste, blood, and blood products in liquid form.

- Needles, scalpel blades, and other sharp material.

- Small contaminated material or broken glass.

1.7.1.10. Group X (cytostatics)

Liquid and solid carcinogenic, mutagenic, or teratogenic products are included in this group, as well as disposable materials contaminated by such products or other highly toxic substances.

1.7.2. Disabling of wastes

Before disposing of waste in the appropriate containers, proper deactivation is recommended in order to minimize risks in the storage and transport thereof. Some of the lines of action to follow with different families of compounds is as follows:

- Acids and bases: Aqueous solutions of acids and bases should be disabled by adjusting the pH to a 6–8 range. Acids can be neutralized by using a solution of sodium bicarbonate or NaOH, and the basic residues with diluted HCl.

- Alkali metals: Before the solid waste of these metals is disabled, it must be cut into pieces as small as possible, and placed in a container with an inert solvent such as toluene or hexane. To disable the waste, EtOH is slowly added, with caution because hydrogen gas is formed as a byproduct. This process is ended by adding water with extreme caution. Perform all these operations in a fume hood.

- Nitriles and thiols: These are deactivated by oxidation with sodium chlorate, the excess of sodium chlorate is destroyed subsequently with sodium thiosulfate.

- Peroxides and other oxidizing agents (Br_2, I_2, etc.): These are deactivated with an aqueous sodium thiosulfate solution.

- Cyanide: This is deactivated by oxidizing with a solution of H_2O_2 in a pH range of 10 to 11.

- Hydrides, borohydrides, and amides: MeOH is added while the container is cooled externally with the residue. Deactivation is ended by adding water with caution.

- Lithium aluminum hydride: This is suspended in ether while ethyl acetate is added dropwise (always in an inert atmosphere and with vigorous stirring). After the reaction, water is slowly added.

- Azides: Deactivated with iodine in the presence of sodium thiosulfate, nitrogen gas is released in this process.

- Halides and alkyl sulfates: Caution should be taken as these reagents are often carcinogenic. They are deactivated by adding them dropwise to a saturated aqueous solution of ammonium hydroxide. They must be cooled externally.

- Acid halides: These are deactivated into the corresponding esters with MeOH at acidic pH, and then neutralized with NaOH.

- Aldehydes: These are deactivated with a saturated aqueous solution of sodium bisulfite, yielding the corresponding bisulfitic combinations.

- Organometallic compounds sensitive to moisture (usually accompanied by organic solvents): Add dropwise to deactivate on *n*-butanol, and the flammable vapors generated are eliminated in the fume hood. The process is finished by adding water with extreme caution.

1.7.3. Disposal of wastes

Once wastes are ranked within the 10 groups listed above and, as far as possible disabled, they can be eliminated by storing them in containers that are designed for that purpose and that should be available at any Organic Chemistry laboratory. These containers should be in perfect condition. They should be closed, and also replaced once they are filled. Do not store excessive waste in the laboratory. The containers should be placed in a ventilated area away from sources of heat or spark.

1.8. References

1. R. H. Hill and D. Finster, *Laboratory Safety for Chemistry Students*, John Wiley, Chichester, UK, 2011.

2. A. K. Furr, *CRC Handbook of Laboratory Safety*, CRC Press, Boca Raton, FL, 2000.

Chapter 2

Lab Notebook

A fundamental aspect of the work performed in an experimental science such as Organic Chemistry is annotation of the results. It is necessary to distinguish between a professional notebook and an experiment report for an experimental course of Organic Chemistry. The two have similarities, but some key differences can be highlighted. Both documents should be prepared in a clear and concise way, containing all the experimental work, the incidents, and any relevant data related to the laboratory. Research-laboratory notebooks, in private companies or academic institutions, are official documents, confidential, belonging to the institution and are invaluable in the subsequent process of publishing scientific articles and patents. In the case of an experiment report for Organic Chemistry teaching, the students must learn the correct experimental procedures to describe the reactions conducted in the experiments with respect to their theoretical background. Furthermore, it is essential to organize, analyze, and draw appropriate conclusions for each experiment.

This chapter begins with some general recommendations and guidelines for developing a good lab notebook and ends with a reminder of the basic calculations that should be included in the notebook, such as how to express the concentration of a solution or the concept of limiting reactant, stoichiometric calculations, and yields.

2.1. Notebook structure

It should be clarified that the structure of a laboratory notebook is not exactly the same for a teaching laboratory experiment as it is for research work. In the case of a research notebook, the aim is to describe research experiments, which in most cases are new experiments. An undergraduate lab notebook describes a documented chemical recipe from a textbook or adapted from a scientific paper. However, both types of laboratory work have many points in common. The easiest way to properly structure a laboratory notebook is to include the following sections:

a) Table of contents: A few pages at the beginning of the notebook are reserved for a table of contents, with the list of experiments and the pages where they are described. This greatly facilitates finding information on a particular experiment.

b) Name of the person or persons who performed the experiment: It is essential, especially in research, to indicate the author(s) of the experiments.

c) Execution date and order number: At the beginning of an experiment it is desirable always to record a number, code, etc. as well as the date of completion. Thus, the notebook becomes a lab diary.

d) Experiment title: It should be clear and descriptive of the experiment, for example:

  • Synthesis of ...
  • Purification of ... by recrystallization ...

e) Experiment goal: This should be as short as possible (one or two paragraphs). It should explain the type of experiment or reaction performed (e.g., nitration reaction through a process of electrophilic aromatic substitution on benzoate), its use in Organic Chemistry (it is very common, rare, used industry, similar to a process occurring in living organisms, etc.), and other general aspects.

f) Reaction scheme: This scheme should be done, and adjusted stoichiometrically (when possible). The structural formulas (developed or semi-developed) of all compounds and chemical reactions involved in the process should be included. It is not recommended to use only the names of the compounds or empirical formulas, as they may lead to confusion.

For example, in the case of isomers, the explanation of the reaction(s) is how the reactions take place, if they are common to other types of substrates, the role of each in the reaction (acts as oxidant, reductant, nucleophile, base, etc.), and the type of the reaction mechanism (S_N1, S_EAr, elimination, etc.). The reaction mechanism should also be explained, with an outline indicating the electron movements by curly arrows and all the key reaction intermediates. The key transition states should also be indicated. Where applicable, outline the regiochemistry and stereochemistry of the reaction and whether competing reactions occur, giving rise to the formation of side products.

g) Identification of reagents and solvents:

  • Identification (relevant data) of the reactants and the products obtained:
    − Physical properties such as the chemical formula, molecular mass, m.p., b.p., density, etc.
    − Danger data and precautions for use (the student should consult the material safety data sheets MSDS).

- Amounts of reagents and solvents used (in grams or moles; when liquid in milliliters).

It is useful to make a table in which all reactants and products, their relevant characteristics (formula, molecular mass, density if they are liquid, and amount used in mg, mmol, and equivalents) are represented:

Table 2.1

Compound	M_w	Density (ml, g, or mg)	Amount	No. mol.	No. eq.
. . .	. . .	. . .	. . .	. . .	. . .
. . .	. . .	. . .	. . .	. . .	. . .

h) Procedure and experimental development: The procedure is the method used for conducting the experiment. This should indicate the assembly (provided it is not too easy), basic operations and techniques used to perform both chemical reactions and the isolation and purification of the products, and a scheme for separating the products yielded in the experiment. Note that one must not copy exactly the recipe from the corresponding original paper; it is the reaction described in the literature but with the details of the present procedure, including slight modifications performed. In case of a new reaction or procedure, all steps must be described in detail.

It should indicate briefly how the experiment was carried out. The actual procedure followed and the modifications made with respect to the original procedure must be described, with a brief justification of any changes were made. First, the product quantities used are noted (in g, mg, ml, or mmol and equivalents); thus, the reagents used in excess or the limiting reagent that affects the reaction yield can be analyzed.

A remarkable aspect to consider is the inclusion of security measures to be considered in terms not only of the danger of reagents and solvents used but also the basic techniques and operations employed.

Then the conditions under which the reaction is performed should be indicated, justifying in a reasonable way the operations performed:

- Solvent and "real quantity" used.
- How the reagents are added, and in which order.
- Temperature and other conditions used.
- Type of stirring: magnetic, mechanical, or manual.
- Material used and its configuration: reflux with or without a drying tube, simple distillation, filtration type, etc.
- Reaction time.
- Which kind of treatment is applied to crude products.

- Isolation and purification of the products yielded.

It is advisable to mention any data of interest observed during the experiment, or any incident occurring, such as:

- Changes in color.
- Special difficulty dissolving some reagents.
- Appearance of precipitates.
- Heat release.
- Generation of gases.
- A procedure in which the reaction is monitored: time, aspect, thin layer chromatography, etc.
- Errors detected in the bibliographic description of the practice and how they have been corrected.

Chromatography TLC plates made during the experiment, either directly or through detailed drawings or photographs, should be included. The purification method should be specified (recrystallization, column chromatography, etc.). When chromatography is used to isolate products, the exact procedure (e.g., CG: amounts used, retention times, peak areas, parameters used, types of columns, temperature, etc.) should be detailed.

Finally, within this section, the amount of each product yielded (weight, number of mol and yield), purity, physical appearance (physical state: liquid, amorphous solid, crystalline solid, color, odor, solubility tests, etc.) should also be indicated, as well as their physical and spectroscopic characteristics (melting point, IR and NMR data, assigning the most remarkable ^{1}H and ^{13}C NMR signals and IR bands). A common way to present the spectroscopic data is as a list or table or on the molecule drawing. It is useful to paste or attach copies of the resulting spectra (indicating also under what conditions the spectra were recorded: solvent, scale, type of equipment used, etc.). It is also useful to indicate the chromatographic R_f values (indicating the eluent composition and type of adsorbent).

Of special interest in these professional notebooks is to describe properly all the changes made in an experiment, in order to describe optimal conditions —for example, to improve yields, or changes to avoid hazards with solvents or reagents, particularly if an experiment is designed to a higher scale up. Also important is to register failed experimental accuracy as a useful tool to detect errors and perform the corresponding changes.

i) Yield: Where appropriate, the reaction or reaction sequence yields should be calculated.

j) Remarks: The trouble or comments considered relevant throughout the experiment development should be commented on. They should indicate whether the reaction proceeded as planned, whether the product was pure and whether spectroscopic or other data were as expected. If something went wrong, it is important to explain the reason why. In the case of a multi-step process the corresponding partial and global yields should be calculated when possible. Unexpected or noticeable incidents must also be reported in the notebook.

k) Conclusions: Present the main conclusions of the experiment: what has been achieved and what has been learned. The conclusions may also be related to other concepts or reactions studied in the course.

l) Bibliography: When using sources from the literature, include them here. Note that bibliographical sources must be cited in sufficient detail that the information can be readily found. For a book citation not only authors and title must be detailed but also the year, editorial information, and page (e.g., J.A. Dobado, F.G. Clavo-Flores, J. Isac-García, *Química Orgánica: ejercicios comentados*, Editorial Garceta, 1st Ed. 2012, pp. 23–34.).

2.2. Experiment report for a practice course

Experiment reports for a practice Organic Chemistry course shows certain parallelism to a professional notebook but with key differences. The main goal of these documents is for students to have a tool to help them to understand clearly what to do in the lab. The experiments that are described in such notebooks are procedures derived from a textbook or a scientific paper and are usually explained by the instructor. Therefore, some previous relevant issues such as hazards related to reagents or techniques, personal protection, or residue handling must be understood clearly by the student before undertaking an experiment, and these issues should be properly reflected in the report.

A possible structure of an experiment report is as follows:

- Date and name of the experiment.

- Main goal of the experiment.

- Scheme of the reaction or procedure.

- Table with reagents and solvents, amounts, properties, and warnings on risks and hazards.

- Procedure.

- Purification method.

- Yield and physical properties.

2.3. Some guidelines for keeping a notebook

The laboratory notebook should be seen as a diary in which each experiment performed and all incidents that occurred are reported.

The guidelines for keeping a good lab notebook include:

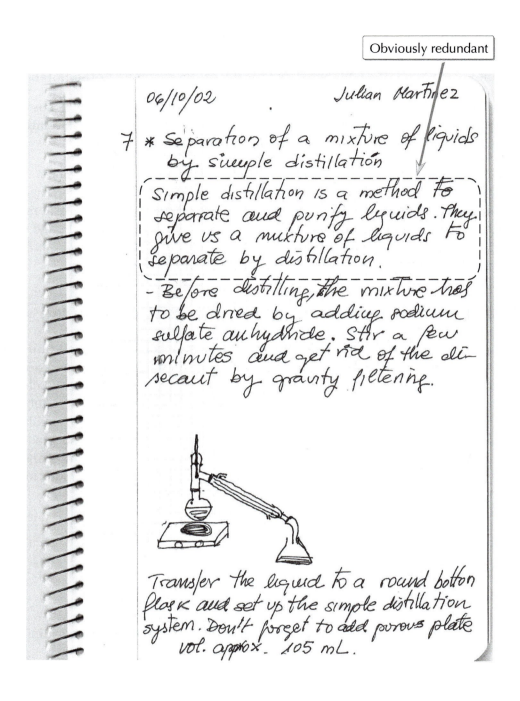

Obviously redundant

06/10/02   Julian Martinez

7 * Separation of a mixture of liquids
     by simple distillation

   Simple distillation is a method to
   separate and purify liquids. They
   give us a mixture of liquids to
   separate by distillation.

   - Before distilling, the mixture has
   to be dried by adding sodium
   sulfate anhydride. Stir a few
   minutes and get rid of the di-
   secant by gravity filtering.

   Transfer the liquid to a round bottom
   flask and set up the simple distillation
   system. Don't forget to add porous plate
        vol. approx. 105 mL.

Figure 2.1: Notebook example.

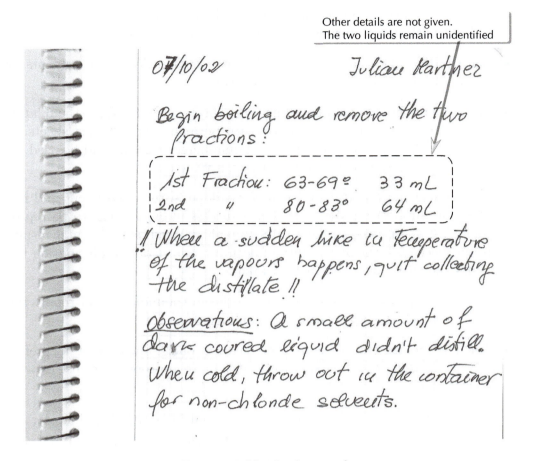

Figure 2.2: Notebook example.

- The student needs to become accustomed to writing preferably with ink, and not with pencil during the course of the experiment.

- The wording and content need to be clear enough for other students or instructors to evaluate the laboratory work.

- The student should avoid handling loose sheets, even to staple them later. It is much safer to use a notebook with permanently attached sheets. A bound notebook, quadrille-ruled, with numbered pages should be used, so that a page cannot be removed without leaving evidence.

- The student must write what really was done in the experiment and not what was supposed to be done. For a successful experiment to be reproducible, it is essential to be faithful to reality. When experimental results are unsuccessful, the availability of accurate and detailed information becomes the only way to correct mistakes.

- It is advisable to write on consecutive sheets (boundless) and report the date on which the experiment was conducted.

- When developing the procedures, try using a concise and clear style, preferably in the impersonal passive voice. It is more appropriate to use expressions such as:

    - Product A was dissolved ... rather than I dissolved product A ...
    - The mixture was separated ... rather than to separate the mixture I followed this or that procedure ...
    - The reaction yield was ... instead of I got a yield ...

- The laboratory notebook is a working tool and therefore often used. When reporting additional notes or comments are needed, corrections, etc. should pose no problem.

For practical purposes, laboratory experiments can be classified into two groups that require different treatment:

- Synthesis of a product.

- Experimental techniques.

2.4. Example of an experiment description

The preparation or synthesis of a compound is to produce a substance with the highest possible purity. When describing, in the laboratory notebook, the synthesis of a product, the following sections should be included:

- Title of the experiment. For example, "Synthesis of 1-bromobutane."

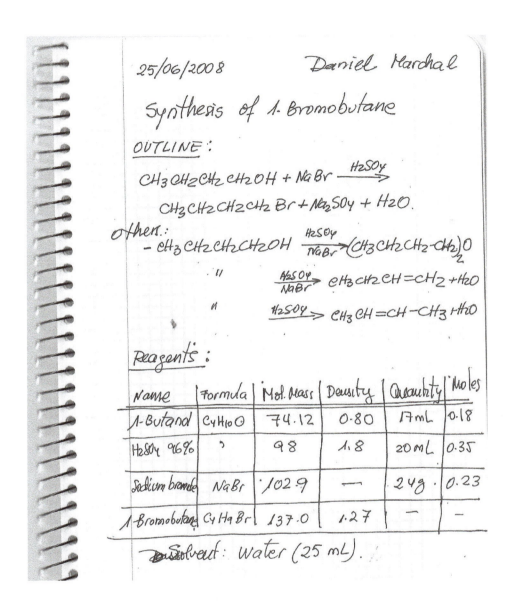

Figure 2.3: Notebook example.

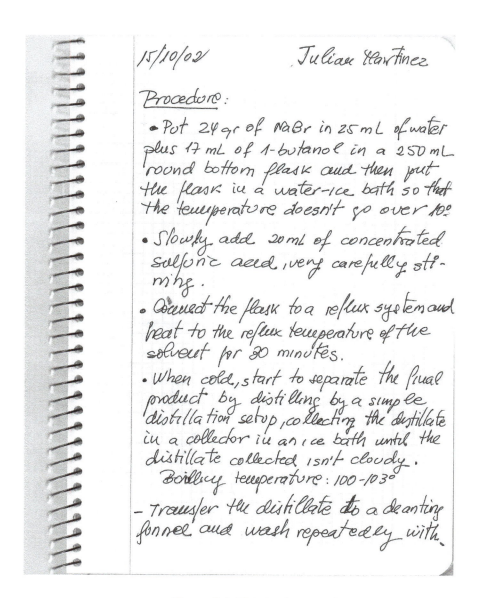

Figure 2.4: Notebook example.

- Outline with the reaction that takes place, reporting molecular formula and molecular mass.

- Scheme with side reactions that occur, if they occur and are known.

- Reagents and catalysts used including:

  − Molecular formula.

  − Molecular mass.

  − Physical properties of reagents: melting point, density, purity.

  − Quantity used in grams and moles and, if appropriate, in milliliters.

- Conditions under which the reaction is performed:

  − Solvent and amount thereof.

- How the addition of the reagents occurs:

  − Mixing all the reagents together at the beginning.

  − Adding dropwise if liquid.

  − In small portions if the reagent is solid.

  − In cold, in hot.

- Temperature at which the reaction occur (specify, if cold, at room temperature, hot, at the boiling point of the solvent, etc.).

- Reaction time.

- The need to carry out the reaction in an inert atmosphere.

- Equipment and assemblies used: commonly used or modified.

- How to follow the progress of the reaction: time, change in color, chromatography, spectroscopic techniques.

- Isolation and purification methods of the product yielded.

- Amount of the final product and calculation of the yield in %.

- Physical properties of pure products:

  − Melting point, m.p., or boiling point, b.p.

  − Optical rotation in the case of chiral molecules.

- Spectroscopic data obtained from the IR, ^{1}H, and ^{13}C NMR, UV/vis.

- Observed product stability and storage conditions.

- Some of the data compiled in the notebook can be tabulated.

2.5. Basic calculations

2.5.1. Concentration units based on mass units

There are different ways to calculate the concentration in a particular mixture or solution, depending on the relative amounts of the components of a mixture or solution using mass units:

- Concentration percentage by mass (m/m = m%): mass of the solute divided by the total mass of the solution after mixing. For example, a 5%-by-mass aqueous solution of sodium hydroxide contains 5 g of sodium hydroxide and 95 g of water in each 100 g of solution.

- Percentage by volume (v/v = v%): volume of the solute divided by the total volume of the solution after mixing. Percentage by volume may be used if both the solute and solvent are liquids, e.g., both EtOH and water are liquids; the concentration of alcohol/water solutions is often given as percentage by volume. For example, a 95% solution of EtOH contains 95 ml of EtOH in each 100 ml of solution.

- Parts per million (ppm): This concentration unit is used for measuring levels of trace components in air, water, biological fluids, etc.

$$\text{ppm} = \frac{\text{grams of solute}}{\text{grams of solution}} \times 10^6 \qquad (2.1)$$

Since the amount of solute relative to the amount of solvent is very small, the density of the solution is, with a good approximation, the same as the density of the solvent. For this reason, parts per million may also be expressed in the following two ways:

- ppmv: mg of solute divided by liter of solution.
- ppmm: mg solute by kilogram of solution.

- Parts per billion (ppb): Strictly speaking ppb is:

$$\text{ppb} = \frac{\mu\text{g of solute}}{\text{liter of solution}} \times 10^6 \qquad (2.2)$$

As in ppm units the amount of solute relative to the amount of solvent is typically very small, the density of the solution is to a first approximation the same as the density of the solvent. For this reason, parts per billion may also be expressed in the following two ways:

- ppbv: mg of solute divided by liter of solution.

$$\text{ppbv} = \frac{\mu\text{g of solute}}{\text{liter of solution}} \qquad (2.3)$$

– ppbm: mg solute by kilogram of solution.

$$ppbm = \frac{\mu g \text{ of solute}}{\text{kilogram of solution}} \qquad (2.4)$$

• Grams per liter: mass of solute in grams in 1 liter of solution.

2.5.2. Concentration units based on moles

• Mol fraction: The mol fraction of a substance in a mixture is expressed as the moles of one component over the total moles of all components in the solution mixture.

$$X_A = \frac{\text{moles of A}}{\text{moles of A} + \text{moles of B} + \text{moles of C} + \cdots} \qquad (2.5)$$

• Molarity (M): It is the number of moles of solute per liter of solution. A 5 M (say "five molar") solution of sodium hydroxide contains 5 mol sodium hydroxide in 1 liter of solution.

• Molality (m): This is the number of moles of solute dissolved in 1 kilogram of solvent.

$$\text{Molality (m)} = \frac{\text{moles of solute}}{\text{kilograms of solvent}} \qquad (2.6)$$

2.5.3. Stoichiometry and balancing reactions

Stoichiometry measures the mass ratios of the chemical elements that are involved in a chemical reaction.

The transformations (chemical reactions) are rearrangements of atoms in molecules and therefore governed by the law of conservation of mass (mass can neither be created nor destroyed during a chemical reaction). Therefore, the total mass of the reactants or starting materials must be equal to the mass of the products.

For all the above, in a chemical equation the same number of atoms of each element on both sides of the arrow should appear. If so it is said that the equation is balanced.

$$2\,H_2 \; + \; O_2 \; \longrightarrow \; 2\,H_2O$$

Reactants = 4H and 2O
Products = 4H and 2O
For adjusting a chemical reaction the steps to follow are:

• Determining which are the reactants and products.

• Writing an unbalanced equation using the formulas of reactants and products, ensuring that no atom is missing.

- Adjusting the reaction determining the stoichiometric coefficients for each molecule of reactant(s) and product(s), giving an equal number of each type of atom on both sides of the reaction. Whenever possible, the stoichiometric coefficients will be given as integers.

For example, the combustion reaction of methane gives carbon dioxide and water. Therefore, we first write the equation unbalanced:

$$CH_4 \ + \ O_2 \ \longrightarrow \ CO_2 \ + \ H_2O$$

The second step would be to adjust the atoms in reactants and products, beginning with the carbon atom, as there is one on both sides of the reaction (one in the reactants and the other in products), the carbon would be adjusted. Following with the hydrogen atom, four in reactive (methane) and only two in the product side (water), so we must double the water:

$$CH_4 \ + \ O_2 \ \longrightarrow \ CO_2 \ + \ 2H_2O$$

Ending up with the oxygen atoms, two atoms are presented at the reagent side and four atoms in the products side, and therefore, adjusting two as the oxygen stoichiometric coefficient in the reactive site balances the reaction:

$$CH_4 \ + \ 2O_2 \ \longrightarrow \ CO_2 \ + \ 2H_2O$$

Verifying the coefficient to confirm that the numbers on both side of the reaction are equal for each atom:
Reactives: 1C + 4H + 4O
Products: 1C + 4H + 4O

2.5.4. Limiting reagent

The molecule in a chemical reaction (reactant or product) that determines the amount of product formed in the chemical transformation is called limiting reagent or limiting reactant. It is used in a balanced reaction to determine the moles of a product obtained from a certain number of moles of a reagent. This is because when we carry out a reaction in an experiment or practice, for practical purposes the reagents are not exactly weighed or measured stoichiometrically. Sometimes, excesses of reagents are used, so that some of the reagents are completely consumed in the reaction while others are spared. The reagent consumed first (stopping the reaction) is called the limiting reagent, since the amount of this reactive used determines the total amount of product formed.

2.5.5. Reaction yield

The reaction yield (absolute yield) of a chemical reaction is the amount of pure and dry product yielded in a reaction. Normally, in order to measure the efficiency of a chemical reaction in organic synthesis, the relative or percentage yield (%) is calculated. Before calculating the yield of a reaction (necessary when preparing the laboratory notebook) it is crucial to know the stoichiometry of a reaction (adjusting the chemical reaction so that there are the same numbers and types of atoms in the reactants and products). Furthermore, we know what the limiting reactant is (expressed in mol, which is given in a smaller amount than the stoichiometry). When the yield of a chemical process is calculated, two situations arise:

- One-step reaction yield: This is expressed as the relative yield (in percentage, %) and results from dividing the moles of the product between the theoretical moles of the product (maximum amount that would result from the product if the entire amount of limiting reagent is consumed in the reaction).

$$A \longrightarrow B$$

$$\text{Yield } (\%) = \frac{\text{mol of the product yielded}}{\text{mol of the product expected}} \times 100 \qquad (2.7)$$

- Multi-step reaction yield: The overall yield of a multi-step reaction composed of various single steps is calculated by multiplying the partial yields for each of the single-step reactions (converting all the percentages to fractions of 100, or to decimals, and multiply them).

$$A \longrightarrow B \longrightarrow C \longrightarrow D \longrightarrow \text{etc.}$$

$$\text{Total yield } (\%) = (\text{yield } 1^{st} \text{ react.} \times \text{yield } 2^{nd} \text{ react.} \times \dots, \text{etc.}) \times 100 \qquad (2.8)$$

For example, in a chemical transformation composed of three reactions and with partial yields each of 25%, 50%, and 75%, respectively, to calculate the total yield the above equation is applied (expressing the partial yields to decimals), resulting the following expression:

$$\text{Total yield } (\%) = (0.25 \times 0.50 \times 0.75) \times 100 = 9.4\% \qquad (2.9)$$

Chapter 3

Laboratory Equipment

3.1. Introduction

Laboratory work requires the use of many tools that enable the efficient execution of basic techniques. A great variety of designs and materials on the market are available, tailored to each type of task. This material can be classified in many different ways. In this chapter, the most common implements (glassware and equipment) used in the Organic Chemistry laboratory are described.

Many of these items are called "glassware." This term comes from the fact that the main parts of such utensils are made of glass, and in these the chemical processes take place, while the remaining parts other than glass in most cases are auxiliary. However, in recent years polymers are replacing glass for certain tasks, so it is also appropriate to describe these new materials.

Because the material available in laboratories is designed to perform specific tasks, some of the material shown in this textbook will be described in Chapter 4, associated with what is known as basic laboratory operations, which comprise a series of unit processes systematically repeated when carrying out an experiment (filter, heat reaction, purified solid or liquid, etc.). Thus, this chapter shows only a preliminary approach to the description of laboratory equipment, as well as the cleaning and conservation procedures. Also, a number of devices with specific applications such as the vacuum pump, hydrogenator, or sodium spinning press are described.

3.2. Manufacture of lab equipment

Laboratory tools are manufactured with the following main materials detailed below:

a) **Glass**: Historically, laboratory equipment has been manufactured with glass. The glass that makes up most of the material is borosilicate laboratory type, since in its manufacturing process certain amounts of boron oxide (B_2O_3), among other compounds, are added to the silica. it is often also called glass mark because of the manufacturer's mark on its surface. Some of the best known brands are Pyrex®, Duran-Shott-Jena®, or Corning®. This glass is currently produced for almost all laboratory glassware. Apart from borosilicate, which is essential, each manufacturer varies the proportion of the other components, based on laboratory studies and for specific uses (see Table 3.1). This gives this type of glass different properties compared to conventional glass such as:

- Scratch resistance.
- Chemical resistance, since it is harshly attacked by the main chemical reagents.
- Resistance to extreme temperature changes without hardly being altered (expansion coefficient of 32×10^{-7} L/°C) in a temperature range between 20 and 300 °C.
- Greater transparency.
- Insulating property (does not conduct electricity).
- Easy to clean.

These characteristics, together with the relatively low manufacturing cost, make it ideal for preparing all kinds of laboratory equipment.

On the other hand, the so-called soda-lime glass cannot be heated, nor can it endure extreme temperature changes. Soda-lime glass is used for everyday items such as bottles, drinking glasses, and flat glass, but its use is largely restricted to laboratory glassware such as Erlenmeyer flasks, beakers, test tubes, and crystallization dishes.

Some laboratory containers are made of amber glass. This glass is manufactured by adding sodium sulfate, iron salts, and carbon. Iron polysulfides are used, depending on the proportion, giving the glass a color ranging from yellow to almost black. Acids as common as the H_2SO_4 or HCl and most organic reagents are marketed in such containers, and they are also used for storing light-sensitive chemicals.

Table 3.1: Approximate composition (%) of different glass types.

Component	Soda-lime glass	Borosilicate glass	
		Pyrex®	Duran®
SiO_2	71–75	80(81)	81
B_2O_3	-	13(12)	13
Na_2O	12–16	-(4)	4
K_2O	-	-(−)	4
CaO	10–15	-(−)	-
PbO	-	-(−)	-
Al_2O_3	-	2(2)	2
NaOH	-	3(−)	-
KOH	-	1(−)	-

b) **Plastic polymers**: In recent years, there has been a dramatic increase in the material made of high strength plastic polymers, including materials of continued use as well as disposable ones. The use of plastic for making these polymeric materials offers a number of advantages, such as durability and mechanical and chemical resistance. With this material, frequent glassware breakages are avoided, increasing the overall safety of the laboratory. Depending on the type of polymer employed, the objects are manufactured compatible with a variety of reagents, including strong acids and bases, general-purpose solvents, and commonly used organic reagents. They are also used to manufacture containers of various kinds, graduated or not, vessels, stoppers, and moving parts such as stopcocks, etc.

The polymeric materials that are commonly used include the following:

- Polypropylene (PP). Translucent or transparent material, resistant to aqueous solutions, acids, bases, and general-purpose organic solvents; thermally stable between -10 and $130\,^{\circ}$C, and thus can be autoclaved, avoiding the formation of meniscus in test tubes and other measurement systems, which facilitates reading. It is incompatible with strong oxidizers. They are used mainly to manufacture beakers, conical funnels, graduated cylinders, Büchner funnels, wash bottles, centrifuge tubes, and desiccators.

- Polymethylpentene (PMP). Transparent, with similar properties to PP, slightly more brittle and thermally stable up to $180\,^{\circ}$C. Incompatible with strong oxidizing agents, hydrocarbons, and some chlorinated solvents (softens, losing mechanical properties). Its manufacturing price is greater than that of PP. PMP is used mainly in the manufacture of beakers, graduated cylinders, and Petri dishes.

- Polyethylene (PE). Transparent, translucent, or opaque. Stable to aqueous, slightly basic or acidic solutions. Compatible with photofinishing reagents and inert to most organic solvents. They are mainly used to manufacture beakers, various containers, wash bottles, and Petri dishes.

- Polycarbonate (PC). Transparent, rigid, and stable in a wide temperature range (from -190 up to $130\,^{\circ}$C). Incompatible with strong acids and bases as well as organic solvents. They are used mainly to manufacture storage containers, Petri dishes, and desiccators.

- Polytetrafluoroethylene teflon (PTFE). Brand named Teflon® is opaque plastic, semi-rigid, and very resistant to strong acids and bases as well as organic compounds. It is thermally stable at a very high temperature range between -200 and $260\,^{\circ}$C. If it were not for the high price, it could replace virtually any glassware. The biggest drawback is its opacity, which prevents a view inside PTFE utensils. It is used mainly to manufacture magnetic stirbars, vials, beakers, flasks, test-tube racks, spatulas, centrifuge tubes, replacement parts such as burette stopcocks, and stoppers.

- Perfluoroalkoxy teflon (PFA). Translucent, flexible, cryogenic, chemically inert, easy to clean, thermally very stable and not contaminant. It is used mainly for manufacturing various containers, Petri dishes, and volumetric flasks.

c) **Other materials**: Different types of utensils in laboratories are also available, manufactured with materials such as:

- Clay and porcelain: Used to make tools with resistance to high temperatures and all kinds of reagents. The crucibles used for heating, melting, burning, or calcined substances are made of these materials and are capsules with hemispherical shape and a lump in the mouth, to facilitate pouring. Spotting tile are made of porcelain, usually white.

- Agate: This microcrystalline quartz presents a high hardness, stability against various reagents, and acid resistance and is used mainly for agate mortars and pestles for grinding a solid with great efficiency.

- Metals (stainless steel, aluminum, etc.): Many tools in the laboratory are made of these materials and are described in the following sections.

3.3. Cleaning of glassware

For proper maintenance, laboratory glassware must be cleaned and dried after the end of each experiment so that dirt does not interfere with the completion of each new experience. Otherwise, there may be disadvantages such as the presence on the glassware of colors that are used to alter the final appearance of the products, lower yields of the products because of the total or partial destruction of reactive substances, or worse, uncontrolled formation of irritating or toxic substances due to chemical incompatibility of reagents with those impurities. This point is most severe in terms of safety.

Waste that can soil/contaminate glassware are diverse in nature and may belong to any of the following groups:

- Inorganic salts.

- Salts with any organic ion such as quaternary ammonium salts.

- Grease from hands.

- Fat from the joints of ground glass.

- Organic substances of varying viscosity.

- Organic solids.

- Resinous polymeric substances, generally brown or dark, which are obtained as byproducts in many reactions.

3.3.1. Conventional cleaning procedures

Washing procedures vary according to the nature of the waste, the degree of adhesion, and the use to be made of the material. For a common reaction and with a mild or moderate amount of dirt, a standard procedure for cleaning glassware would be the following:

- Immediately disassemble all equipment used, to avoid its getting stuck when a reaction is finished.

- Remove waste grease, if any, from joints and ground glass, using a piece of filter paper.

- Rinse with a small amount of solvent such as acetone or CH_2Cl_2 to remove traces of organic matter, especially focusing on the area of the ground glass joints.

Noticeboard 3.3.1

 WARNING!

Organic waste and solvents to be used for washing the material must be disposed of in suitable containers (chlorinated/nonchlorinated) (acid/basic). Never pour them down the drain.

- Wash the material preferably with a liquid soap (solid detergents tend to leave more residues) using a brush of suitable size so that the soapy water can reach all the container areas.

- Rinse with tap water.

- Rinse with distilled water, because tap water has dissolved salts that leave residues after drying.

  —Concerning the graduated container such as volumetric pipettes, burettes, flasks, or test tubes that have been used for measurement of solvents, reagents, or solutions, washing with a compatible solvent followed by washing with distilled water would suffice.

  —Throughout the process minimum safety standards will be considered in order to protect the eyes and hands with glasses and gloves when handling and disposing of hazardous waste in the appropriate containers.

  —In most cases, the protocol mentioned above will be sufficient, although if a standard washing is not effective, it is recommended to follow a procedure that depends on the nature of the residue. For most wastes of inorganic origin, such as metal salts, washing with diluted acid (HCl or

H_2SO_4) is suitable. If the waste persists, try increasing the acid concentration.

—For persistent organic waste, alcohols (EtOH or MeOH) may be employed, as well as DMF or DMSO for washing the material, due to their strong dissolving power, both at r.t. and when heated gently using a hairdryer. In all cases, the students should be aware of the possibility that harmful vapors from solvent or products can be generated.

• Cleaning material

The cleaning brushes are used.

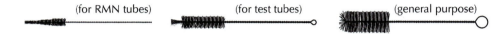

Figure 3.1: Brushes for cleaning labware.

• Cleaning with ultrasonic baths

Ultrasonic baths are used to clean laboratory materials, especially glassware. It is an alternative to conventional cleaning methods. The application of ultrasound to all kinds of soiled material facilitates the removal of particles adhering to surfaces, even in areas of difficult access. It is sufficient for the area to be in contact with the water for effective cleaning.

3.3.2.   Chemical cleaning procedures

Cleaning glassware is easy if done immediately after an experiment. If time elapses, substances in contact with the glass walls can react with the glass or with oxygen from the air, so that as more time passes, the more difficult the subsequent cleaning becomes. If glassware cannot be immediately scrubbed, it is advisable to let it soak in water to prevent dirt from adhering to the glass. There are various soaps and detergents suitable for washing glassware that should be used initially to clean dirty glassware.

When these detergents are ineffective, then solvents such as acetone, toluene, etc. can be employed. The use of organic solvents entails additional risks to health by breathing vapors or making contact with skin or clothing. One of the most widely used solvents for cleaning is acetone, but it poses two problems: it is expensive and highly flammable, so extreme caution is needed, and it should not be handled near sources of ignition such as open flames or very hot surfaces. Always try to use small amounts of acetone (usually 5 ml is enough). The acetone can be used effectively several times before disposal. If the acetone does not work, then try another organic solvent, treating it in the same manner described for acetone. For dirt, stains, and residues that cannot be removed by persistent solvents, acids such as H_2SO_4 or HNO_3 can be used. Wearing goggles, add about 20 drops of concentrated H_2SO_4 and about 5 drops of HNO_3 on the glassware to wash, using extreme caution to avoid any contact with skin or clothes to prevent serious burns and holes in clothing. Once the acids are within the container, proceed to stir gently for a few minutes until the reaction with the contaminant has ceased. Heating in a warm bath can accelerate the cleaning process. After cleaning, discard leftover acids in the appropriate container.

Whatever cleaning method is used with solvents or acids, glassware should be rinsed thoroughly with water and then washed again with soap and water, following the protocol described above.

When the residues adhering to the material become difficult to remove by conventional methods, due to either the nature of the substance or the type of material being handled, more aggressive procedures are required for proper cleaning. The list below with several procedures may be useful:

Noticeboard 3.3.2

 DANGER!

The chemical cleaning methods should be conducted in a fume hood, since they can produce toxic or malodorous gases due to decomposition of organic matter or the oxidizing agent employed.

- Concentrated alkali solutions of alcohols.

  If the waste to be eliminated consists of organic matter, it may be dipped in a solution of MeOH or EtOH saturated with NaOH or KOH. These solutions are very corrosive and attack the skin, so that it is always necessary to work with gloves to avoid burns; they also are potentially flammable, so open flames or heat sources should be avoided. Do not use this procedure for prolonged times or with residues with these characteristics:

    - Material contaminated with substances containing metals.
    - Material contaminated with oxidizing agents.
    - Material composed of crushed glass such as Büchner funnel or some parts in chromatography columns.
    - Graduated pipettes and other volumetric materials, while efficient to remove fat, damage marks to measure the volumes, making it unusable.

- Chromic acid cleaning solution.

  This is perhaps one of the most popular methods for removing organic and inorganic waste in glassware, when normal cleaning methods are not effective. It is effective, for example, in the case of a Büchner funnel or other material with ground glass, or to remove polymeric and resinous substances difficult to clean by other methods. However, it has the serious drawback that Cr(VI) is highly toxic, has carcinogenic effects, and has strong environmental impacts.[1] The cleaning solution is prepared using the following proportions: 6.5 g potassium dichromate, 10 ml of water, and 100 ml of concentrated H_2SO_4. The chromic acid solution may be substituted for ammonium persulfate solutions, having a similar effectiveness for its oxidizing character and being less dangerous.

- Nitro-hydrochloric acid (*Aqua regia*).

  Nitro-hydrochloric acid is a reagent consisting of one part concentrated HNO_3 and three parts concentrated HCl.[2] Nitro-hydrochloric acid has an extraordinary oxidizing character, to the point that it is the only solution capable of dissolving gold. Because of this property, it should be used with extreme care due to its pronounced corrosive nature. Furthermore, when reacting, it generates highly toxic fumes of chlorine and nitrogen oxides. This method should be used only when conventional cleaning methods fail.

- Potassium permanganate.

  Usually, a solution of 10 g $KMnO_4$ per liter of water is used. It can be alkalized before use with Na_2CO_3. After use, the brown spots in the glass can be eliminated with concentrated HCl.

3.3.3. Drying glassware

Glassware must be dried for later use as well as for storage. The easiest way is to air dry by simply placing the items in a drying rack. However, the most efficient method is to use a drying oven. The oven temperature can be set at around 75 °C, to ensure that everything is thoroughly dried. This is a key process, since many reactions may be damaged by the presence of water or other solvents.

> **Noticeboard 3.3.3**
>
> **Practical tips & tricks:**
>
> If dry glassware is needed in a short period of time you can use acetone[a] to drag traces of water before placing the items in the oven. This procedure is recommended particularly for glassware such as reflux condensers, adapters, and other material with curved areas that tend to retain moisture.
>
> ---
> [a]The acetone for washing should be discarded in an appropriate waste container.

---

[1]Some laboratories have banned its use.

[2]It is recommended to add one part of water when stored, to minimize the formation of chlorine gas.

Plastic material, silicone rubber or Teflon, or grease residue material or material that is not completely clean should never be placed in an oven because heat carbonizes residues, making them much harder to remove.

Graduated glassware such as volumetric pipettes, burettes, flasks, or test tubes used for measuring solvents, reagents, or solutions (volumetric) must be dried with a hot air stream (e.g., hair dryer) or air, because the heat of the stove and the subsequent cooling cause expansions/contractions in the glass, respectively, and therefore the measurement accuracy is lost.

Noticeboard 3.3.4

 DANGER!

When removing the material from the stove, wear special gloves, crucible tongs, or any other option to avoid burns to fingers and hands.

3.4. General-purpose glassware

Labware is varied, both in design and composition, because it serves different needs, such as storage of products, make weight and volume measures, performing a reaction, purification of substances, etc. A key part of labware is glass, since it is resistant to most of the reactants, and is transparent (allowing processes to be viewed), and it is an easy material to work with and adapts to any form. Many models, types, and variants of tools are available on the market, making it unfeasible to describe and classify them exhaustively. Many of these items are named for the chemists who designed them.

3.4.1. Assembly of lab equipment

To perform many basic laboratory operations, a series of components must be coupled. This coupling is done by ground-glass joints or combination of threaded adjustments. This allows a hermetic connection between the different parts, preventing the escape of vapors or liquids.

- Ground-glass joints: Ground glass is a technique that gives it a matte finish by a surface treatment with acid or by sandblasting one side of the glass.

  The ground-glass joints, called "conically tapered joints," are the most widespread connections to assemble glassware easily and quickly, so that different parts are sealed without liquids leaking or vapors escaping when heated (fit leak-tight apparatus). For example, a round-bottom flask can be assembled with a reflux condenser. Moreover, these joints are also used for glass stoppers.

One of the glass elements connects an inner surface (or female) of ground glass and the other ground glass surface faces outward (or male), both with the same taper for easy adjustment. The female joint has a flange as discussed below to facilitate gripping.

Ground-glass joints with sharp conical shape. The inner part (male) is shown on the left and the outer part (female) on the right (see Figure 3.2). Ground-glass surfaces are shown with shading. By joining the two parts in the direction of the arrows, they become firmly fastened. Usually some grease is applied to both glass surfaces.

The conical ground joints normally have a taper 1:10 and are marked by a symbol consisting of a capital S, superimposed on a capital T, meaning "Standard Taper."[3] This symbol is followed by a number, a bar, and another number. The first number represents the outer diameter in millimeters at the base (widest part) of the inner join. The second number represents the length of ground-glass bonding in millimeters.

The most common joints in the United States are 14/20 and 24/40. But in Europe, the European standard ISO is binding sizes 10/19, 14/23, 19/26, 24/29, and 29/32. US and European standards (ISO) differ only in length but not in the taper, and can be used in combination. The caps of bottles of chemicals, flasks, funnels, and so on often do not use these standards. There are adapters that allow the assembly between components with different tapers both for expansion and reduction.

Some ground-glass pieces have two glass hooks 180° apart just above the grind that can be used to connect the joint with small metal springs. Although now obsolete, they serve to couple the parts and maintain the joint fixed. By ensuring joints with these springs, overpressure that may occur is prevented, and in these cases the joint opens and releases the pressure. However, caution must be taken in closing the joints because sometimes it is safer for overpressure to be released, resulting in a spill because a firmly attached joint could cause the glassware to explode. For example, the HCl has a vapor pressure at room temperature high enough to blow up standard glass. Silicone is often used to lubricate the stopcocks and ground-glass joints, preventing them from becoming blocked. Although using lubricants keeps joints from leaking, and they can then separate the two parts more easily, the disadvantage being that it can contaminate the reaction products.

The ground joints sometimes are blocked (stiff) after use and the glass is hot and cool. Often, they can be unlocked using solvents in the union with slightly rotating movements of one part against another; heating with a hair dryer, an oven, or using an ultrasonic bath.

---

[3]Taper of 10%, corresponding to a half angle α of the conical ground surface 2°; 51′; 45″ with a tolerance of ± 2′.

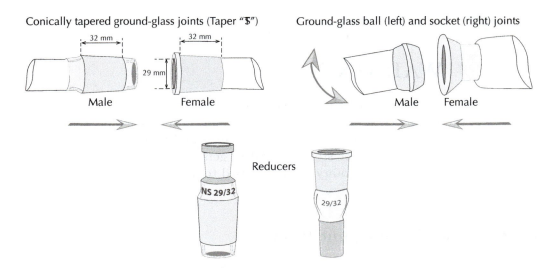

Figure 3.2: Different types of joints and ground-glass adaptors.

- Ground-glass joints with screw connectors.

  There are other interchangeable conical ground joints that improve safety assemblies and provide sealing. There is no need for other elements to ensure the closure (clip), as this kind of joints improves rigidity of the assembly.

- Ball and socket joints.

  These pieces end in male and female hemispherical ground surfaces that fit together. These joints are marked with a code consisting of two numbers separated by a slash. The first number represents the outer ball diameter in millimeters at its base or inner diameter in millimeters at the outermost part of the cavity; in both cases these are the maximum diameters of the joints. The second number represents the inner diameter of the hole in the center of the ball or cavity, leading to the inner diameter of the tube attached to the joint. To keep both pieces together, a clamp allows some movement between the parts assembled by this system.

- Other ground-glass joints.

  Ground glass is also used to manufacture glass desiccators and weighing bottles. The glass desiccators possess the edge of the mouth and a lid with ground glass that makes a perfect lace closure and vacuum of the system.

  A weighing bottle, often used to store or weigh solids, is a cylindrical glass container constructed with a glass lid with ground glass (see Figure 3.2).

Noticeboard 3.4.1

 WARNING!

One of the precautions to consider with glassware manipulation is not to exert unnecessary pressure when assembling or disassembling the material (ground-glass joints, rotary evaporators and flasks, thermometers embedded with rubber stoppers). Glass gets stuck due to heating. In these cases, a gentle rotary motion is used to dislodge the material.

3.4.2. Ground-glass joint flasks

Flasks, vessels designed for mixing or making reactions, are made of borosilicate glass and therefore are resistant not only to heat but also to sudden temperature changes (heat-shock resistant). They may have different shapes such as a flat or round bottom. Following is a list of the most common ones:

- Ground-joint Erlenmeyer flask: Conical in shape with a wider taper at the bottom than at the mouth.

- Round-bottom flask: Can be completely spherical in shape or have a flat bottom.

- Pear-shaped flask: Used in distillations or with a rotary evaporator.

- Volumetric flask: Used to prepare solutions.

- Schlenk flask: Used in air-sensitive chemistry and often connected to Schlenk lines.

All can vary from a few milliliters to several liters, and although the most common round-bottom flasks have one neck, some flasks have twin- or triple-necks for use in more complex assemblies.

3.4.3. Glassware without ground-glass joints

In this section, some of the materials frequently used in the lab are described, including glassware without ground joints (see Figure 3.3).

- Petri dish: This is a box formed by two circular parts that fit together and that close but not tightly. Primarily used in microbiology for cell cultures. In Organic Chemistry it is used because of its low price to store solids (it is also manufactured in plastic).

- Porcelain capsules: They are hemispherical with a peak in the mouth to facilitate pouring and are very useful to heat, dry, or carbonize substances at high temperatures.

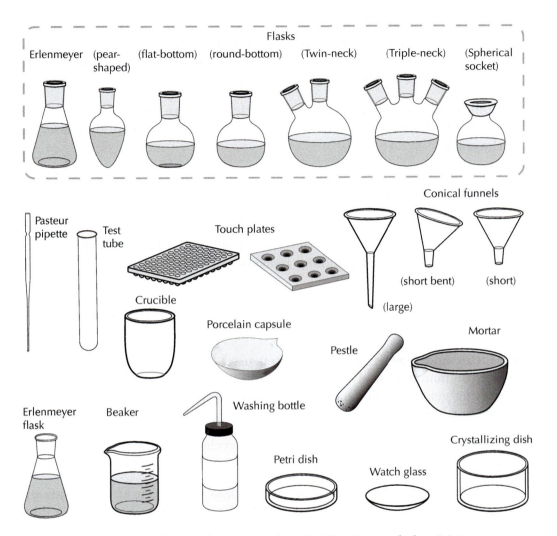

Figure 3.3: Different glassware with and without ground-glass joints.

- Crystallizing dish: These cylindrical vessels with a wide base and low height are commonly used to prepare water, oil or sand baths or serve as containers for different materials. The name comes from their original use to produce a solid by crystallization from a solution.

- Conical funnels: Made of glass or plastic, these are used to transfer solids or liquids. Funnels for solids have the widest tube to facilitate passage of the solid.

- Erlenmeyer flask: This is most common flask in the laboratory and there are also variants having a ground mouth.

- Beaker: They are cylindrical containers having a flat bottom and a spout or lip at the mouth to facilitate the transfer of liquids. They are very common among flasks without grinding. Beakers can hold from a few milliliters to several liters.

- Washing bottles: These plastic or glass bottles have a perforated plug with a thin tube and an angle, which is used to contain solvents.

- Mortar and pestle: A mortar, often used for mixing or shearing solids, may be manufactured with glass, ceramic, or agate. Solids are ground by making gentle circular movements with the pestle resting on the mortar without banging.

- Touch plate: Manufactured usually in porcelain, usually white, these have a series of cavities or wells used for tests and assays on a small scale.

- Pasteur pipette: This is a transparent glass tube open at both ends but with a pronounced narrowing at one. At the upper end, it engages a rubber or latex bulb while the lower end of the tube tapers to a capillary. Used to transfer small amounts of liquids, it is manufactured in plastic in one piece, which is not necessary to couple to a rubber bulb.

- Plugs: These are a variety of very large plugs commonly used in the laboratory, made of glass, rubber, Teflon or other polymers such as PE, which today have replaced cork. Septa caps (singular "septum") are widely used for easily and tightly closing glass, ground, and without ground-glass joints such as Erlenmeyer flasks, round-bottom flasks, bottles, or jars. The cap is adjusted and closed to the flask mouth by folding over the outside of the septum. The main advantage is that it can be punctured with a needle or cannula to introduce a syringe or reactive gases and to work in an inert atmosphere. For flasks with a ground mouth, if a snug fit is required to prevent liquids or vapors from escaping, ground-glass or Teflon stoppers are used.

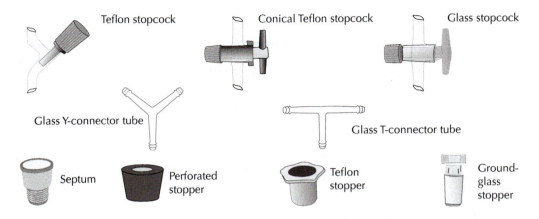

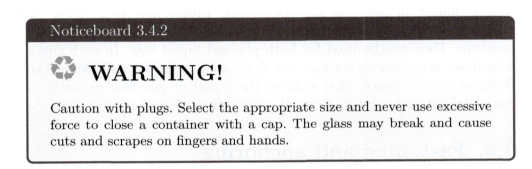

Figure 3.4: Different type of stopcocks, junctions, and plugs.

Noticeboard 3.4.2

♻ **WARNING!**

Caution with plugs. Select the appropriate size and never use excessive force to close a container with a cap. The glass may break and cause cuts and scrapes on fingers and hands.

- Test tube: These are cylindrical glass tubes with one open end (top) and the other hemispherical (bottom). They are manufactured to be heat resistant or not, depending on the type of glass, with many applications in a laboratory. Test-tube holders, designed for storing or handling several test tubes at a time, can be wood, metal, or plastic.

- Watch glass: This is circular in shape and slightly convex and is used to weigh or dry solids.

3.4.4. Stopcocks and connectors

- Stopcocks: Many laboratory utensils have built-in stopcocks to open or close the the passage of liquids or gases. They come with a glass body with interchangeable male and female ground-glass or Teflon joints. Ground-glass joints require good lubrication with silicone grease to prevent stopcock locks.

  Stopcocks are also used for connecting tubes —for example, rubber, Teflon, or silicone —for circulating fluid, to block the passage of fluids, or to control the air input or output in vacuum systems. These valves that connect tubing have a glass segment to either side of the stopcock with a widening conical end, so that the tube fits snugly and does not slip once connected. Some stopcocks are made entirely of plastic.

Figure 3.5: Clamps for flexible tubes.

- Connectors: Flexible tubing is assembled with glass or plastic connectors, which may be straight, T-shaped, or Y-shaped.

3.4.5. Flexible tubing

In laboratories, there are different types of flexible tubing as well as pinchcock tubing clamps for liquid or gas circulation. Tubing may be of latex, silicone, neoprene, PVC, Teflon, etc., depending on the fluid being used and the temperature. They can be used for both gas and liquid flow. In such tubes, it is sometimes necessary to regulate the flow, using a special tool (Mohr clamp or Hoffman screw clamp), that reduces the liquid or gas flow partially or even completely at any point in a circuit consisting of these flexible tubes.

3.5. Fastening and anchoring

To work safely in a laboratory of Organic Chemistry it is essential for any type of container or assembly and its components to be properly assembled and fixed on a support to prevent accidents. The most common devices used in the laboratory for this purpose are described below.

- Ring stand: It is crucial to maintain glassware elements fixed and safe while carrying out an experiment. The supports are formed by a metal rod located upright, usually of stainless steel. These supports are composed of intersecting horizontal and vertical metal rods, and are usually integrated into lab benches or metal supports formed by one or more metal rods, with the corresponding base that allows placing them in different locations. Clamps and clamp holders are attached to these ring stands.

- Clamps and holders: These are complementary fastening devices. The clamps can grasp different glassware such as flasks or more complex assemblies. They consist of a metal rod and two moving metal parts lined with plastic or rubber for easy grip on the glass. Three-finger clamp grips close on themselves by turning a screw, holding the glass without over-tightening and thus avoiding glass breakage. The clamp is attached to a support by a holder, which is a metal piece with two screws fastening the clamp to the metal rod. There are also clamps that carry a holder built into the structure. Ground-glass flasks have a flange. The clamp must be fixed under this flange to ensure adequate attaching.

- Conical joint clips: These are removable metal clips used for tapered ground joints, manufactured of stainless steel and used especially in complex assemblies. There are different clip types depending on the standard size taper.

- Keck clips: Keck clips (patented in 1984 by Hermann Keck) are used to hold together the conical ground-glass pieces. These clamps are joint size color coded with standard tapers of red size 29, green size 24, and yellow size 14. Made of polyacetal (polyoxymethylene POM), they melt at 175 °C and soften at 140 °C. Therefore, these should not be used to join pieces that can reach these temperatures such as in heating under reflux.

- Burette clamps: Burette clamps (also called double "Fisher"-type clamps) were designed to hold a specific burette, with double locking at two nearby points, which prevent the burette from turning (usually when subjected to lab clamp and holder). They also have a very simple system that exerts just enough pressure to secure it firmly while avoiding overpressure that could break glass (as in the case of a three-finger clamp).

- Test tube clamps/tongs: These clamps, used to hold test tubes while being heated or handled, are mainly of two types: wood, similar to clothespins, but with one elongated end, and metal, which are more resistant wear and durable. However, unlike wood, metal is conductive and, if overheated, burns the fingers.

- Crucible tongs: Used to handle crucibles, these are made of metal and also require appropriate gloves when used for this purpose. They are also used to handle the oven-dried material, instead of using a cloth or bare hands.

- Tweezers/forceps: These are used to hold small items or any solid materials and are useful, for example, when preparing a drying tube (guard tube) to manipulate wool inside. There are models on the market with curved tips that facilitate access to holes and are made of stainless steel or plastic.

- Ring clamps: These devices are used to support funnels such as separatory or filter funnels. In some, the end is coupled to a holder to secure them to the metal support. It is appropriate to protect metal rings with a piece of rubber tubing to avoid damaging the funnels. There are rings that are not completely closed to facilitate placement of the funnels.

- Lab jack: These height-adjustable platforms facilitate the distribution and assembly of laboratory experiments, supporting or elevating lab equipment to appropriate heights. They improve overall lab safety, eliminating the need for clips and for placing glass joints under tension.

- Round-bottom flask stand: For the safe handling, of round-bottom flasks, these stands are manufactured in both cork (cork rings) and plastic (polypropylene) materials to prevent these containers from rolling on the lab bench.

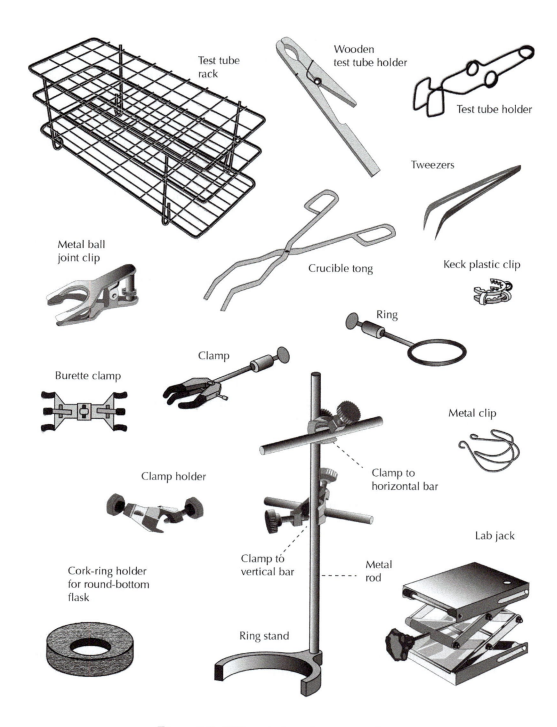

Test tube rack

Wooden test tube holder

Test tube holder

Tweezers

Metal ball joint clip

Crucible tong

Keck plastic clip

Ring

Clamp

Burette clamp

Metal clip

Clamp to horizontal bar

Clamp holder

Clamp to vertical bar

Metal rod

Lab jack

Cork-ring holder for round-bottom flask

Ring stand

Figure 3.6: Different clamping accessories.

- Test-tube rack: These racks are used to hold test tubes, and although traditionally manufactured in wood, today they are manufactured in metal and recently in plastics such as PP, which presents the advantage that the racks can be assembled and put in an autoclave.

3.6. Pressure and vacuum equipment

For a series of laboratory tasks, vacuum and pressure systems are needed. Depending on the required pressure, there are various types.

3.6.1. Water aspirator pump (vacuum aspirator pump)

This is one of the most popular devices to provide a moderate vacuum (up to 50 mm Hg). The vacuum generated depends on the temperature and flow rate at which the water circulates. Water-jet pumps are manufactured of glass or plastic and have a side opening to connect glassware such as a Büchner flask to which a vacuum can be created by a rubber vacuum tube, and then it is firmly fixed to the water supply by means of a screw system or other unscrewed adapters.

Tap water enters through the top of a straight tube in a tight spot and is empted through the drain. This increases the speed of the water, which causes the suction of surrounding air by Venturi effect, generating a continuous vacuum.

Such devices have two drawbacks: the high water consumption, and a change in pressure of water can reverse the water from the tube to the system that is being emptied. The latter problem is avoided by placing a trap to collect the possible water excess. These traps are used as a safety device not only in the water aspirator pumps but also in most vacuum equipment.

3.6.2. Recirculating water vacuum pumps

To avoid the excessive water consumption of water-jet pumps, recirculating water vacuum pumps were designed to create a vacuum with the same principle (Venturi effect), but instead of using tap water, a closed circuit formed by a water tank and an electric motor provides the water flow. These recirculating pumps have the advantage of providing a steady pressure but the disadvantage that the water temperature rises with extended periods of use, thus reducing the vacuum.

3.6.3. Diaphragm vacuum pumps

When lower pressure in the vacuum system is needed (compared to the water jet pumps), diaphragm vacuum pumps can be used. These are electrical devices that generate vacuum by means of pushing elastic walls (membranes) that vary the volume of a chamber. They are specially designed to withstand the attack of volatile organic solvents.

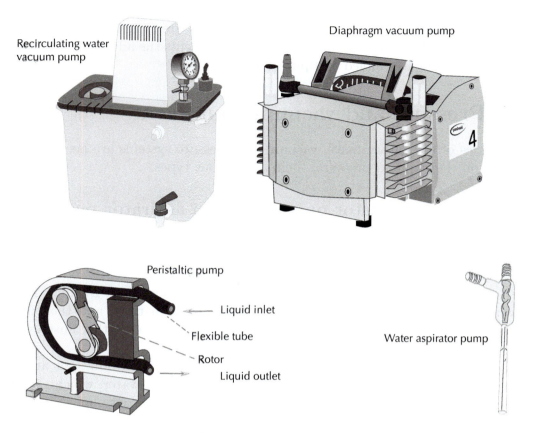

Recirculating water vacuum pump

Diaphragm vacuum pump

Peristaltic pump

Liquid inlet

Flexible tube

Rotor

Liquid outlet

Water aspirator pump

Figure 3.7: Different vacuum and pressure equipment.

3.6.4. Peristaltic pumps

These devices allow pumping a fluid contained in a flexible tube by means of a rotor with a number of moving rollers that push the fluid due to compression. The fluid is isolated from the elements of the pump, so that the fluid (clean or sterile liquid) is not contaminated and, if corrosive, cannot damage the elements of the pump in case of being corrosive. It may also be suitable to generate an air flow at low or moderate pressures.

3.6.5. High vacuum pumps

These electrical devices are similar to diaphragm pumps but can generate a vacuum under 0.1 mm Hg by a mechanism based on oil diffusion (evaporation and condensation cycles). They are typically used for distilling of liquids with high b.p. or generating a vacuum in a multiple Schlenk line.

3.6.6. Centralized vacuum system

Many laboratories now have centralized vacuum systems. Through this central base, a vacuum is generated and distributed throughout the laboratory by means of metal pipes. At each vacuum point, a wrench, manometer, and a trap are installed to prevent aspirated vapors from entering the circuit.

3.6.7. Vacuum gauge and pressure regulator

A pressure gauge is an instrument for measuring the absolute pressure of gases or liquids as well as the differential pressure between two areas or vacuum. The McLeod gauge is a special type of gauge used to measure the vacuum below atmospheric pressure. There are two types —i.e., mechanical or mercury. Mechanical gauges are used in centralized vacuum systems or in some vacuum pumps incorporated as part of the mechanism.

In most cases, for centralized vacuum systems and gas cylinders, it is necessary to reduce the output pressure of gas to a lower value. This is done with a pressure regulator valve, which has two manometers to measure the gas pressure in and out.

Also, the so-called McLeod gauge (see Figure 3.8) accurately measures pressure below atmospheric pressures by a system of thin glass tubes (capillaries) inside mercury, coupled to a movable panel. Mercury is compressed with a sample of the gas system to be measured. The device is placed in a horizontal position and subsequently tilted gently so the mercury in the main bulb can penetrate to the capillaries. When mercury reaches the cut-off line (capacity), the gas inside the left capillary gets trapped and isolated from the rest of the system. This causes an increase in the gas pressure which, when the mercury is pouring, is measured at the scale.

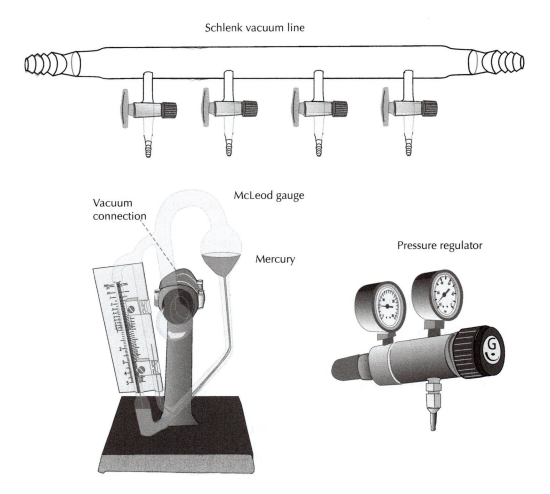

Figure 3.8: A glass McLeod gauge, drained of mercury.

Parr hydrogenator

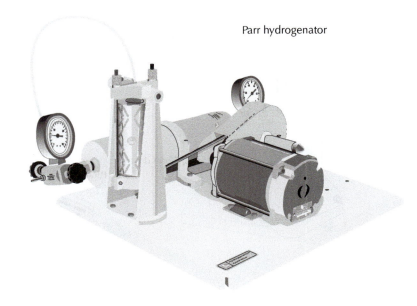

Figure 3.9: Parr hydrogenator apparatus.

3.6.8. Schlenk vacuum line

This device is a glass tube formed by a glass interconnected by a double stopcock system, which has several outputs for several samples. One tube is connected to a source of inert gas, while the remaining ones are connected to a high vacuum pump. A Schlenk line is used for working with compounds sensitive to oxidation by atmospheric oxygen and for removing residual solvent air-sensitive samples. With a set of stopcocks a vacuum or an inert atmosphere can be selected without moving the sample from one line to another.

Between the high vacuum pump and the vacuum device a trap cooled with nitrogen is inserted to prevent volatiles from passing into the pump with the corresponding deterioration.

3.7. Pressure reactors and hydrogenators

For some reactions to be performed at medium or high pressures, some laboratories are equipped with pressure reactors and hydrogenators. Depending on the required pressure, there are different types of reactors and devices. One of the most common reactions in the Organic Chemistry lab that requires such conditions is hydrogenation. The Parr hydrogenator is used for such reactions and consists of a pressure-resistant glass bottle, which is protected by a wire mesh to prevent accidents, connected to a hydrogen cylinder (or gas line) with a pressure gauge for controlling the pressure, and a shaker that moves a cylinder with the product and catalyst. There is also the variant of using a reactor (glass, metal, or synthetic material), where a reaction flask is placed and is equipped also with a magnetic stirrer.

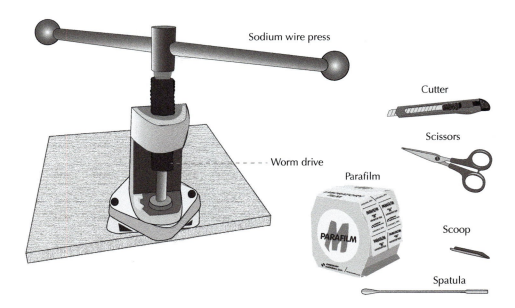

Figure 3.10: Sodium wire press and other common utensils.

3.8. Other materials

- Glass rods: These are solid glass cylinders used to stir solutions, dissolve, and handle solids in the funnel, etc.

- Spatula: Although they may differ greatly in size and shape, all are used for handling small amounts of solids. A special type of metal spatula called a "scoopula"[4] is used for transferring solids to a weighing paper.

- Scissors: There are many types of scissors, with their design depending on their use. Common scissors (office) are used, for example, to cut filter paper.

- Parafilm®: This is a type of thermoplastic film, flexible and self-sealing, that is sold in rolls and is used for closing containers or sealing glass or plastic plugs. It is resistant to most common reagents in a chemistry lab.

- Cutter: This type of knife has a blade with a very sharp edge and a handle of plastic or metal. It is often used for cutting TLC plates, rubber or silicone tubing, cork, etc.

3.8.1. Sodium wire press

Sodium is a ductile, malleable metal, with a low melting point. Its highly reactive nature makes handling it relatively dangerous. It reacts violently with water and

---

[4]Name registered by Thermo Fisher Scientific (from the term "spatula-like scoop"), which is used widely to describe this type of spatula.

with low-molecular-weight alcohols, producing hydrogen. It is easily oxidized with air by covering a layer quickly with oxide or sodium carbonate, which is passivated. Its reactivity in organic reactions can be increased by adding it to the reaction in the form of a fine wire (to increase its surface area). It can be safely handled if cut into small pieces and compressed in a sodium press and collected in a container filled with an inert liquid sodium (hexane or toluene). Other than their use as a reagent, sodium wires are often used to maintain anhydrous solvents after distillation.

Sodium can be weighed in a container with a dry inert liquid, and the sodium is added in small pieces or in wire if available. Sodium wastes in the laboratory can be discarded by reaction with EtOH (never add water) in a fume hood and away from heat sources.

Chapter 4

Basic Laboratory Operations

4.1. Introduction

Under the name of basic laboratory operations are grouped methods and general procedures that enable a successful experiment to be performed. When a reaction is carried out, the protocol begins with the choice of the adequate reagents and solvents that must be weighed or measured. It continues with the reaction, for which the solid or liquid reagents have to be transferred, mixed, stirred, heated, or cooled for certain periods of time and at certain temperatures, and then it will be necessary to isolate from the reaction crude the product(s) and proceed with their purification. This will require specific operations depending on the nature and properties of these products, such as liquid-liquid extraction depending on the different physicochemical properties of the components of a mixture, distillation properties in all its variants, filtration, recrystallization, drying, and especially the use of the different chromatographic techniques.

Finally, we should characterize the final products of the reaction by determining their physical properties, such as melting point (m.p.) or boiling point (b.p.), and their spectroscopic properties resulting from the interaction of electromagnetic radiation with matter.

Together with the basic operations, a number of auxiliary tasks are necessary, such as cleaning and maintenance of laboratory equipment or as preparing solvents. All these procedures are closely linked to the use of specific materials and devices. In the following sections, the most common basic operations in the Organic Chemistry laboratory are described.

4.2. Weighing of solids and liquids

One of the first techniques that students should become familiar with in the laboratory is that of weighing substances to carry out chemical reactions with the required amounts. Furthermore, it is essential to properly calculate yields

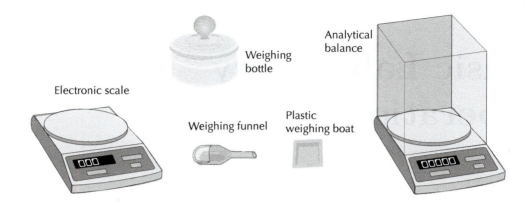

Figure 4.1: Different types of balances.

and to weigh the amount of solute required for a solution of a predetermined concentration.

In the Organic Chemistry laboratory, the most commonly used balance types for weighing substances are the following (see Figure 4.1):

- Top-loading balance: This is used to determine or weigh substances with an average accuracy from medium or low. It has a sensitivity of 0.1 g or 0.01 g, typically with weighing capacities of 100 to 200 g at an accuracy of 0.001 g. There are also some that may have weighing capacities of up to 2 kg and with up to 0.01 g accuracy. The weighing error with this type of balance is acceptable in Organic Chemistry experiments where the amounts of substances are usually not less than 1 g. They are significantly cheaper than analytical balances.

- Analytical balance: This is a very precise instrument (usually digital) for weighing with a small margin of error. It can weigh the mass of a substance in grams, with an uncertainty of 0.00001 g (0.01 mg). For most experiments in a practical Organic Chemistry course, such precision is not required except in microscale reactions, in the preparation of samples for spectroscopy, i.e., UV/Vis, or in determining the optical rotation of a particular substance.

a) Weighing solid substances: To weigh a solid, first use a weighing dish on a balance pan and then set to zero by pressing the corresponding tare button on the scale; next, with a spatula, the substance is added to the value of the required mass. The following must be taken into account:

- No solid reagent should ever be weighed directly on a balance pan as contact with chemicals may deteriorate the pan with time. Once used, the balance should always be left completely clean of reagents (a small brush can be used for this purpose).

- For solids and relatively inert compounds, a piece of glossy paper can be used to prevent solid residues from being retained, and the remaining substance is transferred with the help of a spatula. The paper size must be appropriate to the size of the balance pan. If too large, it may touch some part of the balance or be moved by small air currents generated by the movement of the hands when weighing, causing significant errors in weighing.

- This procedure is not suitable for weighing corrosive or oxidizing substances. Typical cases are potassium permanganate (attacking cellulose paper) and sodium that reacts with air. For these cases, a tared beaker or flask can be used, because their wide mouth allows substance to be added or removed comfortably. Another possibility is to use a Petri dish or a watch glass. Many laboratories have weighing pans made of glass or inert polymers that are useful for this purpose. Weighing funnels (weight substances) can also be used to weigh a solid (see Figure 4.1). These are made of glass or polypropylene and allow the solid to be dragged with a solvent directly through the nozzle onto a flask. In the case of sodium, the flask or vessel used must contain an inert liquid that prevents contact of the metal with the air.

b) Weighing liquids or syrups:[1] This is done with the help of a dropper and a container (flask, Erlenmeyer, etc.) to avoid unnecessary water transfers. The container may be the same as the one in which the reaction is to be performed. The empty container with the dropper inside is tared and then the sample is taken with the dropper. The contents are added to the container, and the amount is found by keeping the dropper within the container to the desired mass. Then add part of the solvent to be used in the reaction, and using this, the content of the dropper is cleaned of remaining reagent.

Noticeboard 4.2.1

 Practical tips & tricks:

- To prevent round-bottom flasks from rolling on the balance plate, a rubber cone from the vacuum filtration equipment can be used as a support.

- Record the tare value of the flask with ground mouth: it can be written in pencil on the inside. The writing remains, even after washing the material with soap.

- The sodium in a container with liquid should be weighed as quickly as possible to prevent passivation of metal.

---

[1] In Organic Chemistry, syrup refers to a thick liquid. An example is honey composed basically of a mixture of various kinds of sugars that occasionally crystallize.

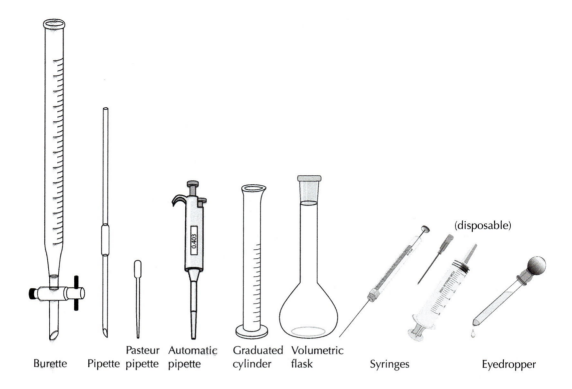

Figure 4.2: Different materials used for measuring liquids.

4.3. Measurement of liquid volumes or solutions

The most frequent case of measuring volumes in the Organic Chemistry laboratory arises when a certain amount of a reaction solvent is needed, to prepare eluents in chromatography, or to prepare solutions of a given concentration. There are different types of materials used for this purpose, which vary depending on the amounts handled and the precision required. The equipment used for measuring volumes of liquid is called volumetric ware (see Figure 4.2).

The most common methods and apparatuses for measuring volumes are described as follows:

a) Graduated pipette: is a transparent glass tube with one conical end that allows fairly accurate volume measurements. There are two types of pipettes: Mohr, backward or drain-out pipettes; and serological, forward or blow-out pipettes. To fill, the pipette is sucked (using propipetter) at the top. Never pipette by mouth, given the volatility and toxicity of many of the compounds used in an Organic Chemistry laboratory. The most common pipettes are Mohr pipettes. These have a series of marks drawn on the glass indicating the volume. The liquid is sucked and the liquid level adjusted once at zero, and then the liquid is dropped to the division that indicates the volume needed. Blow-out pipettes (double capacity) are fixed volume and have a double calibration mark. First, the liquid is drawn to the top mark and then carefully dropped to the bottom mark. Although they have the disadvantage

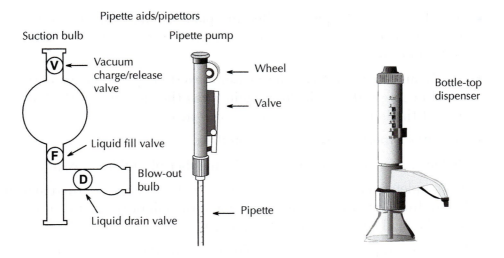

Figure 4.3: Different models of pipette fillers and dispensers.

of measuring a fixed volume of liquid, they are much more precise than Mohr pipettes, because the volume measure is not changed if it breaks or deforms the conical tip. The error limits associated with pipettes by volume are 2, 5, 10, 30, 50, 100, and 200 ml. Their associated errors are ± 0.006, ± 0.01, ± 0.02, ± 0.03, ± 0.05, ± 0.08, and ± 0.10, respectively.

b) Pipette fillers: To fill the pipette to the desired mark by pulling, devices called safety pipette fillers are used. The most common types are a hollow rubber sphere with a three valves "V," "F," "D" (see Figure 4.3). First, valve "V" and the rubber ball are pressed simultaneously, so that a depression is made. Subsequently, with the "F" valve, the liquid is filled up to the right mark, and finally the valve "D" is pressed, whereby the fluid flows out. Also, a hollow plastic tube is used that has a plunger that moves through a sprocket. The bottom tip has a rubber piece that is adjusted to the mouth of the pipette. Turning the gear sucks up the liquid. When the liquid reaches the desired height, the contents can be released into another container by pressing the valve side. In these propipetter models, it is important that the rubber parts are not damaged, either the rubber joint of valve or the rubber piece that fits into the pipette, because they are useless if they cannot hold a vacuum.

c) Use of pipettes: Proper use of a pipette involves the following procedure:

- Insert the pipette (with the conical tip down) into the container to extract a given sample volume.

- Place the pipette filler on the flat end of the pipette and lift the liquid above the upper calibration mark.

- Quickly, remove the pipette filler and place the index finger to seal the top and to prevent tipping, and then proceed to graduate to the top calibration mark.[2]

- Gently and slowly release the pressure exerted by the finger until the liquid begins to descend. Press again when the meniscus of the liquid is zero. If the liquid descends too far, start again.

- Move the pipette to the destination flask.

- Reduce finger pressure to release the number of milliliters required.

- To completely empty a graduated pipette, the finger is removed and the liquid drops by gravity, but do not force the fall of the last drops, they must be in the conical tip of the pipette.

- In graduated pipettes different volumes of liquid can be measured, with a graduated scale.

- The volumetric pipette has a single calibration mark, which can only measure a fixed volume.

d) Automatic micropipette: This is a device for measuring and transferring small volumes, typically from 1 to 1000 μl. Micropipettes are classified into three groups: P1000, useful with volumes of 200–1000 μl; P200, useful with volumes of 20–200 μl; and P20, useful with volumes of 0.5–20 μl.

The liquid is drawn up and expelled through hollow plastic parts (disposable tips) that are thrown away. These are embedded in a tube called a plastic shaft, which is attached to the handle or body of the pipette containing a plunger within a cylinder that causes suction when moved. The body serves as a handle and the top has a plunger button connected to the plunger, which controls the suction and expulsion of the liquid when the thumb exerts pressure. This button has two stops. The body has a button to disconnect the tip from the plastic shaft with a tip ejector arm. There are automatic pipettes of fixed and variable volume. The variable volume in the body has a mechanism for adjusting the volume and a digital display window indicating the selected volume.

- Procedure for pipetting:
    - Set the disposable tip to the plastic shaft by a slight turn without pressure between the pipette and tip.
    - Adjust the volume.
    - Using the thumb, press the plunger button to the first stop and dip the tip into the liquid of 2–5 mm, keeping the pipette vertical.
    - Raise the thumb and let the plunger button return to its initial position.

---

[2]This can also be performed by pressing the emptying point of pipette filler without removing it from the pipette.

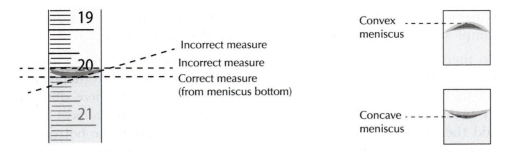

Figure 4.4: Correct meniscus reading in graduated tubes.

— Support the pipette on the inner wall of the container at an angle of between 10 and 45° with the container; slowly press the plunger button to the first position, and afterward to the second, and without releasing the button, empty the pipette by pressing to the second stop.

— Remove the pipette from the container.

— Remove the tip by pressing the ejector button.

Noticeboard 4.3.1

 Practical tips & tricks:

Bring to volume: The dropper is never inserted through the neck of the flask but should be supported on the mouth of the flask at an angle of approximately 75° to the neck. Thus, if by error the glass tube dropper escaped from the dropper bulb, the tube would not fall into the flask. When the level gets closer to the calibration mark, the process has to be done drop by drop. Remember that the volumetric flask has exactly the indicated volume when the meniscus of the liquid is tangential to the calibration mark. In order to verify this tangent (do not make a parallax mistake), the flask must be placed on a flat surface and the eyes at the height of the calibration mark. If an error is made to bring to volume and the meniscus of the liquid exceeds the calibration mark, the procedure must be started again. To avoid further mistakes, do not remove liquid from a flask if it has not yet been brought to volume correctly.

---

e) Volumetric flask: This type of flask, a necked one, narrow, and flat bottomed, is used to accurately measure a given volume of liquid. The neck carries a mark indicating where the liquid level can be read, corresponding to the volume capacity of the flask. The liquid forms a curvature in the neck called the meniscus (see Figure 4.4). When the tangent of the meniscus, produced by the liquid surface and the wall, reaches the level of the mark, the volume indicated by the manufacturer of the flask has been reached. Its main use is to prepare mixtures or solutions of known concentration —for example, samples in UV/Vis spectroscopy and optical-rotatory power.

To prepare a solution in a volumetric flask, proceed as follows: Weigh or measure the volume corresponding to the amount of solute (depending on whether it is solid or liquid), and transfer to a beaker; there dilute little by little with the solvent. Then pour into the flask until almost full, and add the solvent slowly using a Pasteur pipette to match the bottom of the meniscus line with the calibration mark. Next, the flask is capped and shaken to thoroughly mix the solution.

f) Automatic dispensers: These are designed to fit bottles of reagents and solvents, so they are usually sold in different sizes and with different dispensing capacities (> 0.1 ml). The device for exerting pressure is housed in a plunger that is activated by uploading and lowering it (similar to a syringe). The plunger should move smoothly to prevent the entry of air. Also, before dispensing liquid the device should be purged (the plunger without having air inside). Otherwise, the amount of liquid dispensed will not be accurate. To purge the device, the plunger must be refilled several times, dispensing the liquid into another container (which is discarded) until the pump is completely full of liquid. The sample is dispensed in the final container where it will be used (round-bottom flask, Erlenmeyer flask, beaker, etc., and if necessary it should be weighed). Finally, ensure that the last drop in the dispenser is transferred to the final container. This device is particularly suitable for handling strong acids such as H_2SO_4 or HNO_3.

g) Graduated cylinder: The transparent graduated cylinder is made of glass or solvent-resistant plastics such as polypropylene (PP) or polymethylpentene (PMP), allowing relatively accurate volume measurement. The bottom is closed and has a base that supports it. The top is open and usually has a peak that can easily pour the liquid. They may have a capacity ranging from 5 to 2000 ml. When greater accuracy is desired, other material is used, such as volumetric flasks and pipettes.

h) Syringes: These consist of a plunger inserted in a graduated tube with a small opening in one end through which the contents of the tube are ejected and to which a needle can be attached. They are used to inject small amounts of gases or liquids in inaccessible areas or to take samples of the components in these areas. To fill, the plunger is pulled then placed with the needle pointing upwards, the plunger is depressed to expel any remaining air bubbles, and then the liquid is ejected by pressing the plunger.

The disposable ones are usually made of PP. With these syringes, the volume measured is approximate. Also, they are available in borosilicate glass. The most common sizes have a capacity of between 1 and 100 ml. In this case, the volume can be measured with more precision. Finally, precision syringes can measure volumes from 5 μl.

Glass syringes should be washed immediately after use with a suitable solvent to remove any remaining liquid and then washed with deionized water and finally with acetone.

i) Burettes: These are graduated tubes (with subdivisions that, depending on the total volume, reach tenths of a milliliter or less) with a uniform internal diameter and fitted with a stopcock at the bottom. They are used to measure variable quantities of fluids, mainly in volumetric analysis, or dispense a reagent until the reaction end point is reached. Error tolerances of burettes are associated with this volume; for burettes of 10, 25, and 50 ml their associated errors are ± 0.02, ± 0.03, and ± 0.05, respectively.

Liquid should never be measured on the top of the burette, since it is designed to dispense liquid through the bottom. The amount of liquid dispensed is calculated by measuring the difference between the initial and final value.

Noticeboard 4.3.2

 Practical tips & tricks:

Never use an oven to dry precision material, such as pipettes and graduated flasks.

Stopcocks are usually made of glass in Geissler burettes (which are attached by bases). To avoid this drawback, Teflon® can be used, as it is inert and resistant. Another simpler type of stopcock that is often used in Mohr burettes is the Bunsen stopcock, composed of a glass bead in a rubber tube, which, when the tube is pressed, allows the liquid to pass.

The volumes measured with burettes are accompanied by systematic errors, due to the drops affixed to the bottom or small air bubbles located behind the stopcock. Ensure that the drain is not too fast. To avoid errors by default, the end point with a drop hanging from it has to be collected by tapping gently with the receiving vessel (Erlenmeyer). A drop has an approximate volume of 0.05 ml, and missing a drop can have a significant effect on the measurement of small volumes.

4.4. Transfer of solids and liquids

Once weighed, a solid can be transferred to the appropriate container directly from a weighing bottle, boat, canoe, dish, or paper.

When there are no weighing bottles to facilitate the transfer of solids to other containers, they can be easily transferred in a simple piece of weighing glossy paper. Glossy paper containing the solid is folded along the diagonal, then the

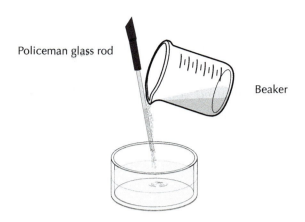

Figure 4.5: Policeman rod used for transferring liquids.

opposite ends of the diagonal between the index finger (inside) and thumb (out) are attached, and finally proceed to transfer.

To transfer a liquid from one container to another, the following devices can be used:

a) Conical funnels: These offer the most common and safe procedure, as spills and splashes are avoided while transferring liquids from bottles or between laboratory bottles and containers (graduated cylinder, beakers, flasks, etc.).

b) Policeman rod: This glass rod has a flexible scraper (natural rubber) or spatula at one end.[3] This implement is used when pouring a solution from one container to another —for example, from a beaker to a separatory funnel (see Figure 4.5).

c) Liquid-transfer pumps: In the case of using large vessels, both manual and electric pumps are often used, but these must be adapted to the type of liquid (corrosive, organic nature, etc.) to be handled.

4.5. Techniques used in chemical reactions

4.5.1. Temperature measurement (thermometers)

In many laboratory processes, the temperature must be measured in both cold and hot reactions. There are different types of thermometers for every need and temperature range. The most common consists of a graduated glass tube containing liquid.[4] In some cases thermometers are supplied with a ground-glass piece to adjust to the distillation head or in other cases with the appropriate adapter (typically a screw cap with a rubber ring). It should be noted that

---

[3]The name comes from the rubber piece, the main function of which is to prevent scratching the glass container with the rod.

[4]Due to the harmful nature of mercury this has been replaced today by a colored liquid.

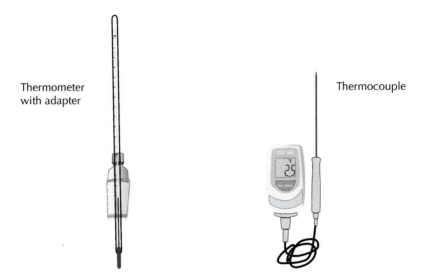

Figure 4.6: Different temperature measurement devices.

the liquid in thermometers does not present discontinuities. These are typically caused when the thermometer reaches temperatures above their use limit. Make sure that the temperature range to be measured is found always within the range of the scale of the thermometer.[5] Today, it is increasingly common to use electronic devices for measuring temperature or thermocouple (see Figure 4.6). These can be connected to magnetic stirrers to allow the temperature of the hot plate to be regulated with more reliability.

4.5.2. Heating techniques

The vast majority of organic substances are flammable and many also volatile, so that the heat of a chemical reaction or dissolution carries risks if the proper precautions are not taken. Therefore, today almost no heating techniques involving direct exposure to flame are used in the Organic Chemistry laboratory (for danger when handling volatile or flammable organic solvents), as in the case of Bunsen burners. Currently, they have been replaced by electrical devices with thermostatic heaters.

a) Bunsen burner: This is used to heat test tubes or to sterilize other types of samples or instruments.[6] A flame is generated at the end of a metal tube by a continuous stream of a gas (butane, propane, or such) that is mixed with the air coming through a hole. The air inlet can be adjusted with a stopcock or by manipulating a slit, so as to regulate the composition of the gas/air mixture and the type of flame. The color of the flame indicates its temperature. A

---

[5]These models are not of clinical thermometer use so they do not have to be shaken to lower the temperature.

[6]In Organic Chemistry these are also used to mold glass rods and make capillaries.

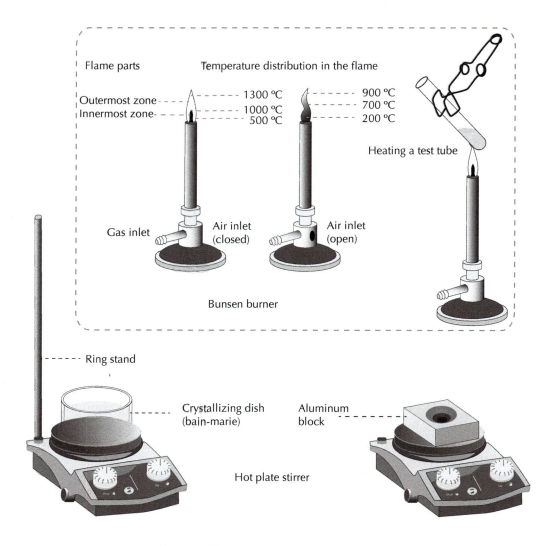

Figure 4.7: Bunsen burner and hot plate.

yellow (dirty flame) indicates low temperature, approximately 300 °C. With increasing proportions of air/gas, the flame turns blue, and, reaching higher temperatures (approximately 700 °C), becomes almost invisible. Although not recommended because of danger, test tubes, for example, are still heated in laboratory tests (see Figure 4.7). The test tubes are supported by a metal or wood clamp, and the approaching flame is slightly inclined (45°) and oriented at a safe direction with occasional shaking.[7]

b) Hot plate: The item most used for heating a reaction is a hot plate with an internal temperature regulator. The majority of hot plates today also include magnetic stirring plates, which can simultaneously be heated and stirred. The container or the apparatus used is placed directly on the heat source of the

---

[7]So that if splashes occur damage is minimized.

plate or can be inserted into a water, oil or sand bath, or instead placed on an aluminum block. In every case, the temperature should be controlled with a thermometer inserted in the bath or on the aluminum block.

Noticeboard 4.5.1

 DANGER!

Never heat an open container with a flammable solvent on a hot plate. Vapors may cause fire on contact with the hot zone or electrical components of the plate.

---

c) Water bath: The simplest bath, which is still used, is a double-boiler water bath, or bain-Marie.[8] This bath is highly effective when temperatures below 80 °C are required. Reaction temperatures between 70 and 80 °C are reached. The water is placed in a crystallization dish and a container with the reaction set in the bath. However, the only drawback is the relatively quick evaporation of the water. Rapid evaporation can be prevented by fitting a piece of aluminum foil over the bath surface.

d) Oil bath: Fluids transfer heat in an efficient and homogeneous way, and therefore they are used in heat reactions. These fluids must be thermally stable and non-flammable within the temperature range used. It is important to know its flash point.[9] The fluid is added to a container, usually a glass crystallization dish, which is placed on a heat source (the same hot plate). The crystallization dish has the advantage of having a short wall, which allows a view of the reaction. Mineral oils or silicones are used as heating fluids.

Noticeboard 4.5.2

 DANGER!

Mineral oil is flammable. Particular care is needed to avoid dripping oil onto the hot plate. Also, if the water is mixed with oil, there is a risk of splashing when the temperature rises above 100 °C. The danger is that water, being denser than the oil, sinks to the bottom and produces spray at high temperatures.

---

[8]The term comes from Mary, the sister of Moses, who was considered the first alchemist.
[9]The minimum temperature at which a mixture of fluid vapors ignites in air.

e) Sand bath: A crystallization dish can be filled with sand instead of fluid. To heat small containers at temperatures not exceeding 200 °C, sand baths are prepared by adding sand to the crystallization dish to a height of about 3 cm.

- Aluminum heats up very quickly.
- It can reach high temperatures.
- The blocks can be cooled quickly if held with tongs and immersed in water.

Table 4.1 lists some materials used in the heat baths, the temperature range commonly used, and its flash point.

Table 4.1: Range of working temperatures reached by various substances used in electrical baths.

Material	Flash point (°C)	Temp. of work (°C)	Observations
Water	-	30–70	—Cheap —Non-flammable, non-toxic —No waste —Evaporates quickly
Mineral oil	113	30–90	—Cheap —Darkens and polymerizes with time —Dangerous[a] —Low flash point
Paraffin	-[b]	55–180	—Cheap —Does not degrade over time —Dangerous[a]
Oil silicone	150–350	30–230	—Wide temperature range —Relatively stable with use —Dangerous[a]
Sand	-	30–500	—Cheap —It is not flammable —For practical purposes has no temperature usage limit

[a] It can produce projections, if water is spilled accidentally when hot.
[b] Varies by source.

f) Aluminum heat blocks: This is an accessory that is placed on the hot plate, producing an effective heat transfer between it and the reaction container (similar to a heating mantle). There are various designs, some hemispherical to fit round-bottom flasks and, further improving stability, some with different orifices in the block to fit different types of flasks. In this case, it is possible to heat several flasks or vials at a time. Aluminum heat blocks have a number of advantages:

g) Thermostat bath: This container for water or oil can be heated by a resistance, and stable temperature can be maintained for long periods of time.

Noticeboard 4.5.3

 DANGER!

Never heat fluid bath to a temperature higher than 20 °C below its flash point.

---

h) Heating mantles: These have a metal housing with a cavity in which there is a bed of glass fiber or ceramic material that fits the shape of the container, which can be kept in direct contact with the glass container (round-bottom flask; see Figure 4.8). It has an electrical resistance that controls the temperature by a power regulator connected to a thermostat. Heating mantles are provided with magnetic stirrers and can be heated while stirring. The most common are specially designed for heating round-bottom flasks from 50 ml to 5 l or even higher. This heating system is suitable for flasks above 500 ml capacity, particularly for distilling solvents in large quantities, where required temperatures are relatively high, in a stable manner. For smaller flasks the heating plates are used more frequently.

The fastest way to regulate the rate of warming a heating mantle is by changing the distance between the cavity and the flask, rather than by manipulating the thermostat. This separation can be changed quickly by moving the flask up and down by means of a suitable support mounted outside the mantle. Although heating mantles are very easy and safe to use, extreme caution must be taken to avoid at all costs dropping any chemicals (including solvents) on them, for two reasons:

- The temperature reaching the resistance can cause the thermal decomposition of substances generating toxic fumes or fire.
- Device malfunction may occur and even a short circuit.

Less common in the Organic Chemistry laboratory are heating mantles designed to heat beakers. Unlike the previous type, these have a cylindrical shape instead of a hemispherical one.

i) Microwave equipment: An alternative to using the above devices is to use microwave equipment, which can provide thermal energy to a reaction with the added benefit of saving time and energy.

This way of heating uses the property of some molecules to transform electromagnetic energy (near 900 MHz) into heat. This causes the rotation of

Figure 4.8: Heating mantle.

the dipoles within the liquid, causing the polar molecules to align and then relax in the oscillating field of electromagnetic radiation. When the rotation energy of these dipoles dissipates, the liquid heats, so that, when heating occurs inside the liquid, no heat is transferred from the container and thus the liquid is at a higher temperature than the container. However, the disadvantage is that, in order to reproduce the results properly, sophisticated equipment is needed, unaffordable in most teaching laboratories.

j) Oven: Used mainly for melting metals and the synthesis of ceramic materials, ovens are also used to activate molecular sieves and to produce anhydrous salts.

4.5.3. Cooling techniques

Sometimes it is necessary to cool a reaction below r.t. —for example, in highly exothermic reactions where it is necessary to eliminate or alleviate the detached heat or because the reactants or products are thermally unstable. In other cases, a solution needs to be cooled simply to promote the crystallization of a given compound or in some cases to facilitate condensation of a liquid. Procedures used in the laboratory for cooling include:

a) Ice bath: This is the easiest option. A crystallization dish is filled with ice, which must be crushed and mixed with water to facilitate contact with the flask to be cooled.

b) Freezing mixtures: These tend to be binary blends that acquired a lower temperature than the components separately. The simplest of all is an ice/salt mixture. In this case, it is difficult to set the temperature, since it depends on the ratio of ice to salt. Table 4.2 lists different refrigerant mixtures classified by temperature.

Table 4.2: Range of working temperatures reached by different freezing mixtures used in the laboratory.

T (°C)	Freezing mixture[a,b]	T (°C)	Freezing mixture[a,b]
13	p-Xylene/CO_2(s)	−56	n-Octane/CO_2(s)
12	Dioxane/CO_2(s)	−60	Isopropyl ether/CO_2(s)
6	Cyclohexane/CO_2(s)	−61	Chloroform/CO_2(s)
5	Benzene/CO_2(s)	−72	EtOH/CO_2(s)
2	Formamide/CO_2(s)	−77	Butyl acetate/CO_2(s)
		−78	Acetone/CO_2(s)
0	Ice	−83	Propyl amine/CO_2(s)
−5 to −20	Ice/NaCl	−83.6	Ethyl acetate/N_2(l)
−10.5	Ethylene glycol/CO_2(s)	−89	n-Butanol/N_2(l)
−12	Cycloheptane/CO_2(s)	−94	Hexane/N_2(l)
−15	Benzyl alcohol/CO_2(s)	−94.6	Acetone/N_2(l)
−22	Tetrachloroethylene/CO_2(s)	−95.1	Toluene/N_2(l)
−22.8	CCl_4/CO_2(s)	−98	MeOH/N_2(l)
−25	1,3-Dichlorobenzene/CO_2(s)	−100	Diethyl ether/CO_2(s)
−29	o-Xylene/CO_2(s)	−104	Cyclohexane/N_2(l)
−32	m-Toluidine/CO_2(s)	−116	EtOH/N_2(l)
−38	Heptan-3-one/CO_2(s)	−116	Diethyl ether/N_2(l)
−41	Acetonitrile/CO_2(s)	−131	n-Pentane/N_2(l)
−42	Pyridine/CO_2(s)	−160	Isopentane/N_2(l)
−47	m-Xylene/CO_2(s)	−196	N_2(l)

[a] CO_2(s) = dry ice.
[b] N_2(l) = liquid nitrogen.

If the value of the reaction temperature is not critical, the cooling mixtures can be replaced by others that yield a similar temperature and where the solvents used are less toxic or cheaper.

For the reactions usually performed in a practice laboratory, baths with temperatures above 0 °C can be replaced by water/ice, monitoring the temperature with a thermometer. For temperatures from 0 to −20 °C range, the ideal solution, due to its low price and availability, is to use ice or ice/salt mixtures.

If temperature must be maintained for long periods, it is desirable to isolate the vessel containing the freezing mixture and the flask in which the reaction is performed.

The freezing mixtures, especially those involving more or less toxic or hazardous solvents, are replaced by recirculating chillers.

c) Thermostatic cooling bath: The ideal procedure to maintain the reaction at a low temperature is to use a device called a cooling bath, usually constituted by a metal tank filled with a liquid that does not freeze at the operating temperature, where the container is immersed with the reaction.

d) Wort chillers: These have a cooling unit to produce cold but with an external accessory (cooling coil, similar to a cold finger) that can be used to generate localized cold in other types of reaction vessels.

e) Recirculating chillers: Referred to as closed loop coolers, these are widely used for driving liquid at low temperatures in the usual water condensers. Their most common use is in rotary evaporators and reflux of low-boiling-point solvents, and it is usually filled with a freezing mixture similar to antifreeze used in automobiles.

4.5.4. Stirring, mixing, and grinding

Most organic reactions require stirring for various reasons, such as:

- To dissolve one of the components.

- To mix reagents properly when they are poured within the reaction.

- To make mixtures of immiscible solvents.

- To avoid clumping by the appearance of precipitates.

- To generate homogeneous boiling.

- To facilitate the course of reactions in heterogeneous phase due to the presence of:

  - Solid-phase reagents.
  - Solid-phase catalysts.
  - Exchange resins.

The most common stirring methods in the laboratory are the following:

a) Magnetic stirring: This has become the most popular stirring system in the Organic Chemistry laboratory. The vessel (Erlenmeyer flask, beaker, etc.) is placed and centered above the hot plate stirrer. A magnetic stir bar, which is coated with an inert material (polypropylene or Teflon®), is added inside the container. The stir bars are manufactured with different forms according to the application (see Figure 4.9). To prevent accidents caused by the vibration of the stirring, firmly fix the flask with clamps and regulate the speed of rotation appropriately. If this is excessive, splashing may occur or even a blow that could break the glass flask, or, if it is too slow, stirring may be ineffective, especially in the case of reactions that are in heterogeneous phase or highly viscous.

b) Mechanical stirring: This is used when large quantities of reagents (>1,000 ml) are handled or in cases of viscous mixtures where magnetic stirring is ineffective.

Figure 4.9: Various models of magnetic stir bars.

The assembly used consists of a flask, usually with several necks to adapt a number of other elements. A glass or a metal rod with a system of metal blades, glass, ceramic, Teflon®, etc. is inserted through the middle neck of the flask and is driven by an adjustable-speed motor. The flask is adapted with an joint stirrer bearing that allows the rod to freely rotate and that prevents solvent vapors from escaping the bath, whether hot or cold.

However, these mechanical stirring devices are rarely used in a teaching laboratory of introduction to experimental Organic Chemistry.

Another type of material used for mixing solids is mortars (see Figure 3.3, p. 57).

4.6. Reflux

The great majority of chemical reactions are performed in solution. Most organic solvents are volatile and many flammable, so that if a reaction is warmed in an open container, the solvent evaporates shortly afterward, making contact with the heat source and causing the consequent danger of fire. To heat a reaction safely in the necessary time, the technique called reflux is used (see Figure 4.10). A round-bottom flask with ground-glass stoppers coupled to a reflux condenser is employed. The solution with the reactants is placed in the round-bottom flask, which is set on a heat source. After heating begins, when the solvent starts to evaporate, the vapors ascend to the condenser, where they cool, condense, and fall back to the flask, with two effects:

- The volume of the solution does not vary.

- The reaction temperature is kept constant, corresponding to the b.p. of the solvent at the pressure at which the experiment is performed.

Thus, the reaction can be heated safely for long periods of time. The boiling should be homogeneous to avoid splashing; for this, magnetic stirring is used.

4.6.1. Condenser types

There are two groups of condensers that can be used in this technique:

a) Air-cooled condensers: These are the simplest and consist of a hollow tube that is coupled to the flask using a conical joint and are cooled by ambient air. Others have a female ground joint for coupling in order to mount other devices via adapters (thermometer, addition funnels, inert gas, etc.; see Figure 4.11).

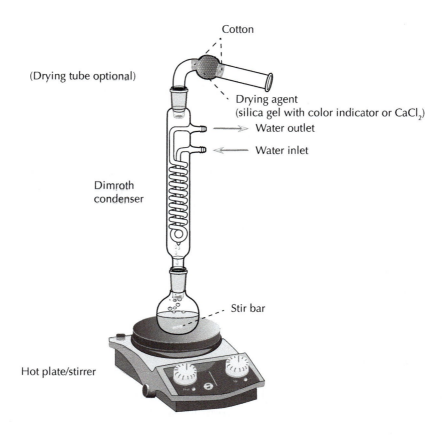

Figure 4.10: Basic reflux setup.

b) Water-cooled condensers: The most efficient and safe, these condensers have
   a double wall or indoor coil through which water circulates to cool the ascending vapors. To circulate the water, these condensers have connectors:
   i.e., one water input and the corresponding water output to attach silicone
   rubber laboratory tubing. The connection to the water circuit is such that
   water input is from the closest to the mouth of the flask area. The connection
   to the water circuit is conducted in countercurrent with the rising vapors (see
   Figure 4.12).

   Other types of coolants can be used instead of water. Allowing the condenser
   to cool below 0 °C, these can be cooled in an external cooling pump. Different
   types of water-cooled condensers are available on the market. Below, some
   of the most common types are described.

   - West, Liebig, and short condensers: These condensers are formed by a
     straight tube surrounded by a chamber through which a coolant liq-
     uid passes. There are two water intakes (bottom inlet and top outlet).
     Coolant circulation causes the ascending vapor to cool and condense.
     The only difference between these three types of condensers are the di-
     mensions and the short condenser where the ground joints are integrated
     into the tube by decreasing the length of the device (see Figure 4.12).

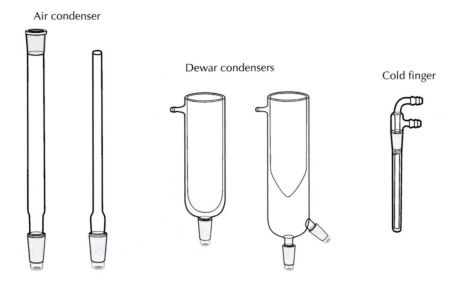

Figure 4.11: Other types of condensers.

These condensers are usually used in simple distillation equipment, the most common being the Liebig condenser.

- Allihn condenser: To improve the efficiency of the Liebig condenser, the inner tube surface is increased in spherical form, only half of the tube in the Allihn-Dronbitter condenser or in the whole tube in the Allihn condenser (see Figure 4.12). The Allihn condenser is optimum for laboratory-scale reflux.

- Davies condenser (double surface): This has a larger contact surface than in the Liebig condenser of equal size. It is especially suitable for reflux of low b.p. liquid due to double surface, inside and outside.

- Graham and Hartzler condensers: These have a spiral tube that runs the entire length of the condenser, resulting in a condensation surface three-fold that of a Liebig condenser of similar dimensions (see Figure 4.12). They are extremely efficient in the reflux position, where countercurrent cooling can be arranged.

  There are two possible configurations for a Graham condenser. In the first, rarely used, the spiral leads the coolant, and condensation takes place on the outside of the spiral. This configuration maximizes the flow capacity, because vapors can flow over and around the coil. In the second configuration, the outer tube contains the coolant, and condensation occurs within the coil. This configuration maximizes the collection of condensate, since all vapors flow through the entire length of the spiral and thus have prolonged contact with the coolant.

  The Hartzler condenser is similar to the Graham one but with a mixed design, with a bulbous portion as in the Allihn type.

- Spiral and Dimroth condensers: These are similar to the Graham condenser, but with a double inner coil to the coolant, for the coolant inlet and outlet being at the top. Vapors pass through the outer tube from the bottom up. Dimroth condensers are more effective than conventional spiral condensers and thus are one of the most widely used models. This type of condenser has been adapted to most existing rotary evaporators (see Section 4.10, p. 117).

- Triple jacket: The triple-jacketed design results in a highly efficient condenser.

- Hopkins Friedrichs condenser: Also known as spiral or coil condensers. This consists of a large inner helix-type cold finger set inside a larger diameter cylindrical casing, which increases the effective area of condensation. The refrigerant flows through the inner tube, and the turbulent flow increases heat transfer and efficiency. It has a common feature of presenting a ground-glass mouth that can be fitted with different components, as it takes an inert atmosphere.

c) Other types of condensers (cold finger and Dewar flask, see Figure 4.11): In addition to the air or water condensers, other types of devices to generate/store cold exist on the market. Cold finger is used to generate cold in a given area or localized surface and is typically used in a sublimation apparatus or as a compact version of a reflux condenser to carry out chemical reactions or distillations. The Dewar flask is an insulated container used especially to store liquefied gases (e.g., dry ice, mixtures of acetone/dry ice, etc.), having a double wall with a vacuum between them and silvered surfaces facing the vacuum.

4.6.2. Solvent choice in a reflux reaction

A critical aspect when carrying out a reaction is the choice of solvent. Depending on the type of reaction, the reagents used, and the temperature at which the reaction is carried out, different solvents can be used. Sometimes one of the reagents, because of its properties, has a dual role, acting also as a solvent. However, most commonly, a solvent should be chosen, or even mixtures are used. In refluxing reactions, the solvent selection is critical, since it must meet certain conditions and have characteristics such as:

- Adapting its properties to the type of reaction (polarity, b.p., protic or non-protic character).

- Dissolving the reactants and products at b.p.

- Being inert to the reactants and products at b.p.

- Being easy to remove once the reaction is finished.

In Table 4.3 the most significant properties of the solvents commonly used in the Organic Chemistry laboratory are indicated [1].

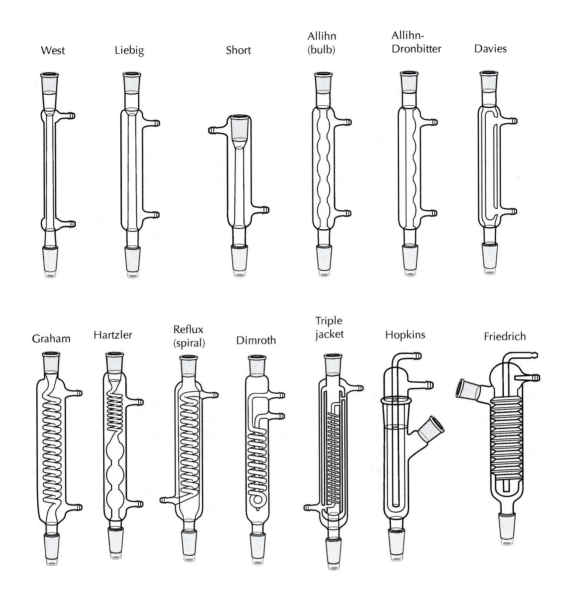

Figure 4.12: Different types of water-cooler condensers.

Table 4.3: Physical properties of the main solvents used in the Organic Chemistry laboratory.[a,b]

Solvent	m.p. (°C)	b.p. (°C)	D_4^{20} (mg/ml)	n_D^{20}	ϵ	R_D	μ (Debye)
Ethyl acetate	−84	77	0.901	1.3724	6.02	22.3	1.88
Acetone	−95	56	0.788	1.3587	20.7	16.2	2.85
Acetonitrile	−44	82	0.782	1.3441	37.5	11.1	3.45
Acetic acid	17	118	1.049	1.3716	6.15	12.9	1.68
Water	0	100	0.998	1.3330	80.1	3.7	1.82
Bromobenzene	−31	156	1.495	1.5580	5.17	33.7	1.55
Cyclohexane	6	81	0.778	1.4262	2.02	27.7	0.00
Chlorobenzene	−46	132	1.106	1.5248	5.62	31.2	1.54
Chloroform	−64	61	1.489	1.4458	4.81	21	1.15
Dichloromethane	−95	40	1.326	1.4241	8.93	16	1.55
Diethyl ether	−117	35	0.713	1.3524	4.33	22.1	1.30
DMF	−60	152	0.945	1.4305	36.7	19.9	3.86
1,2-Dichloroethane	−36	84	1.253	1.4448	10.36	21	1.86
DMSO	19	189	1.096	1.4783	46.7	20.1	3.90
DME	−68	85	0.863	1.3796	7.2	24.1	1.71
1,4-Dioxane	12	101	1.034	1.4224	2.25	21.6	0.45
EtOH	−114	78	0.789	1.3614	24.5	12.8	1.69
HMPA	7	235	1.027	1.4588	30.0	47.7	5.54
Isopropanol	−90	82	0.786	1.3772	17.9	17.5	1.66
MeOH	−98	65	0.791	1.3284	32.7	8.2	1.70
tert-Butyl alcohol	26	82	0.786	1.3877	10.9	22.2	1.66
Nitrobenzene	6	211	1.204	1.5562	34.82	32.7	4.02
Nitromethane	−28	101	1.137	1.3817	35.87	12.5	3.54
o-Xylene	−25	144	0.880	1.5054	2.57	35.8	0.62
Pyridine	−42	115	0.983	1.5102	12.4	24.1	2.37
Carbon disulfide	−112	46	1.274	1.6295	2.6	21.3	0.00
CCl$_4$	−23	77	1.594	1.4601	2.24	25.8	0.00
THF	−109	66	0.888	1.4072	7.58	19.9	1.75
Toluene	−95	111	0.867	1.4969	2.38	31.1	0.43
Trichloroethylene	−86	87	1.465	1.4767	3.4	25.5	0.81

[a] Melting and boiling points, density and refractive index at 20 °C, dielectric constant, molar refraction, and dipole moment 20 °C.
[b] Data taken from ref. [1].

According to their different nature and structure, solvents can be classified into the following classes:

a) Polar protic solvents: These have hydrogens bonded to oxygen, so the general formula is ROH. This group includes water, alcohols, and carboxylic acids.

- Water (b.p. 100 °C): Most organic compounds are insoluble in water. By contrast, it is a very good solvent for ionic substances or very polar compounds. In many reactions, especially hydrolysis, mixtures of water may be used with other solvents, giving homogeneous solutions with low-molecular-weight alcohols, acetone, THF, DMSO, DMF, or acetic acid or heterogeneous mixtures such as toluene. In all cases, water retains its properties. Water is incompatible with substances yielding rapid hydrolysis reactions such as hydrides, organometallic compounds, some halides, or carbanions.

- Alcohols and MeOH (b.p. 68 °C), EtOH (b.p. 78 °C), propan-1-ol (b.p. 97 °C), butan-1-ol (b.p. 118 °C), or *tert*-butanol (b.p. 82 °C) are used in reactions where it is necessary to dissolve highly polar compounds, provided they do not react with the starting materials or with the final products. In conventional esterification reactions (acid + alcohol), they are used as the reagent and solvent. They are incompatible with strong bases due to the formation of alkoxides, and strong acids can give dehydration reactions (especially in heat), among other substances, which are incompatible with both substances. Special mention for the reaction with concentrated HNO_3 is due because these substances generate alkyl nitrates $(R-O-NO_2)$ by a violent and potentially explosive reaction.

- Carboxylic acids: Acetic acid is the most commonly used of these (b.p. 118 °C). They dissolve major organic substances, being miscible with water in all proportions, and with a $pK_a = 4.7$. Thus, it is a weak acid and can be used when these conditions are compatible with the reactants and products. Carboxylic acids are unable to protonate alkenes but are useful to stabilize polar reaction intermediates.

b) Aprotic polar solvents: These have a permanent dipole moment although they have no -OH groups. Such solvents are especially useful in S_N2 reactions because they stabilize the charge separation in the transition states. This group includes the following ones:

- Acetone (b.p. 56 °C): This is a good solvent for compounds of medium to high polarity and is easily removed by evaporation. It is miscible in all proportions with water. Some salts such as NaI or CsF are solubilized. It is typically used in nucleophilic displacement reactions, where the nucleophile is part of a soluble salt and the leaving group forms part of insoluble salts, and thus a measure of the reaction progress is that under reflux a precipitate appears. It is incompatible with strong bases, since stabilized carbanions are generated.

- DMSO (b.p. 189 °C): This dissolves most organic and inorganic substances. As in the case of acetone, it is miscible in all proportions with water and is an excellent solvent for S_N2 reactions. It is also incompatible with strong bases. It has a nucleophilic character. One drawback is that it cannot be removed by reduced pressure (rotary evaporator) due to the high b.p.; therefore, the reactions in which it is used are usually processed by adding ample water and extracting the reaction products with an organic solvent immiscible with water (see Section 4.6, p. 89).

- DMF (b.p. 152 °C): Its properties are very similar to DMSO, but without a nucleophilic character. It is able to dissolve most of organic substances and many inorganic compounds. It does not react with strong bases, including hydrides. Given its high b.p., it is not usually removed by evaporation; otherwise abundant water is added and the product extracted with a mixture of toluene/diethyl ether (3:1).

c) Ethers of medium to low polarity: These are immiscible with water.

- Diethyl ether (b.p. 35 °C): This is a solvent widely used in Organic Chemistry. Having a very low b.p. and being highly flammable, it is only occasionally used in reflux reactions. It solvates metal ions, although not in ionic compounds, and therefore it is used in organometallic reactions.

- *tert*-Butylmethyl ether (b.p. 55 °C): Less volatile than diethyl ether, it presents the same applications and is safer.

- THF (b.p. 67 °C): This is one of the most commonly used solvents in Organic Chemistry. It has a great ability to dissolve organic compounds and to solubilize many inorganic substances, although ionic salts are mostly insoluble. It can easily be removed by evaporation. It is miscible with water, so that in many reactions it acts as a cosolvent to solubilize insoluble or partially water-soluble compounds. Lewis acids are well solvated by THF (and other polar substances), which is widely used in reactions including organometallic compounds.

d) Acetonitrile (b.p. 82 °C): This is an aprotic and very polar solvent that has characteristics and the ability to dissolve similar to alcohols.

e) Ethyl acetate (b.p. 77 °C): This is not a common solvent used in a reflux reaction. Although incompatible with acids or bases, especially when in contact for long periods of time, it is widely used in recrystallization.

f) Pyridine (b.p. 115 °C): This aromatic amine is quite nucleophilic. Given its basic character, it is employed in reactions that require the presence of a base. It is toxic and gives off an unpleasant odor, and therefore its use is commonly avoided when possible.

g) Chlorinated solvents:

- CH_2Cl_2 (b.p. 40 °C), being water immiscible, is one of the most widely used solvents in Organic Chemistry. It is incompatible with strong bases. Its b.p. is somewhat low, so its use is limited in reflux reactions.

- CCl_4 (b.p. 78 °C): It can dissolve most organic substances and is miscible with organic solvents, including EtOH. It has the advantage of being non-flammable.

h) Non-polar or low-polarity solvents: These have a low dielectric constant and are immiscible with water.

- Aliphatic hydrocarbons: Most notable are the commercial hexane (mixture of isomers of C_6H_{14} formula and b.p. 69 °C) and petroleum ether (hydrocarbon mixture including b.p. in a range of 30–60 °C). Quite inert, they do not react with acids or bases and they are very apolar, dissolving organic molecules of very low polarity. Their use as a solvent for reactions at reflux is quite limited.

- Aromatic hydrocarbons: The most common are benzene (b.p. 80 °C) and toluene (b.p. 111 °C). Benzene is almost forbidden as a solvent because of its carcinogenicity. It has been replaced by toluene for sharing similar properties as solvents.

4.6.3. Reflux under anhydrous conditions

Many reactions carried out under reflux are incompatible not only with water but also with ambient moisture. As described in Section 4.6.4 (see p. 98), the condenser would be open at its upper end, meaning that the reaction crude is exposed to ambient humidity during the reaction time. To avoid this disadvantage, drying tubes are used. These pieces of glassware are bent glass tubes with a single ground joint for coupling to the condenser (see Figures 4.13 and 4.10) or a trap safety. A certain amount of solid that acts as a drying agent to prevent moisture from entering the assembly is inserted inside. Furthermore, no seal is needed to prevent overpressure generated inside while the reaction is heated. Alternatively, a syringe can be used with the plunger withdrawn and filled with desiccant between two cotton balls or glass wool. The condenser is covered with a septum, and the syringe with the desiccant provided with a needle prick in the septum, taking care in this case to make another empty needle puncture in order to prevent overpressure.

- Preparing a drying tube: A drying tube is prepared by inserting a wad of cotton through the opposite side to ground-glass joint to prevent the drying agent from falling into the apparatus. Then the drying tube is filled with drying agent, and a cotton plug is inserted to keep the drying agent from spilling.[10]

  The setup should not be very compact, because if the air does not freely circulate, the drying tube acts as a plug and increases the pressure inside the apparatus with the consequent danger of overpressure.

  The types of drying agents most used are:

  - Anhydrous calcium chloride ($CaCl_2$), traditionally used as a drying agent, has great capacity to retain water to the point that if left in

---

[10] When the drying tube (filled with desiccant) is not in use, it is convenient to store with two stoppers or sealed with Parafilm® at their ends.

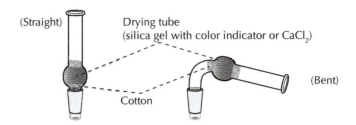

Figure 4.13: Different types of drying tubes.

contact with air it is deliquescent.[11] Thus, the drying tube has to be prepared again, virtually every time it is used.

– Drierite®, the trade name of a solid form of granules composed of 97% anhydrous calcium sulfate and 3% cobalt chloride on the solid surface, has a bluish color that turns pink with hydratation, indicating that the drying agent is no longer effective. This is because cobalt chloride $(CoCl_2)$ is anhydrous blue and when hydrated turns violet $(CoCl_2 \cdot 6H_2O)$.

– Silica gel with a moisture indicator, has an operation mode similar to that of Drierite®, but the solid that retains moisture is silica gel, and it also comes in granular form.

The drying tubes filled with Drierite® or silica gel with a moisture indicator have the advantages that the color indicates when drying efficiency is decreasing and that they can be recycled. Once hydrated, they are placed in the oven to regain the intense blue color characteristic of the anhydrous desiccant, which can be used repeatedly.

4.6.4. How to assemble a reflux apparatus

• First, a clamp nut is fixed to a support or to the metal bars of the lab table or fume hood.

• The solution with the reactants is transferred to a container with a ground-glass mouth, usually a round-bottom flask, using a funnel or conical solid funnel to prevent soiling the ground-glass mouth. If soiled, it must be cleaned carefully using a piece of filter paper.

• It is very important to maintain homogeneous boiling without liquid splashing. For this, it is necessary to use magnetic stirring over hot plates that include this device and to place a magnetic stir bar inside the flask. If the hot plate does not have magnetic stirring, then, before heating, a small piece of ceramic material (boiling chip, boiling stone, or anti-bumping granule) can be placed in the flask to serve the same function of maintaining homogeneous boiling without splashing.

---

[11]Property of bodies, to absorb atmospheric moisture to form an aqueous solution.

> **Noticeboard 4.6.1**
>
> **DANGER!**
>
> Whenever a container is heated, it must never be completely closed, or the volume of liquid must not exceed 50% of the total volume of the container chosen.

- It is placed on a heat source (heating plate or heating mantle), adjusting the mouth of the flask with a keck clip.

- A condenser is coupled to this container. The water condenser called a Dimroth condenser is most commonly used and it consists of a tube fit to an inner grinding coil through which water circulates. The refrigerant is fixed with another clamp to the corresponding support, and tubing is connected to the water circuit.

- The reflux condenser can be replaced by a straight condenser or another condenser, although this lowers the efficiency. Then, if the reaction conditions require a drying tube, it is coupled to prevent moisture from entering the reaction.

> **Noticeboard 4.6.2**
>
> **Practical tips & tricks:**
>
> Before heating:
> - Make sure that everything is securely fastened.
> - Check that the stir bar or the boiling chip has been placed in the flask.
> - Check that the ground glass joints are tightly sealed (using silicon) to prevent solvent vapors from escaping.
> - Check that the water flows through the reflux condenser.

- Once the assembly is finished, the reaction is heated. When the flask is heated, solvent vapors are generated. These vapors, upon contact with the walls of the inner tube cooled with water, lose their heat, condense as a liquid, and fall back into the flask, without the consequent loss of solvent. Note that a "reflux ring" forms at a certain height of the condenser. Thus, the reaction is refluxing. Moreover, in this way, not only does the reaction run at a constant temperature without loss of solvent but also the temperature reached coincides with the b.p. of the solvent (see Figure 4.10).

- When the reaction is finished, do not remove the system until the reflux has ceased.

4.6.5. Reflux with addition of reagents

In many reactions, the reagents need to be added gradually within a given time interval. For the simplest reactions performed at r.t., it suffices to uncover the flask, and add a solid with a spatula, or, if it is a reagent solution, it can be added, for example, using a dropper. In other cases, a syringe can be used with the reagent or the corresponding solution provided with a needle and the liquid placed into the container by puncturing a septum.

However, this is often not possible if the reaction is carried out, for example, under reflux. It would force the device to be assembled hot, each time a reagent addition is needed. Also, the reflux cannot be opened or closed if the reagents or products are sensitive to moisture or oxygen.

For this purpose, the "dropping funnel" is used (see Figure 4.14). These are vessels with two ground-glass settings: one male, for coupling to the mouth of a flask equipped with a stopcock that regulates the drip, and a female on top to cover it with a glass stopper or Teflon®. In many cases, these funnels have a scale on the body to control the flow more easily. A variant of these devices is called "dropping funnel with a pressure-equalizing tube." These have a hollow glass tube that connects the top of the funnel with the space between the stopcock and the ground joint. Thus the pressure between the upper and lower parts of the funnel are balanced, facilitating the dripping inside the flask. This device is particularly suitable for reactions in closed flasks. For reflux apparatuses, the flask to use has at least two ground mouths, to which the condenser and addition funnel are connected, or, if it has a single mouth, a Claisen adapter is used.

4.7. Isolation and purification of solids

4.7.1. Filtration

This is the process by which a solid in suspension in a liquid is separated. In the Organic Chemistry laboratory, two main types of filtration are employed, gravity filtration and vacuum filtration.

a) Gravity or simple filtration: This is used to separate a solid, which normally will separate from a liquid of interest. The material needed to perform this filtration is a conical funnel, a metal ring to attach the funnel, a fluted filter paper, and an Erlenmeyer flask to collect the liquid after filtration. The solid is retained in the filter paper fitted into the conical funnel, through which the solution is recovered.

The fluted filter paper is prepared from a square piece of filter paper by cutting it to fit a circle. Then the paper is folded as shown in Figure 4.15. The purpose of the pleats is to facilitate passage of liquid through the paper

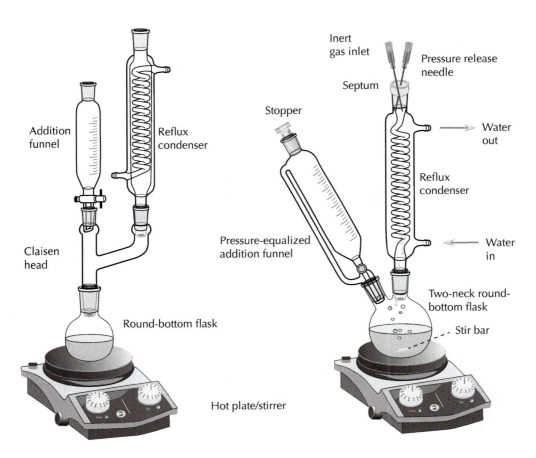

Figure 4.14: Reflux with the addition of reagents. Addition funnel with a pressure-equalizing tube, a two-neck flask, and inert gas (right) and with a neck flask and a Claisen head (left).

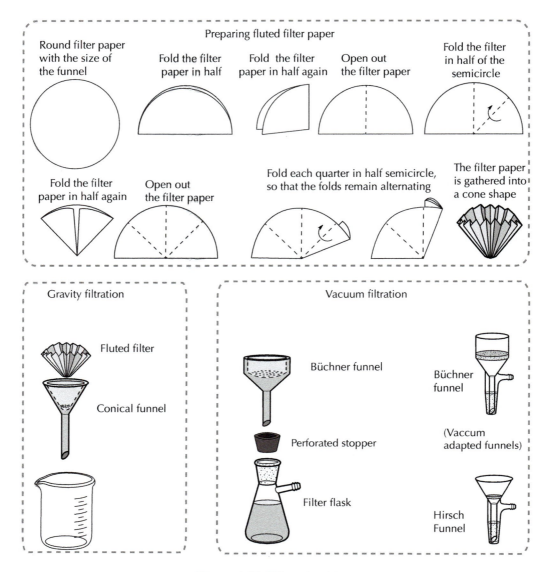

Figure 4.15: Filtration kit.

by increasing the effective area of the filter paper. The filter size must not protrude above the edge of the conical funnel.

Procedure for filtering:

- The metal ring is fixed to the corresponding support and the conical funnel is introduced.
- The height of the ring is regulated to the outlet pipe of the funnel and placed 1 or 2 centimeters below the mouth of the collector.
- The fluted filter paper is prepared and placed in the conical funnel.
- The filter is wet with the same solvent mixture to be filtered.
- The liquid is poured into the filter without allowing it to overflow. This can be done using a policeman rod (see Figure 4.5).
- The liquid is allowed to fall by gravity onto the collector.
- Finally, a small amount of solvent is added to wash the remaining solid in the filter.

b) Vacuum filtration: It is used mainly for separating a solid suspended in a solvent when a solid is desired. The material needed for a vacuum filtration is the following:

- Büchner funnel: This is a cylindrical porcelain or plastic sieve having small holes on a flat bottom (see Figure 4.15).
- Filtration flask: This resembles an Erlenmeyer in its conical shape, but on top protrudes a tube topped by a tapper for connection to a vacuum. A disk of filter paper is placed in the Büchner.
- Rubber cone adapter: This is placed between the Büchner and filtration flask so that they fit tightly when the vacuum is made.
- Vacuum trap: This device is inserted between the vacuum source and filtration flask to prevent liquid from remaining in the filtration flask from passing to the device that generates the vacuum (water aspirator pump, etc.) and the reverse process in water pumps; that is, if a vacuum fluctuation occurs water passes through the water pump. This floods the filtration flask and later the Büchner with the solid.

The filtration flask should be fixed to a support by means of clamps and connectors with a support rod to prevent leakage.

Filtration method:

- A disk of filter paper is cut, making sure the filter is the right size without touching the walls of the Büchner[12] or leaving any holes uncovered.

---

[12]The Büchner funnel has a filter plate with large holes, so that circular filter paper should be placed to remain totally flat and cover all these holes. The Büchner funnel can be used as a template, marking the perimeter with a pencil on a piece of filter paper for precise cutting.

- Once the filter paper disc is placed, the Büchner is embedded in the rubber cone and placed on the filtration flask, securing them with a clamp and connector with a support rod.
- A vacuum trap is prepared and is fixed with a clamp and connector with a support rod at a suitable distance so that they can connect the vacuum tubing to the vacuum source and this to the filtration flask.
- The paper disk is moistened with a few drops of the same solvent mixture to be filtered so that it is completely stuck to the bottom of the Büchner.
- The mixture to be filtered is poured into the Büchner, and just as in gravity filtration, the solid is extracted with the help of a glass rod.
- All the liquid is poured out, using more solvent if necessary to move the remaining solid into the vessel containing the mixture, checking at all times that the filter has no trace of solid.
- Once the filtration is complete, the filtration flask tubing is disconnected from the vacuum, and then the vacuum is closed. It is very important to carry out the operation in order, because if done in reverse order the tap water can enter the filtration flask due to low pressure, and the filtration will have to be repeated.

Vacuum filtration can also be performed for disposable solids —for example, when gravity filtration is slow.

When the solids to be filtered have a very small particle size, the filter paper might not have a small enough pore size. In these cases, cylindrical funnels with a filter plate or conical funnels, called Hirsch funnels, are used.

4.7.2. Centrifugation

Centrifugation is a technique that, by rotating the sample at high speed, separates solids suspended in a liquid (or liquids of different densities). The solids are housed in special centrifuge test tubes or in plastic containers (such as Eppendorf tubes). Rotation at very high speeds (strong centrifugal force) ensures that the denser components of the mixture settle to the bottom of the tube, leaving the less dense components closer to the mouth.

The centrifuge tubes are placed in compartments called rotors, which must be filled with the same amount of sample so that they have the same weight[13] (see Figure 4.16). There are many types of centrifuges (table, floor, microhematocrit, etc.) adapted to different types of applications and samples. There is a wide range of centrifuge tubes, both plastic and glass, with different sizes and designs, often with a thickness greater than that of test tubes and generally with a capacity of 1–25 ml. They are made of different materials such as glass or plastic polymers.

---

[13]This is to balance the rotor, because unbalance at high speeds may cause damage in the centrifuge shaft.

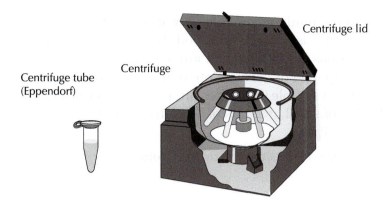

Figure 4.16: Centrifuge.

Centrifugation generates much higher gravity forces in certain periods of time, allowing the separation of precipitates, suspended cells in biological fluids, etc., more efficiently than with conventional filtration. Although not a filtration technique, it can be used to separate impurities that are suspended within a liquid.

4.7.3. Recrystallization

In a typical laboratory experiment, a solid that is separated from a reaction crude is usually accompanied by impurities, so that purification is necessary. By the technique called recrystallization, many solids can be purified using pure solvents or solvent mixtures. Recrystallization is based on the different solubility that a solid substance presents in a solvent at r.t. or when the solvent is hot. The recrystallization process is carried out with product loss, so the overall reaction yield is affected.

The crucial point of the crystallization process is the choice of the solvent that must meet the following properties:

- Total solubility of the substance to be purified at high temperatures.

- Low capacity of dissolving the impurities that contaminate the product in any temperature range.

- Absence of chemical reaction with the product to be purified.

- Generation of good crystals of the product to be purified.

- Easy removal.

For practical purposes, two types of recrystallization can be distinguished: in water and in organic solvents.

1. Recrystallization from water: Many organic compounds are insoluble in water at r.t., but they are hot. For this purpose a suspension of the solid is

prepared in the minimum amount of water in a beaker or Erlenmeyer flask, and the mixture is brought to a boil. If the solid does not dissolve under these conditions, small amounts are added and the water is boiled again until the compound dissolves. Note that suspended particles corresponding to part of the insoluble impurities often remain. If the solid is dark in color, the addition of small amounts of activated carbon will retain most of these colored impurities.

The procedure for recrystallization from water is as follows (see Figure 4.17):

- Transfer the solid for recrystallization to an Erlenmeyer or beaker.

- Dissolve the substance in the minimum amount of hot solvent.

- If the starting crystals exhibit intense color due to the presence of impurities, add some activated carbon to remove them, since the colored impurities are retained in the activated carbon.

- Heat the mixture to boiling on a hot plate, checking that the product to recrystallize has been completely dissolved. Magnetic stirring should be performed (or add a boiling chip) to avoid sudden boiling and the formation of solid splashes.

- Disconnect the hot plate.

- Filter the hot solution by gravity using a conical funnel and fluted filter to remove insoluble impurities as well as activated carbon (discarding the solid residue, composed of activated carbon and insoluble impurities, on the fluted filter). Tweezers, tongs, or a piece of filter paper should be used when manipulating an Erlenmeyer flask or beaker to avoid burns when handling hot glass containers.

- When the product solidifies in the funnel, add hot water.

- As the solution cools, the corresponding product crystals will form (external cooling helps this process).

- Finally, after the filtrate completely cools, these crystals are filtered under vacuum and washed (with the cold solvent used for recrystallization) in a Büchner to remove adhering filtrate and then dried to remove traces of solvent.

In the case of incomplete crystallization, concentrate the filtrate (by heating and by evaporating half of the solvent), and repeat the process.

2. Recrystallization in organic solvents: When a volatile organic solvent is used instead of water, the heating part of the solution is performed with a reflux assembly to prevent flammable volatile organic solvent vapors from causing fires.

The procedure for recrystallization is as follows:

1A) Dissolve the solid in water (beaker)

1B) Alternative: dissolve the solid in a volatile organic solvent (reflux)

2) Add activated carbon

3) Hot gravity filtration

4) Crystallization (external cool)

5) Vacuum filtration

Figure 4.17: Steps in recrystallization.

- Set a round-bottom flask with a clamp and a connector on a hot plate.
- Transfer the solid to be recrystallized using a solid funnel or a piece of coated paper.
- Add the solvent or solvent mixture and a boiling chip or a stir bar (if a magnetic stirring plate is available).
- Attach a reflux condenser to the flask, connect the tubing, and open the water or cooling circuit.
- Heat the mixture to reflux to dissolve the solid.
- Turn off the heating plate and allow to cool in order to stop the reflux.
- While the flask is still warm, gravity filter the contents of the flask, helping with tongs.
- Cool the filtrate to r.t. and let stand until the end of the crystallization of the solid.

Sometimes crystallization is facilitated by seeding a few crystals of the product or scratching the bottom of the container with a glass rod. In both cases crystallization nuclei are generated, accelerating the process.

The main reasons for failure with this technique are:

- The wrong solvent is chosen for the recrystallization.
- The wrong amount of solvent is used to dissolve the recrystallization solid.
- Lack of precipitation when the solution is cooled.
- An oily substance formed instead of a solid precipitate.

4.7.4. Sublimation

Sublimation or volatilization is a change of state from solid to gas without passing through the liquid state. A typical example is dry ice, which can sublimate at r.t. and at atmospheric pressure. This technique is used for purification of solids that exhibit exceptionally high vapor pressures and m.p., making these solids transform directly to the gas phase. Other substances are subject to sublimation, such as iodine, sulfur, naphthalene, etc. This purification technique is suitable for poorly soluble solids and cannot be isolated in a pure state by successive recrystallizations. Normally, it is performed using vacuum distillation equipment (see Figure 4.18) for very pure yield. Unlike recrystallization, this technique is not commonly to used purify solids.

For certain substances, sublimation at atmospheric pressure can be performed as follows:

- The solid that is to be converted into steam is heated in a beaker.

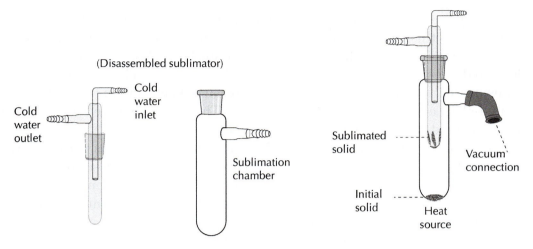

Figure 4.18: Sublimator.

- A cold surface (cold finger) is set over the beaker.

- When the steam meets the cold surface above, and the temperature is lower, there is a return to the initial solid-state crystals.

4.7.5. Drying solids

The crystalline products resulting from filtration are not dry. Usually they are impregnated with the solvent in which crystallization occurred and may trap moisture. For an accurate calculation of the yields of the reactions and a precise determination of physical constants such as melting point, it is necessary for the products to be dry. After filtration, they are usually left in the Büchner with a stream of air for a while, but they still tend to retain solvent residues.

Figure 4.19: Desiccator for drying solids.

To dry a solid efficiently, a desiccator is used (see Figure 4.19). This is a tool made of thick glass or polycarbonate, which comprises a cylindrical vessel with a flange inside on which a perforated porcelain, glass, or plastic plate is deposited, so that the container has two different areas. The desiccator has a hemispherical cover made of glass, the two parts being in contact through a ground-glass edge that ensures a tight seal. The polycarbonate has a rubber union. The lid has a stopcock for connection to a vacuum.

The solid in a Petri dish, a crystallization dish, or a beaker is placed on the perforated plate of the container. In these latter two cases it is desirable to cover them by adjusting the mouth of the container with a piece of aluminum foil and puncture it with, for example, a needle to prevent the solid from escaping the container when the desiccator opens or closes.

A desiccant is placed under the perforated plate. To facilitate the cleaning and maintenance of the dryer, the desiccant should be in a container such as a crystallization dish. Drying agents that can be used include $CaCl_2$, Drierite®, silica gel, NaOH, H_2SO_4 (concentrated), and phosphorus pentoxide.

The choice of desiccant depends on the nature of the solid (if sensitive to acids or bases) and the moisture content or the type of solvent that is impregnated.

The desiccator lid is opened and closed by moving it laterally with respect to the body of the desiccator. In the case of glass desiccators, they must be lubricated with silicone grease to facilitate opening and closing of the ground joints.

Noticeboard 4.7.1

 Practical tips & tricks:

To get a solid from a desiccator with a vacuum, do not open the stopcock directly to allow the passage of air, as whirlwinds will form and disperse the solid throughout the desiccator with splashes. For the proper opening of the vacuum stopcock use a piece of filter paper in the nozzle because it causes the air to enter very slowly. By vacuum, the paper will be attached to the output of the stopcock and will release from it when the outside pressure equals the interior pressure.

Perform the drying process as follows:

- Slowly open the vacuum stopcock.

- Open the desiccator lid laterally, shifting the lid over the body.

- Check the condition of the desiccant, and replace it if it is not in good condition.

- Insert the solid into a container.

- Close the lid by moving it laterally on the desiccator.

- Connect the stopcock to a proper vacuum tubing for a few minutes.

- Close the valve and disconnect the tubing.

- Keep the product in the desiccator as long as necessary until it loses moisture. To determine this, check the solid crystals, which should not be matted. If moisture persists, repeat the entire sequence.

4.8. Liquid-liquid extraction

The liquid-liquid extraction (typically extraction) is one of the most common basic operations in the Organic Chemistry laboratory and allows the isolation and purification of a product resulting from a chemical reaction.

Extraction can be defined as the transfer of a substance X from a liquid phase A to another liquid phase B. Both solvents should be immiscible, thus forming two distinct phases. X sharing between phases A and B is given by the Nernst equation:

$$K_D = \frac{C_{B(X)}}{C_{A(X)}} \tag{4.1}$$

where $C_{B(X)}$ and $C_{A(X)}$ are the concentrations of X at B and A, respectively, and K_D is the partition coefficient, which depends on the temperature.

This operation is typically performed between an aqueous solution (aqueous layer) and a water-immiscible solvent (organic layer) with the aid of a separatory funnel. The relative position of the two layers (upper, lower) depends on the relationship of densities. Chlorinated solvents (CH_2Cl_2, $CHCl_3$, CCl_4, etc.) always remain in the lower layer, as it is denser than water.[14] However, other organic solvents typically have lower densities than water (diethyl ether, ethyl acetate, toluene, hexane, etc.) and therefore always remain in the upper layer. Clearly, water-miscible solvents are not useful for this process —for example, acetone, MeOH, EtOH, etc.

a) Separatory funnel: Liquid-liquid extraction is conducted on a laboratory scale with a separatory funnel, a conical or pear-shaped container with a ground-glass stopper on top and a stopcock that connects the container with an outlet tube terminating in a chamfer.

For a successful extraction, the following steps are necessary (see Figure 4.20):

- The separatory funnel is set up on a stand with a ring clamp and a container or collector; usually an Erlenmeyer is placed below it.

- The separatory funnel height is adjusted so that the outlet tube is left a few centimeters from the collector.

---

[14] *tert*-Butyl chloride, with a density of 0.789 g/mol (less than water), is an exception.

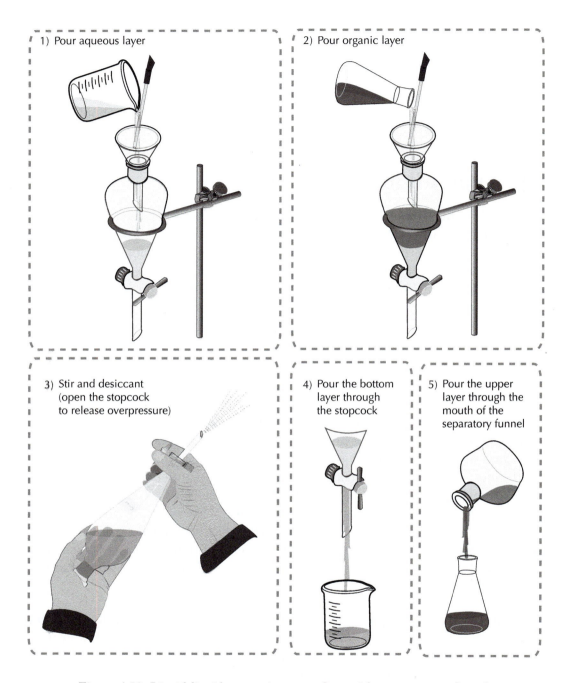

Figure 4.20: Liquid-liquid extraction procedure with a separatory funnel.

- Check that the stopper is closed before adding any liquid into the funnel.[15]

- With the use of a stemmed funnel the two immiscible layers are poured into the separatory funnel.

- Check that the stopper is properly adjusted and closed correctly without any leak.

- Remove the ring clamp and hold firmly on the side of the stopper with the left hand (for right-handed) or right (for left-handed) (see Figure 4.20).

- With the free hand hold the funnel between the fingers in the area of the stopcock, so that the stopcock can be opened and closed comfortably with the tip of the index finger and thumb.

- Then shake the funnel vigorously with both hands.

- It is important to occasionally open the stopcock to remove the excess pressure that sometimes builds up inside (do not direct the exiting gases at yourself or another person).

- After shaking, the funnel is placed back into the ring clamp, and the stopper is removed.

- Let stand until the two layers are separated (decanted).

- The layer remaining at the bottom is emptied by opening the stopper to the limit of the two layers.

- Finally, the top layer is poured out by taking the funnel in hand by its neck.

b) Sometimes the separation between the two layers in the extraction process presents a number of difficulties that slow down the process, such as the formation of interfaces, foams, or emulsions. There is no standard procedure to resolve such problems. Sometimes, a simple rapid rotation of the funnel is enough to break them. In other cases it is helpful to add a few crystals of salt (NaCl) or a concentrated brine solution. But overall, waiting is the most effective solution.

When extraction takes place, the two layers are saturated with respect to the other solvent: water with organic solvent and the organic layer with water. Therefore, the water must be removed from the organic solvent in order to obtain the pure product (see Section 4.8.1, p. 114).

However, the extraction is not always used simply to separate compounds, distributing them among organic and aqueous layers. Sometimes, it is useful to force the extraction accompanied by a chemical reaction (most commonly through an acid-base reaction or by reactions forming metal complexes),

---

[15]There are plugs, both glass and Teflon®, having a slot connecting with the hole usually found in a ground-glass funnel. Therefore, check that the two match before shaking.

enabling the purification of the compound if it is contaminated with by-products with different chemical properties. For example, in the case of aldehydes, such contaminants can be removed from an organic layer by washing with an aqueous bisulfite solution; alkenes also may be purified with silver salts or carboxylic acids with basic solutions, etc. This purification process of liquid is called "washing of solutions" and is performed in the same way as a liquid-liquid extraction.

4.8.1. Liquid drying

The synthesis and isolation procedures of organic compounds generally require the availability of a solution of these compounds in an organic solvent. For example, in the case of extraction with aqueous solutions of an organic layer, a transfer of the water to the organic solution occurs due to the partial miscibility of the organic layer and water. It can also happen that some reaction takes place in the aqueous solution and requires removal of the product with an organic solvent for proper isolation and characterization. The usual procedure for liquid drying is to treat the organic solution with a drying agent. Desiccants are anhydrous inorganic salts that take up water until they become hydrated. The most commonly used salts are listed in Table 4.4.

Table 4.4: Desiccants commonly used.

Compound (anhydrous)	Capacity	Drying speed	Drying aplications
$CaCl_2$	High	Medium	Drying of hydrocarbons
$CaSO_4$ (Drierite®)	Low	Fast	General purposes
$MgSO_4$	High	Fast	Not applicable to very sensitive acid compounds
K_2CO_3	Medium	Medium	Not in acid compounds
Na_2SO_4	High	Slow	General purposes

For drying a liquid, a desiccant is left in contact with the solution in an Erlenmeyer flask, and after a few minutes it is separated by gravity filtration.

Molecular sieves are synthetic materials that mimic the structures of natural zeolites. They consist of aluminum silicates and alkali metal or alkaline earth. They have a high absorption capacity for small molecules and a certain selectivity depending on the pore size (see Table 4.5). Molecular sieves are activated when water is removed from the structure holes. Removing the hydration water causes no significant structural changes. Molecular sieves can be used as drying agents due to these properties or to remove other types of small molecules that contaminate the solvent. When the sieve no longer absorbs more molecules, it is said to be saturated. A sieve can be regenerated to active again by subjecting it simultaneously to heating and the passage of a gaseous flow of air or nitrogen.

The steps for drying a liquid are the following:

Table 4.5: Types of molecular sieves, selectivity, and applications.

Pore size (Å)	Does absorb	Does not absorb	Application (drying)
3	NH_3, H_2O	C_2H_6	Polar liquids
4	H_2O, CO_2, SO_2, H_2S, C_2H_4, C_2H_6, C_3H_6, EtOH	C_3H_8 and higher hydrocarbons	Non-polar liquids and gases
5	Normal hydrocarbons (linear) to n-C_4H_{10}, alcohols to C_4H_9OH, mercaptans to C_4H_9SH	Isocompounds or rings larger than C_4	
8	Branched and aromatic hydrocarbons		Gases
10	Di-N-butylamine	Tri-N-butylamine	HMPA

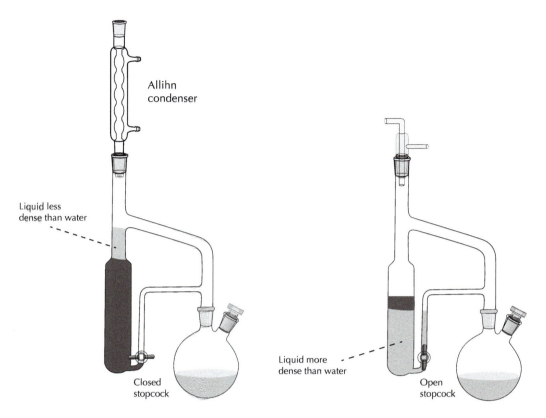

Figure 4.21: Continuous liquid-liquid extraction.

- The desiccant is added to an Erlenmeyer flask with the solution.

- Swirl the Erlenmeyer flask by hand for a few seconds, cover, and let stand.

- After a few minutes, the liquid is separated from the solid by gravity filtration.

4.8.2. Continuous liquid-liquid extraction

With a liquid-liquid extraction it is not possible to isolate an organic product even using the strategies outlined above, because the partition coefficient is very unfavorable for the desired product. The solution would need a large number of liquid-liquid extractions, but this is not practical. In such cases, continuous liquid-liquid extraction is used, considering the varying density of the organic solvent compared to water (see Figure 4.21).

4.9. Solid-liquid extraction (Soxhlet)

This is used for isolating compounds present in a solid, using a liquid solvent. For example, the extraction of active principles of plants in Natural Product Chemistry. Although a compound can be extracted from a solid with a liquid by a simple extraction and subsequent filtration to be effective, this process would require many tedious extraction and filtration steps. Therefore, this equipment was designed so that, through a continuous extraction a compound within a solid can be extracted in a very efficient way. With the Soxhlet equipment, multiple extractions are performed automatically (continuously) using the same solvent, which evaporates and condenses, recirculating in the equipment (recycled).

The Soxhlet equipment consists of the following elements (see Figure 4.22):

- Reflux water-cooled condenser

- Soxhlet cartridge with the solid to be extracted

- Soxhlet column extraction where the cartridge is inserted with the solid to be extracted and has a siphon through which the solvent falls after extraction to the flask

- Round-bottom flask three or four times the size of the solvent chamber

Procedure 4.9.1:

The solid to be extracted (seeds, plant leaves, etc.) is placed in a cartridge made with filter paper (ready-to-use cartridges are available on the market). Once the solid has been introduced, it is placed inside the extraction column. The flask is filled with a solvent to approximately half, the boiling chip is added, and heating is begun. By heating the solvent (used to remove the solid inside the Soxhlet cartridge), it evaporates and the vapors rises through the extraction column to the area where the condenser is cooled, condensing the vapors that fall back drop by drop onto the cartridge, and thus the active ingredient of the solid is extracted. Through the siphon and when a sufficient level of solvent is reached, the soluble component that is intended to separate from the solid drops by gravity to the solvent chamber. When reheated, the pure solvent is evaporated and falls again over the sample and more of the compound is extracted. The process is automatically repeated until that extraction is considered complete.

4.10. Removal of solvents under reduced pressure (rotary evaporator)

A common operation in Organic Chemistry laboratory is the elimination of a volatile organic solvent coming from a reaction mixture or a process such as liquid-liquid extraction. Although this operation, in principle, could be performed by simple distillation, the fastest and most convenient method is the use of a rotary evaporator for distillation under reduced pressure (see Figure 4.23). Basically, this consists of an electric motor, which causes rotation of a tube

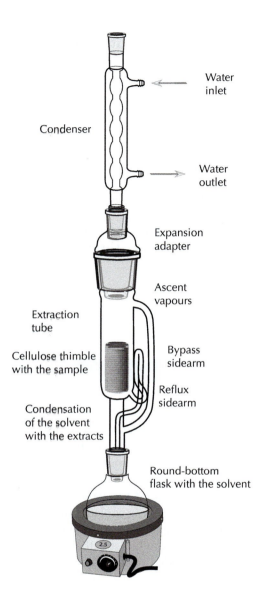

Figure 4.22: Soxhlet extractor.

with a ground-glass joint guide tube, to which a round-bottom flask containing the solution is coupled. This flask is partially immersed in a water bath, maintaining the rotation. The bath temperature should not exceed 35–40 °C for manipulation of the most common organic solvents. Coupled to the system is a condenser that circulates a liquid (water or antifreeze). This produces the condensation of the solvent, which is recovered in a collector. The set is a closed system connected to a vacuum system (vacuum pump, water pump, or vacuum circuit).

There are solutions that when placed in a rotary evaporator have a strong tendency to form foams or to produce splashes that move up the guide tube to be deposited in the indoor coil. This drawback is avoided using the so-called foam brake. This is a glass adapter located between the guide tube and flask

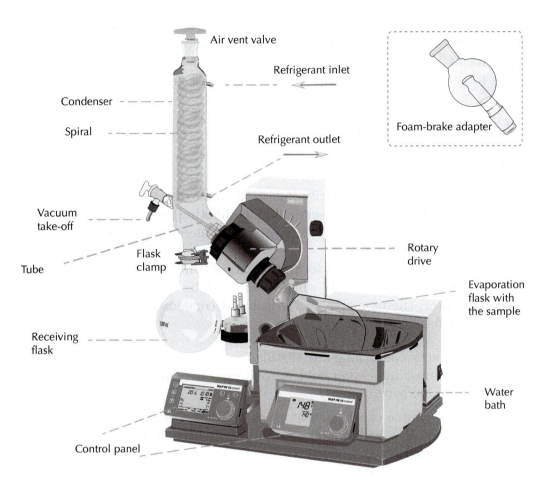

Figure 4.23: Rotary evaporator.

containing the solvent to be removed. It has two ground-glass joints: the female on top, connected to the guide tube, and the male at the bottom, connected to the flask. The body consists of a glass sphere with elbow pipes or another device to prevent any material in the flask from ascending the guide tube.

4.11. Distillation

4.11.1. Simple distillation

This is a technique used to purify liquids by removing non-volatile impurities or separating mixtures of immiscible liquids differing in b.p. by at least 25 °C, which must also present b.p. <150 °C at atmospheric pressure. The process is governed by Raoult's law.[16]

The most usual setup to perform distillation is illustrated in Figure 4.24. It takes a ground-glass joint flask, which should not be more than half full, a distillation adapter, thermometer, water-jacketed condenser, distillation adapter, and collector flask.

- Some points concerning the distillation setup: The assembly must have clips, clamps, and clamp holders connected to a support stand; make sure all joints are snug.

    The heat source can be a heat plate or heating mantle, and for homogeneous boiling, a boiling stone should be used or, if available for the heat plate, an appropriately sized stir bar for magnetic stirring the flask liquid.

    The distillation adapter (three-way adapter) allows the thermometer to be placed and the vapor to be diverted to the water-jacketed condenser using standard connections sealed by ground-glass joints. The thermometer should be placed with its bulb below the lower level of the lateral line of the distillation connector so that it is in the vapor stream of the distillation.

    Water-jacketed condenser (or Liebig condenser) is the simplest design for water-cooled condensers (see Figure 4.12). Although the condensing surface is relatively small, these condensers are effective for a variety of applications in addition to simple distillation. The cooling circuit must be connected so that the water input is the nearest to the distillation bottom and the exit closest to the distillation adapter with the thermometer.

- Simple distillation setup:

    - First, the clamp and clamp holder are fixed to a support stand (or lab frame) at the lab desk or fume hood.

---

[16]The relationship between the vapor pressure of each component in an ideal solution depends on the vapor pressure of each individual component and the mol fraction of each component in the solution.

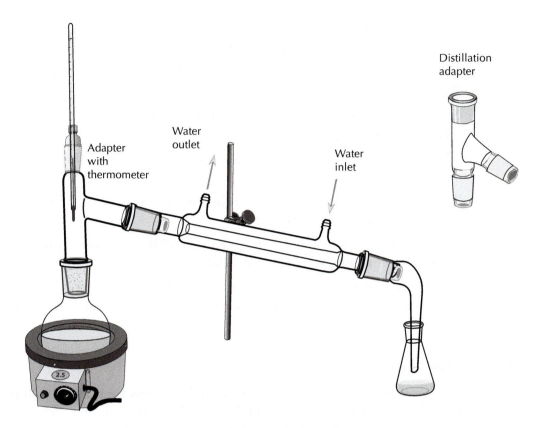

Figure 4.24: Simple distillation equipment.

- Then the round-bottom flask is placed on the hot plate, adjusting a clamp at its neck to the support stand.

- Later, a distillation adapter (three-way adapter) is connected to the flask.

- Next, the thermometer is placed with its adapter on the top of the distillation adapter.

- Then another clamp and clamp holder are adjusted to the support stand in order to fix the condenser.

- The water-jacketed condenser is adjusted to the second clamp and fixed to the distillation adapter.

- Another distillation adapter is mounted with a receiver container (beaker or conical), where the distillate is collected.

- Finally, the rubber tubing is connected to a water-jacketed condenser to provide the water (inlet from water supply, lower end, and outlet to sink, upper end).

4.11.2. Fractional distillation

This is used to separate liquid components that differ by less than 25 °C in b.p. Each of the components to separate are called fractions. It is similar (see Figure 4.25) to the simple distillation setup in which, between the round-bottom flask and the distillation adapter, a rectifying column was inserted (this column may have a different design, like a Vigreux column, filler column, etc.; see Figure 4.26). When the mixture is heated, the vapor rises and goes around the hollows of the rectifying column, enriching it with the more volatile component, while the liquid flows down, passing through the column into the round-bottom flask, enriching the less volatile component and causing separation. The overall process is equivalent to performing multiple simple distillations of the sample.

- Main types of rectifying columns:

    - Vigreux column: This is a modification of the air condenser. It has a number of deep, pointed indentations in its sidewall where the vapors condense while passing through the column and that serve to increase the contact surface between the vapor and condensate.

    - Snyder column: This is a type of very effective column containing a number of segments that in turn have a floating glass sphere to increase the mixture of vapor and condensate.

    - Packed columns: These are air condensers that are fillers with small pieces of glass or ceramic Raschig rings.[17] To keep the filler inside

---

[17]Pieces of tube with similar length and diameter; their name comes from the inventor, the German chemist Friedrich Raschig.

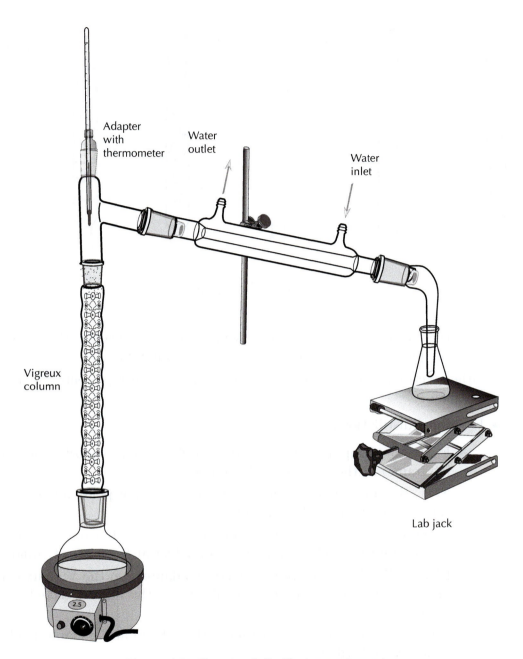

Figure 4.25: Fractional distillation equipment.

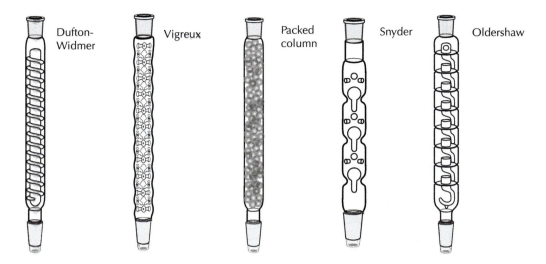

Figure 4.26: Different types of rectifying columns.

these columns, there is usually some kind of constriction that allows the passage of vapors but prevents their falling into the flask.

Other rectifying columns such as the Overshaw are represented in Figure 4.26.

- Fractional distillation setup:

  - First, the clamp and clamp holder are fixed to a support stand (or lab frame) at the lab desk or fume hood.

  - Next, the round-bottom flask is placed on the hot plate, adjusting a clamp at its neck to the support stand.

  - Then assemble the Vigreux column to the round-bottom flask neck and secure with a clamp and a clamp holder at the support stand.

  - Place a distillation adapter (three-way adapter) on the Vigreux column, and adjust a clamp horizontally to set the water-jacketed condenser.

  - Fix the water-jacketed condenser with a clamp and a clamp holder as well as a distillation adapter.

  - Fix the distillation adapter to the water-jacketed condenser.

  - Also, set clips to adjust distillation adapters.

  - Next, place the thermometer with its adapter into the top of the three-way adapter.

  - Adjust the distillation adapter with another clip.

  - Place an Erlenmeyer as a collector under the distillation adapter, adjusting the height with a clamp and a clamp holder.

  - Finally, connect the rubber tubing of the water-jacketed condenser to water (inlet from water supply, lower end, and outlet to sink, upper end).

4.11.3. Vacuum distillation

Also called distillation under reduced pressure, vacuum distillation is used to purify or separate liquids or syrups having a boiling point above 150 °C or a lower boiling point but thermally unstable. The vacuum can be created using a water pump or vacuum pump.

For this type of distillation a setup similar to a simple distillation assembly is used, but with some differences. For example, a special distillation adapter having a vacuum inlet (vacuum adapter) is used. On the other hand, it is usual not to employ a boiling chip for homogeneous boiling because when the vacuum is made, the air that has entrapped the boiling chip can come off quickly and compromise the vacuum. Instead, it uses magnetic stirring. In some cases, as when the amount of liquid to be distilled is small, it is common to use a special assembly (Claisen) for the vacuum, requiring a capillary as a stirring system and vacuum control.

- Precautions:

    - The glass material used in vacuum distillation should be carefully examined to prevent scratches or cracks because they present a risk of implosion once assembled and operating. Especially in the case of using very low pressures, it can be protected by wrapping the assembly in some kind of mesh or adhesive tape, which, in case of an accident, would prevent glass fragments from scattering.

    - Grease is carefully applied to ground-glass joints in order to ensure the best fit and avoid a loss of vacuum.

The vacuum distillation devices can have different arrangements and components, but the most popular one is similar to the simple distillation assembly, except for the vacuum adapter connected to the vacuum line.

- For vacuum distillation, proceed as follows:

    - First, clamps and clamp holders are fixed to the support stand of the lab table or to the fume hood.

    - The ground-glass joints of all parts of the assembly are lubricated with silicone grease.

    - Then the round-bottom flask is placed on the hot plate and adjusted by its neck with a clamp.

    - The liquid to distill is mixed in with a stir bar.

    - Then the distillation adapter (three-way adaptor) is assembled on the round-bottom flask.

    - For liquids that tend to form foam or to splash, it is advisable to insert a Claisen adapter between the flask and the condenser.

    - Then the thermometer is located with its adapter on top of the distillation adapter.

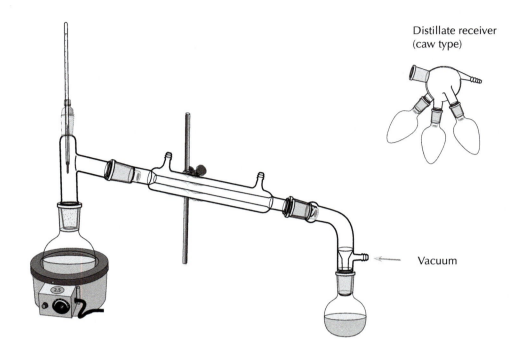

Distillate receiver
(caw type)

Vacuum ←

Figure 4.27: Vacuum distillation equipment.

- A water-jacketed condenser is mounted and secured with a clamp and clamp holder to the support stand and with a clip to the distillation adapter.

- Also, the vacuum tubing of water-jacketed condenser is connected to water inlets and outlets (recirculation or tap/sink).

- The distillation adapter is mounted and coupled to a ground-glass-joint flask.

- If necessary, the collecting flask is placed in an ice bath to facilitate the condensation of the distillate.

- The vacuum line is connected.

- The pressure gauge of the vacuum pump is opened.

- In some cases, it is appropriate to cover the flask and the distillation adapter (or Claisen adapter) with aluminum foil to prevent heat loss.

  To distill a mixture of two or more components, a special distillation receiver adapter termed "caw type" is used. Usually it has two or three settings (male), which are connected to each round ground-glass-joint flask. Turning this adapter, the different fractions are collected (see Figure 4.27).

4.11.4. Steam distillation

Steam distillation is applied to separate a substance from a mixture. It is particularly suitable for compounds that are poorly soluble in water, that have a very high b.p., and that decompose when distilling. This technique is also employed to purify substances contaminated with large amounts of resinous impurities and to separate high-b.p. solvents from solids that do not drag. A device that drives a steam current (typically water vapor) to drag that substance is used, as shown in Figure 4.28. According to the way in which the vapors are generated, the following two types of steam distillations are distinguished:

- Distillation with an external source of steam: This is performed with an assembly such as that shown in Figure 4.28 (similar to a reaction with gaseous reactants) except that in this case the gas bubbled over the solution of the product is water vapor produced in an auxiliary assembly.

- Distillation with an internal source of steam: In this case, a mixture of a product is dissolved in an organic solvent, and water is distilled. This is performed as simple distillation, and an addition funnel can be placed in the assembly with water gradually added (see Figure 4.28).

4.11.5. Azeotropic distillation (Dean-Stark)

An azeotrope is a mixture of two or more liquid components under constant boiling, and distillation processes are performed as if they were a pure compound (see Table 4.6). Azeotropic distillation is a useful procedure for removing a liquid from a crude reaction by a co-distillation with an immiscible organic solvent. This technique is often used in equilibrium reactions where water is formed as byproducts of the reaction. Water removal shifts the equilibrium of the reaction toward the product side. If the reaction is carried out, for example, with toluene, which is less dense than water, the steam in the reflux condenser will consist of an azeotropic mixture of toluene and water. When this mixture is condensed, it falls into the so-called Dean-Stark, forming two layers: the top layer will consist of toluene and the bottom layer water. When the liquid level in the Dean-Stark trap reaches the top of the side arm, the toluene flows back into the reaction flask. The water can be removed through a stopcock in the bottom of the Dean-Stark trap (see Figure 4.29).

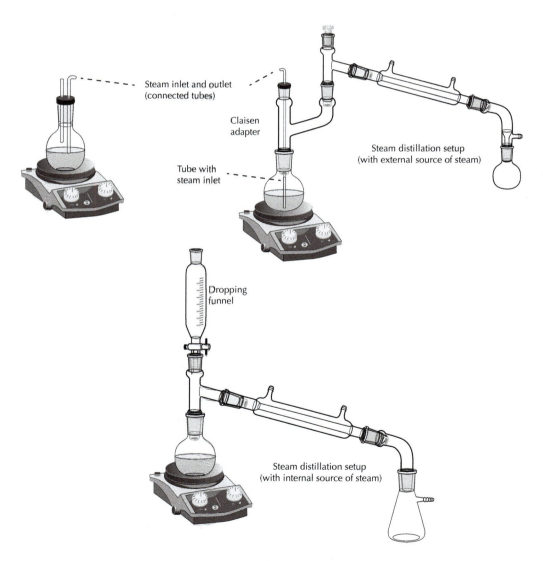

Figure 4.28: Steam distillation equipment.

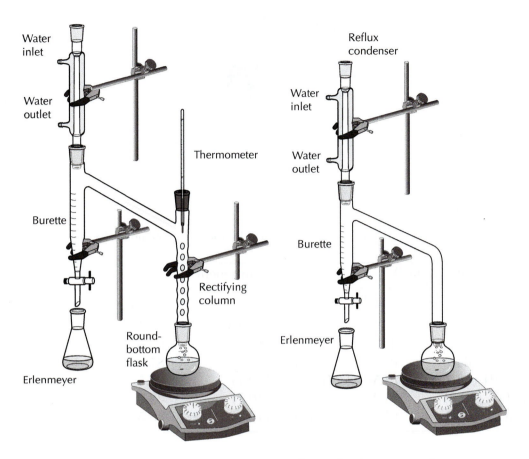

Figure 4.29: Azeotropic distillation (Dean-Stark equipment).

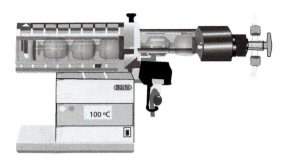

Figure 4.30: Kugelrohr distillation apparatus.

Table 4.6: Different azeotropes formed by the most common solvents.

Component A		Component B		Azeotrope
B.p. (°C)	% (weigh)	B.p. (°C)	% (weigh)	B.p. (°C)
H_2O (100)	1.3	Diethyl ether (34.5)	98.7	34.2
H_2O (100)	1.4	Pentane (36.1)	98.6	34.6
MeOH (64.7)	12.1	Ketone (56.1)	87.9	55.5
MeOH (64.7)	72.5	Toluene (110.7)	27.5	63.5
EtOH (78.3)	68.0	Toluene (110.7)	32.0	76.7
H_2O (100)	13.5	Toluene (110.7)	86.5	84.1

4.11.6. Kugelrohr distillation

This is a device that allows the distillation of small amounts of liquids, including solvents with a low b.p., and the drying of samples. The system has a number of glass spheres that are connected. The balls rotate and simultaneously move in front of a heat source, and the assembly is connected to a vacuum pump. Samples can be observed directly through the transparent glass to see how the liquid passes from one ball to another (see Figure 4.30). The device also serves to dry solid samples, in which case a desiccator is used.

4.12. Reactions in an inert atmosphere

There are reagents or products that are sensitive to moisture, to oxygen in the air, and even to the CO_2. In these cases, an effective protection of the chemical reactions is necessary, not only because the reaction yields may be affected by the partial or total destruction of any sensitive material during the reaction but also because some of these unwanted transformations may become violent, resulting in explosions or fires. Reactions of this type are among the most complex processes that may occur in an Organic Chemistry laboratory. This is why these reactions are discussed after the main basic laboratory operations are known.

With relatively simple guidelines, it is possible to maintain an inert atmosphere within an assembly using conventional laboratory equipment. There are several aspects that should be taken into account:

a) The material must be completely dry.

b) The solvents must be anhydrous.

c) Inside the assembly the air must be replaced by an inert gas such as nitrogen or argon, for which the reaction should be performed under a constant flow of inert gas or at least displace the air with an inert gas while maintaining slight pressure of the inert gas in the interior.

4.12.1. Precautions with the material

The material should be dry and should be used immediately after removal from the oven. It may even be desirable to use a hot air dryer on the various components while they are assembled to prevent any accumulation of moisture.

- Solvents: First, it should be borne in mind that these types of reaction solvents must be anhydrous, —i.e., completely moisture free. For this, the solvents must be of high quality and need special treatment for the complete elimination of water traces. Although many suppliers sell solvents packaged as anhydrous, the removal of the seals and the exposure, although minimal, to the atmosphere may invalidate their use for some particularly sensitive reactions. In these cases, the solvent must be anhydrous immediately before use.

4.12.2. Using anhydrous solvents

There are basically two methods for producing anhydrous solvents:

a) Individualized chemical treatment. Drying of the most common solvents:

- THF (tetrahydrofuran): First, it is treated with calcium hydride or molecular sieves 4 Åand then heated to reflux under an inert atmosphere in the presence of sodium wire and benzophenone for several hours until the solvent becomes deep blue. This indicates that the solvent is dry. At this point, the required amount of solvent is taken. If the mixture becomes orange, a new amount of dry solvent should be prepared.

- Diethyl ether: It is dried in the same manner as the THF.
  For such solvents an assembly is often used in which reflux is performed with the drying agent until the colorimetric indicator shows that the solvent is anhydrous (see Figure 4.31). From this point the solvent can start being collected in a flask that acts as a reservoir, triggering a set of stopcocks. If not used, the solvent may be returned to the flask where

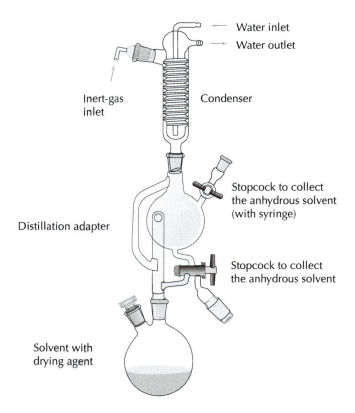

Figure 4.31: Equipment for drying solvents.

the reflux is performed to keep it in contact with the drying agent. The whole process is conducted in an inert atmosphere.

Noticeboard 4.12.1

 DANGER!

To remove residual sodium of this procedure, EtOH is slowly added. This process is slow and facilitated by the occasional shaking of the flask. When no more hydrogen is released, a small amount of hydroalcoholic mixture 1:1 can be added. If there is no more reaction, water is added until the formation of two distinct layers that must be separated and poured into the appropriate waste container.

- CH_2Cl_2: It is pretreated with calcium hydride and then distilled on the same compound. There are no colorimetric indicators to show that the solvent is dry.

- MeOH: For most uses, it is dried over 3 Å molecular sieves overnight and then distilled. Alternatively, MeOH can be dried with magnesium methoxide. To a volume of 1 l, 100 ml of dry MeOH made by the above process are refluxed with magnesium turnings (5 g) and iodine (0.5 g) until all the magnesium has reacted. MeOH is added to a volume of 1 l and is refluxed for 2–3 h in an inert atmosphere.

- Acetonitrile (MeCN): It is stirred with 4 Å molecular sieves and distilled over calcium hydride.

Noticeboard 4.12.2

 DANGER!

The process of removing magnesium or calcium hydride is similar to that for sodium. When the hydrogen release is stopped by adding EtOH, water can be added carefully until no bubble formation is observed. If an aqueous and an organic layers form, it must be separated and poured into a suitable waste container. If the water and the solvent are miscible, more water is added and poured into the waste container.

- DMF (N,N-dimethylformamide): It has the disadvantage that it decomposes slowly at r.t. and more rapidly at reflux, giving dimethylamine and carbon monoxide. The decomposition is catalyzed by traces of acids and bases or basic drying agents such as calcium hydride. It can be dried with barium oxide or 4 Å molecular sieves at r.t. overnight, whereupon the drying agent is removed by decantation and vacuum distillation ($\sim$ 20 mm Hg is sufficient to reduce the boiling point of DMF vacuum).

To store anhydrous solvents for a period of time it is advisable keep them in a container over 4 Å molecular sieves. Any reaction or distillation should be undertaken in an inert atmosphere.

b) Solvent dispensers: It is increasingly common in the Organic Chemistry laboratories to have dispensers of anhydrous solvents. High-quality solvents are used that are stored in metal security containers and that pass through columns with a specific filler that holds water or other waste molecules such as alcohols or amines, before leaving through the corresponding valve. This is done by entrainment of the solvent by a stream of inert gas. The valves allow the filling of all flasks without air contact.

4.12.3. Origin of inert gas

a) Gas cylinder or centralized circuit: Inert gases may be stored in a gas cylinder or may be available in a centralized system. In any case, the gas pressure is regulated with a flow-controlling valve that allows fluid-pressure reduction and the maintenance of the outflow pressure at a constant fixed or selected value. To control the flow of inert gas in an assembly, a bubbler is used at the end of the assembly. This is a glass device (see Figure 4.32) with glycerin

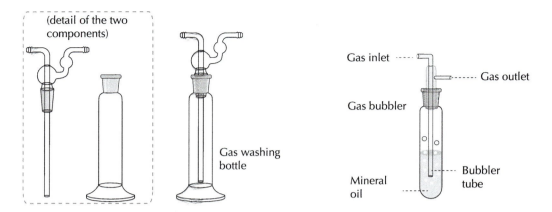

Figure 4.32: Gas bubbler.

or a mineral oil that lets the gas flow. The flow rate is regulated depending on the gas bubbling in the fluid.

b) Using a balloon: A rubber balloon filled with inert gas and attached to a syringe with a needle inserted through a septum, so that a slight pressure is maintained within the inert gas.

4.12.4. Transferring air-sensitive or moisture-sensitive reagents and solvents

In reactions in an inert atmosphere it is common to handle reagents that are sensitive to moisture or oxygen. These kinds of reagents can be transferred in several ways to the containers in which the reaction is conducted. The following describes some of these ways. First, it is noted that in the case of liquid reactants in solution, the bottles are sealed by a double seal formed by a metal capsule similar to that of bottles of soft drinks with a hole in the center that exposes a small surface of a polymeric material such as polypropylene or similar one that can be pierced by needles several times without the reagent contacting with the air. For conventional bottles, a septum can be used as a cap.[18]

A simple syringe with a needle can be used to take the quantities needed. To facilitate the filling of the syringe and prevent taking in moisture or air, another suitable needle is attached to a flexible tube with nitrogen or argon. Increasing the inert gas pressure drives the liquid filling the needle.

To transfer larger amounts, a cannula can be used, this being a steel tube with a needle at both ends. The container with the reagent or solvent is connected to a flexible tube with inert gas, with the cannula at one end and the other connected to a flask with a septum and a gas outlet, such as needle or other bubbler. Increasing inert gas pressure in the reagent container or the sol-

---

[18]Septum closure can be sealed with Parafilm®.

vent forces it to pass through the cannula to the flask. Refilling ceases when inert gas flow is stopped.

4.13. Reaction and removing of gaseous products

In many organic reactions, one of the reactants may be a gaseous product (ozone, ammonia, CO_2, HCl, etc.), in which case the reagent must be generated before use, and afterward the reagent must be bubbled in the reaction flask.

In other organic reactions, harmful gas is released. So as not to be expelled into the atmosphere, a reflux is usually set up or an assembly of the reaction so that the gases remain trapped in a solvent or in a solution that allows neutralization. In both cases, a bubbler is used (see Figure 4.32), which in some cases is provided with a foot. Depending on the reaction type, the bubbler is placed in the assembly: before (gas-generating reaction) or after (gas-releasing reaction).

4.14. Chromatographic techniques

Chromatography is a separation technique with extraordinarily versatile variants. In all chromatographic separations, there are two phases: a mobile phase and a stationary phase, which move relative to one another while maintaining contact. The sample is introduced into the mobile phase, and the sample components are distributed between the stationary and mobile phases. The components of the mixture to be separated spend different times in each phase, and therefore separation occurs. If a component spends most of the time in the mobile phase, the product moves quickly, whereas if it spends the main time in the stationary phase, the product is retained and its output is much slower.

The most commonly used in Organic Chemistry is liquid-solid chromatography in its two variants: thin-layer chromatography (TLC) and column chromatography (CC).

4.14.1. Thin-layer chromatography (TLC)

This is one of the most popular solid-liquid chromatography variants,[19] and its most common uses are:

- To identify the components of a mixture.

- To identify two substances.

- To follow the course of a reaction.

---

[19]Izmailov and Scraiber, in 1938, used glass plates to spread very thin layers of alumina and then applied plant extracts, thereby giving rise to the first form of TLC. Egon Stanl (1956) gave the name of "thin-layer chromatography" (TLC) and standardized the procedures, equipments and adsorbents, popularizing this simple, efficient, and inexpensive technique.

- To determine the appropriate conditions for the separation of mixtures by CC.

- To follow the development of a CC.

a) Adsorption process

The sample is applied to the layer and is adsorbed on the material surface by the action of electrostatic forces (van der Waals forces, hydrogen bonds, inductive effects, etc.). Later, when the plate is exposed to a flow of solvent by capillarity, the substance undergoes a competition between the active site of the adsorbent and the solvent.

b) Adsorbents

The most common adsorbents in TLC are:

- Silica gel (SiO_2) (used in 80% of separations).
- Alumina (Al_2O_3) (acid, neutral, or basic).
- Siliceous earth or Kieselguhr.
- Cellulose (native or microcrystalline).
- Polyamide.

In an Organic Chemistry laboratory, the silica gel is used in most cases as a support. These adsorbents are characterized by the particle size (pore volume, pore diameter, surface area, homogeneity) and purity. The adsorbent is placed on a plate, which acts as an inert support and which is usually made of glass, aluminum, and polyester.

Although the use of commercial chromatofolios is increasingly widespread, TLC plates can be prepared using glass slides. They can be prepared by placing two slides together, clean and dry, in a suspension of silica gel 35% AcOEt containing $CaSO_4$ as adhesive. For the immersion in the suspension of adsorbent, tweezers are used. Silica gel forms a film on each of the faces of the slide that is exposed to the suspension. Evaporation of the solvent leaves the plate ready for use.

c) Determining R_f value.

For help in identifying products that are in a mixture, the so-called ratio to front (R_f) value is often used, calculated according to Eq. 4.2.

$$R_f = \frac{\text{Distance traveled by the compound}}{\text{Distance traveled by the solvent}} \tag{4.2}$$

4.14.1.1. Description of materials and methods used in this technique

The materials used in this technique are listed below:

- Chromatographic plates

- Capillary and Pasteur pipettes

- Eluents

- Tank elution

- Developer

a) Chromatographic plates: These are used as the adsorbent-carrier sheets of glass, plastic, or aluminum. The sizes (TLC foils) of conventional TLC plate are 20×20, 10×20 and 5×2 (in cm). There are plates containing a fluorescent indicator (F_{254} or F_{366}).[20]

b) Capillaries and Pasteur pipettes: The sample is applied dissolved onto the plate whether the target is preparative or analytical. For preparative purposes, the sample is dissolved and applied with a Pasteur pipette; this is called band application. If the aim is to perform the technique for analytical purposes, the sample dissolved is applied using a glass capillary. In this case, it is called point or spot application. The sample should be applied at a distance from the edge of the plate. The height and the precise location where the sample is placed can be marked with a pencil to avoid damaging the adsorbent.

c) Choice of eluent: Choosing a solvent in which the components of the mixture present an R_f average around 0.3–0.5 is recommended. The search for a suitable eluent requires testing several solvents of different polarities or mixtures. Table 4.3 lists a number of organic solvents sorted by their dipolar moment. When a compound is eluted at an R_f less than 0.2 or greater than 0.7, what looks like a single compound may actually be a mixture. In these cases, a change should be made to a more or less polar solvent, respectively.

For less polar compounds, moving from the origin, apolar eluent such as hexane should be used. In the case of compounds of medium polarity, hexane/ethyl acetate and hexane/diethyl ether mixtures in different proportions are recommended. More polar products are retained in the adsorbent, and they require more polar solvent mixtures such as CH_2Cl_2/MeOH in different ratios.

One way to check whether the eluent is appropriate involves taking a sample of the mixture to be chromatographed, dissolving it in the chosen solvent mixture, pricking a TLC plate with a capillary, and observing the result of diffusion of the sample (see Figure 4.33). The solvent choice is correct when the product spot forms a circle of approximately half the size of the solvent front on the plate (see Figure 4.33 (ii)). Otherwise, when the product spot and solvent front are very close to each other, or when the product spot is

---

[20]The number displayed as a subscript indicates the excitation wavelength (in nm) of the indicator.

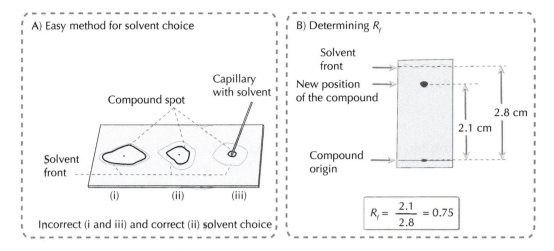

Figure 4.33: Determination of R_f and method to choose a solvent.

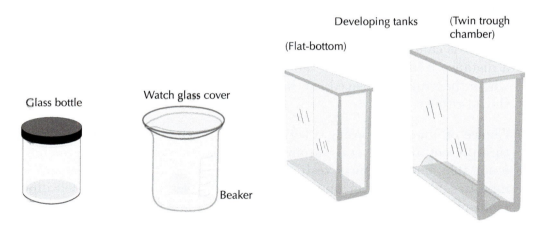

Figure 4.34: Sprayer and tanks for developing TLC plates.

close to the origin, the solvent choice is not correct (see Figure 4.33 (i) and (iii)).

d) Elution tank (chromatography tanks): This is a closed container used for the development of TLC plates. Its atmosphere is saturated with eluent vapors. There are various types of tanks on the market (see Figure 4.34). Flat-bottom tanks with different designs, dual compartment tanks (with advantages the former of less solvent consumption and no waste, allowing the environment to be saturated with different vapor-eluting solvent), etc. Closed bottles and even a beaker with a watch glass are also used. The tank is filled with the eluent to a height that does not reach the area where the products are located. Usually it is a good practice to place in the elution tank a piece of filter paper that reaches the mouth of the tank so that the eluent rises by capillarity, and the tank becomes saturated with eluent vapors. This facilitates proper development of the plate. When the solvent front has reached the upper edge of the plate, this tank is removed and then revealed.

e) Detecting or visualizing (revealed): If the sample is not colored, a method that allows visualization of the components present is required. This procedure is known as revelation. These methods are:

- Using a UV lamp with 254 nm filter: When using plates with a UV indicator, if the substances have at least one chromophore, darker spots are observed in light green and sometimes bright spots, mainly blue ones.

> **Noticeboard 4.14.1**
>
> **DANGER!**
>
> Never look directly into the UV light, because it can cause permanent damage to the eyes.

- Chemical methods: There are many developers for TLC, the most common listed below [2]:
  - H_2SO_4/EtOH 50%. Universal and inexpensive developer. Once the TLC plate is sprayed, it has to be heated either with a hot air dryer or on a hot plate to see the spots, presenting these dark colors.
  - $KMnO_4$ (3 g)/K_2CO_3 (10 g)/water (300 ml). This is considered a universal developer, incompatible only with eluents containing amines. The TLC plate takes the typical color from permanganate stains shown with yellow and brown hues.
  - Phosphomolybdic acid (10 g)/EtOH (100 ml). Expensive and applicable to the majority of organic molecules.
  - Ninhydrin (0.1 g)/acetic acid (2 ml)/acetone (100 ml). It is used mainly for amino acids and nitrogen compounds.
  - Iodine. It is one of the oldest developers that is still used for a variety of organic molecules. After the TLC plate stays in a tank a few minutes, a few iodine crystals appear, and brown tones become visible but disappear when the plate is removed. Therefore, it is advisable to mark the spots with a pencil in order to keep a reference for later.

Developers can be used by spraying the TLC plate or by immersion in a tank with the developer solution. In the case of spray, both plates are used that allow concentration of the developer sprayed onto the TLC plate and are often placed in a fume hood. Thus the aerosol remains can be extracted from the workplace (see Figure 4.35).

f) Method for analytical TLC (see Figure 4.36):

- Prepare a solution with a small amount of the sample.

Figure 4.35: TLC chromatography.

- For 20×20 cm TLC foils, cut using a cutter in plates of 1/3 TLC foils high and wide depending on the number of samples that will be applied.

- Draw a pencil line 0.5 cm from the edge of the plate, and score the places where the sample with the capillary will be applied.

- Dip the capillary into the sample solution, and transfer the contents to a point marked on the plate. Dry if necessary with an air dryer to prevent the dissolution from spreading along the surface of the plate before putting it into the tank.

- Place the plate in the tank, and let the eluent to rise to the top.

- Take it out of the tank and reveal it.

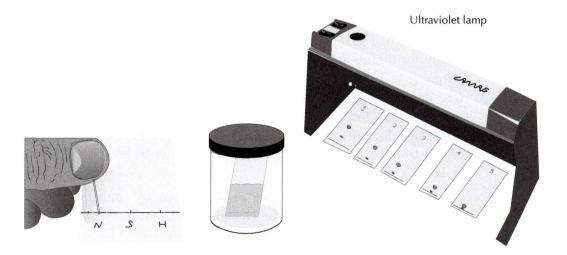

Figure 4.36: TLC application, developing tank, and revealing with a UV lamp.

4.14.2. Column chromatography (CC)

A method frequently used for purifying substances and separating mixtures at a preparative scale. Use a chromatography column as the main device. The column is filled with an adsorbent that acts as a stationary phase, usually silica gel or alumina. The sample is introduced into the column and a solvent or solvent mixture (eluent) is passed to separate the impurities of desired product or the components of a mixture, due to the different retention time on the stationary phase to separate. The eluent can be run throughout the column either by gravity or by exerting pressure with an inert gas or by an air pump. This type of CC is called flash chromatography. In the latter case, the column must have a ground-glass adjustment[21] at the top to which it is connected to an adapter having a side tube with a tapper joined to a flexible tube that is supplied with pressurized gas. Flash chromatography is very efficient and has the advantage that it is much faster than the gravity chromatography. The stationary phase used in this technique is specific to it, as it has a smaller size than a conventional chromatography particle. This is because finer particles of absorbent involve a longer effective path for the eluent, which is beneficial the higher the flow rate of the mobile phase, as in flash chromatography.

a) Method for performing a pressure CC with silica gel:

- Selection of the mobile phase: Solvent choice is crucial for good separation. If the solvent to be used is very polar, elution is fast and there is little separation between the product and impurities or between components of the mixture. If instead the solvent is apolar, the compounds will be retained in the column. Therefore, several TLC plates should be developed with silica gel to determine the solvent (or solvent mixture) most suitable to keep the observed spots as far apart as possible.

- Stationary phase: The amount of silica gel employed depends on the amount of sample to be purified and on the R_f of the spots on TLC. As a guide, for a sample presenting a pair of spots with R_f about 0.5 and with appreciable separation between them in TLC, the amount of silica gel (type 40–75 μm) per unit mass of sample would be 50:1. The column is fixed with a clamp and a clamp holder to a lab stand, and the stationary phase is introduced by forming a slurry with the same eluent to be employed after a powder funnel is used. If the column has porous glass at the base plate, it can be added directly; otherwise a compact cotton ball is placed using a glass rod to retain the silica gel on the column, and the eluent passes through. It is important that the ground-glass adjustment of the column gel does not become soiled with silica gel, and the possible column residues should be dragged with solvent. Some sand can be put on the edge of the silica gel to prevent

---

[21]Do not forget to fix the ground-glass joint with a Keck clip or metallic clip.

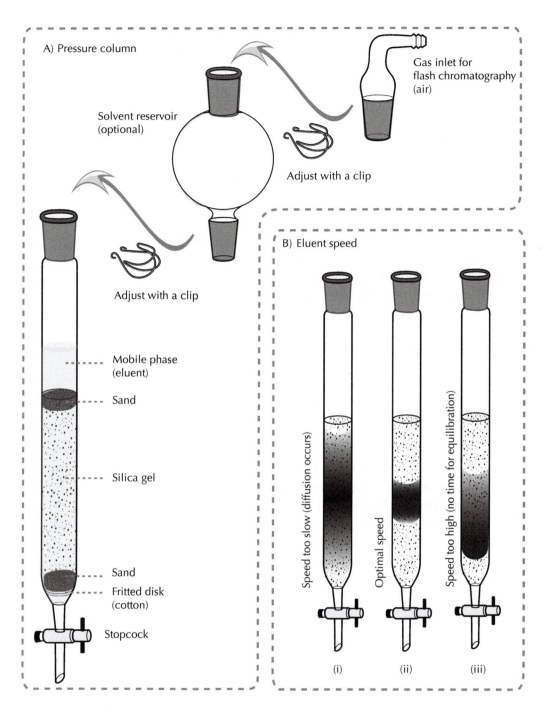

Figure 4.37: Column chromatography (CC).

removal when the eluent is added. Column size must be large enough to leave a free volume to allow adding the eluent.

- Sample application: Usually the sample is placed in the column, adjusting the eluent level just to the silica gel limit. If the sample is soluble in the eluent, it is dissolved in the minimum amount and transferred to the column with a dropper, on silica gel, avoiding splashing the walls of the column. In the case that the solvent is insoluble, in a round-bottom flask a solution of the sample to which silica gel is added is prepared. The solvent is evaporated under reduced pressure on a rotary evaporator placing a break foam between the guide tube and the flask to prevent silica gel from missing the guide tube of the flask. A similar effect to break foams can be achieved if a spatula is placed in the flask, taking care not to prevent the rotation of the flask. After the solvent evaporates, the silica gel with the product adsorbed on the surface thereof is introduced into the column.

- Addition of eluent: Once the sample has been placed in the holder, the eluent is added and an adapter with a side outlet is connected to the pressure device or pump with a flexible tubing. It is secured with a clip and pressure is applied. An additional container can be used for the solvent, placing it between the column and the pressure tap (see Figure 4.37).

- Elution rate: For each chromatographic separation, there is an optimum value of the elution rate. If the rate at which the eluent flows is too slow, the sample will diffuse excessively into the column, the run time will lengthen the separation time, and the efficiency of the separation will decrease. If instead the eluent flow is too fast, not enough sample will be provided to interact with the column packing, and the separation time will also be inefficient. Moreover, for small-diameter columns, the eluent flow should be much slower than for the large diameter, in order to achieve an efficient separation.

- Collecting eluent and isolating the products: The usual practice is to collect the eluent in a test tube, to be placed neatly in a rack. To find out which are the product fractions or products that have been purified, an analysis of these by TLC is made. Once the product is identified, the corresponding fractions are collected in a round-bottom flask and the solvent is removed (rotary evaporator).

4.14.3. Ion-exchange chromatography

Ion-exchange chromatography (or ion chromatography) allows the separation of ions and molecules with a net charge. Ionic resins are used as stationary phase. Depending on the load of the sample to separate or purify, two types of anion and cation exchange are distinguished. Buffered aqueous solutions are used as a mobile phase.

4.14.4. Size-exclusion chromatography

In this type of chromatography, the stationary phase is composed of cross-linked polymers that form a porous three-dimensional interconnected network that allows or prevents the passage of molecules based on their size, promoting separation. The stationary phase is usually introduced into a column. It is particularly suitable for separating high-molecular-weight compounds such as polypeptides, proteins, and nucleic acids.

4.15. References

1.   W. M. Haynes, D. R. Lide, and T. J. Bruno, *CRC Handbook of Chemistry and Physics 2012-2013*, CRC Press, Boca Raton, FL, 2012.

2.   B. Fried and J. Sherma, *Thin-Layer Chromatography, Third Edition, Revised and Expanded*, Marcel Dekker, New York, NY, 2003.

Chapter 5

Determining Physical and Spectroscopic Properties

5.1. Introduction

The organic compounds, once they have been isolated and purified, are characterized according to their physical and spectroscopic properties. These properties will indicate unambiguously their structure and purity. This chapter describes the lab equipment that allows the determination of the most common physical properties (including spectroscopic or spectrometric) and how to prepare a sample of a substance for analyzing these properties. Moreover, the basics of the different methods and equipment necessary to make such determinations are also described. Also, a description is provided on how to prepare the sample for applying spectroscopic and spectrometric techniques that today constitute the quickest and most reliable methods for structure determination of organic compounds: ultraviolet-visible spectrophotometry (UV/vis), infrared (IR), mass spectrometry (MS), and/or ^{1}H proton and ^{13}C carbon nuclear magnetic resonance (NMR). A description of the instruments needed to carry out such spectroscopic and spectrometric techniques as well as the instrumental background is not included because it is beyond the scope of this book.

5.2. Determining physical properties

5.2.1. Melting point

The melting point (m.p.) of a substance is the temperature at which a solid changes to the liquid state (melts). This physical property is easy to measure, is characteristic of a specific substance, and serves as a purity index. The m.p. is hardly affected by changes in pressure. The fusion of a substance is usually given in a range of a few °C and, with increasing amounts of impurities in a solid, the m.p. decreases and the interval in which the solid changes to liquid lengthens. The m.p.s of pure common organic compounds have values below 300 °C and are listed in databases and manuals. For the m.p. of a compound to be correctly determined, it is necessary to start from a dried sample that is free of solvent residues.

5.2.1.1. Sample preparation

The first step in determining the m.p. of a solid is to place a few crystals of a solid into a glass capillary closed at one end (m.p. capillary tubes are available on the market). For this, a small amount of solid is placed on a watch glass or in a Petri dish, and the capillary tube is filled with the solid 2 or 3 mm in height. Then the capillary with the solid is reversed and is dropped inside a piece of hollow glass rod supporting it at the work table. Thus, the solid is compacted on the closed end of the capillary. Once the sample is prepared in this way, it is placed in the corresponding m.p. apparatus for measurement.

5.2.1.2. Apparatus for measuring the melting point

Various types of devices are available on the market for measuring the melting point (see Figure 5.1). The simplest of these is the Thiele tube. This tube is a heat-resistant glass containing a fluid (typically a mineral oil) with a stopper together with thermometer that has an opening through which an m.p. capillary is introduced with the sample. The Thiele tube is heated with a gas burner until the fusion of the solid is seen in the capillary. The thermometer and capillary should not be in contact with the glass tube. The Thiele tube has been replaced in most laboratories by electrical m.p. devices, which are safer and more accurate. There are different designs, types, and brands on the market for m.p. devices, most of them digital, but they all work similarly. The simplest model has a hole through which the capillary is inserted, leaving the hole in front of a viewfinder. The rate at which the temperature increases is adjustable and the m.p. is determined when the fusion of the solid is seen in the capillary thorough the viewfinder. Instead of having a fluid to transmit heat, some devices have a metal block.

There is also automatic equipment that allows a direct reading of the m.p. without the need of an observer to see when the solid melts. More specific is the use of a microscope with a metal block, a resistor, a potentiometer, and a thermometer. The sample is placed between two glass plates and gradually heated. This procedure is more precise and allows the observer to see whether there are other significant changes in the sample, such as softening or changes in color or homogeneity.

5.2.2. Boiling point

The boiling point (b.p.) of a substance is the temperature at which the vapor pressure of the liquid equals the atmospheric pressure, and therefore the value depends on the conditions (pressure) under which the measurement is made. The b.p. increases with the molecular mass of the substance and the types of intermolecular forces present. For an organic substance that is liquid at r.t., the

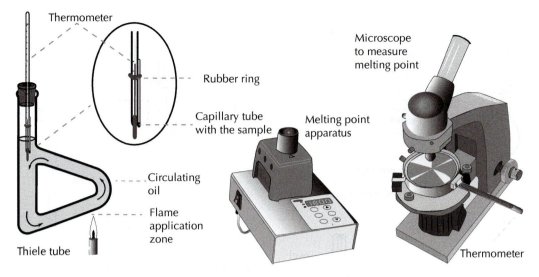

Figure 5.1: Different devices for measuring the melting point (m.p.).

b.p. is a measure that characterizes the substance and is indicative of its purity. Depending on the available amount of a substance, different procedures exist to determine the b.p., but in essence they involve distillation.

5.2.3. Flash point

The flash point is the lowest temperature at which a volatile substance evaporates to form an ignitable mixture with air in the presence of an igneous source and continues burning after the trigger source is removed. This parameter is related to the degree of danger of a volatile substance. Tests to determine this property often use a small flame for ignition. The liquid is heated slowly from a temperature lower that the flash point, with increasing temperature steps and applying a test flame to the vapor chamber. The flash point is the temperature at which a flash is observed when applying the flame or ignition source.

5.2.4. Density

The density of a compound is defined as the mass of the substance per volume unit. In the International System of Units, the density values are expressed in $kg \cdot m^{-3}$, although the most common way to express it is in $g \cdot cm^{-3}$.

The density of a substance as a solid, liquid, or gas is determined with different devices:

- Hydrometers: Also called hygrometers, these determine the density of liquids. They are built in a hollow glass cylinder with a heavy bulb on the end so it can float. A scale according to their level of sinking gives the density of the liquid. Depending on the type of solution or liquid used to measure the density, specific names are used (alcohol meter, milk density meter, wine meter, etc.).

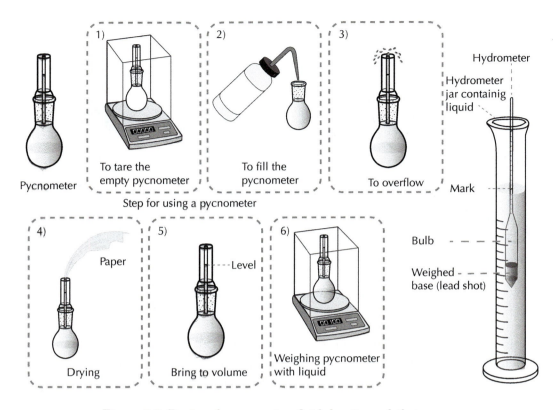

Figure 5.2: Devices for measuring fluid density and their uses.

- Pycnometers: These are used to determine the density of liquids (see Figure 5.2). Pycnometers maintain a fixed volume by placing different liquids inside, this allowing the densities of two liquids to be compared by weighing the pycnometer with each of the liquids. Liquid density is determined by following the steps shown in Figure 5.2. The empty pycnometer (W_p) (step 1) is weighed and then filled with the liquid of unknown density, making sure that the liquid, by capillary action, rises and overflows (steps 2 and 3). The remaining liquid is absorbed with a piece of paper until the liquid level reaches the calibration mark (steps 4 and 5). Both the container and the cap should be thoroughly dried to remove any drops of liquid that had overflowed. The liquid is weighed again (now flush, step 6) and the value (W_x) is noted. When the volume of the pycnometer (V_p) is known, the following calculation is made:

$$D_x = \frac{(W_x - W_p)}{V_p} \tag{5.1}$$

5.2.5. Solubility and partition coefficient

IUPAC defines the solubility of a substance (solute) as the analytical composition of a saturated solution, expressed in terms of the ratio of the solute in a given solvent. The solubility can be expressed as a concentration, molality, mole fraction, mole ratio, etc. The solubility of a substance depends heavily on temperature. This fact is used for purifying compounds by recrystallization (see Section 4.7.3 on p. 105).

In the case of two immiscible solvents capable of dissolving a substance at least partially, the solute is distributed between the two. The partition coefficient (K) is called the distribution coefficient (D), while the partition coefficient (P) is the solute concentration ratio between the two solvents.

$$K = \frac{[\text{Substance}]_1}{[\text{Substance}]_2} \tag{5.2}$$

where $[\text{Substance}]_1$ is the concentration of the substance in the first solvent, and $[\text{Substance}]_2$ is the concentration of the substance in the second solvent.

5.2.5.1. Partition coefficient octanol-water (LogP)

The octanol/water partition (K_{OW}) of a substance, also called the partition coefficient (P_{OW}), is the ratio of concentrations of a substance in a mixture of two solvents, octanol and water. Both solvents are immiscible and therefore form two phases.

$$K_{OW} = K_{\frac{\text{octanol}}{\text{water}}} = \frac{[\text{Solute}]_{\text{octanol}}}{[\text{Solute}]_{\text{water}}} \tag{5.3}$$

The behavior of such drugs or contaminants is a noteworthy concept to discuss because it is indicative of the tendency of a substance to be solubilized in tissues from an aqueous medium. Instead of the values of K and P, their logarithms, log K_{OW} or log P_{OW}, are often used.

The octanol/water partition coefficient can be determined experimentally by liquid-liquid extraction and by high-performance liquid chromatography (HPLC) and can also be theoretically calculated from parameters related to the structure of molecules.

5.2.6. Refractive index (index of refraction)

The refractive index or index of refraction (n) (measuring speed of propagation of light in a substance) is generally used in laboratories and in industry to verify the composition and purity of a liquid. It is a useful and inexpensive technique in certain cases. One of the most common applications in the industry has been, for example, for determining the concentration of sugar or salt in foodstuffs. A refractometer is used for measuring the refractive index.

There are different refractometer models (see Figure 5.3), but two of the most widely used are the Abbe refractometer (invented by Ernst Abbe in 1874) and the portable refractometer. Both allow the measurement of the refractive index of a liquid on a surface by depositing a drop of the sample and adjusting certain settings. Abbe refractometers can also use two refractive methodologies to take measures: total refraction or critical-angle refraction.

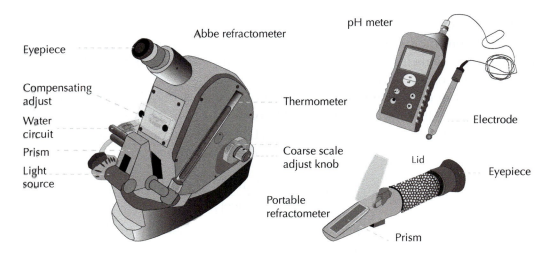

Figure 5.3: Two refractometer types and a pH meter.

Although Abbe refractometers at first used a monochromatic light (sodium D line, 589 nm), today they run on natural light or white light. They must be calibrated at least once with distilled water.

These devices have two scales: One that directly indicates the refractive index (1300–1700) and one that indicates the percentage of dry matter of sugary liquids (0–85%).

The Abbe refractometer can be used only for substances with a refractive index that is less than that of flint glass ($n = 1.7$).

As the refractive index varies with temperature, accurate refractometers have a thermometer together with a water-circulation system fitted with a thermostat.

The portable refractometer is often used to measure refractive indices of fruit juice in the field and works directly with natural light. With a scale ranging from 0 (distilled water), it is able to give the sugar concentration. A drop of juice is deposited on the prism, the lid is closed, and the measurement is made directly on the scale, while aiming the prism toward the light. This usually involves an automatic temperature compensation to eliminate that factor (0.00045 refractive index units per degree Celsius, around 20 °C).

5.2.7. Specific rotation

The deflection plane of polarized light is the property of having molecules called "optically active." The technique to measure the deflection angle is called "polarimetry." The polarimeter is an instrument to measure this magnitude when polarized light passes through a solution of a given sample. It is measured in degrees (specific optical rotation) and is calculated with the following equation:

$$[\alpha]_D^{25} = \frac{\alpha_{\text{observed}}}{c \times l} \tag{5.4}$$

Specific rotation $[\alpha]_D^{25}$ is defined as the angle of optical rotation observed (α), when a light beam $\lambda = 589$ nm (sodium D line) passes through a sample in a cell of step 0.1 m, concentration 1 g·dl^{-1} (or 1 g·100 ml^{-1} and cell 10 cm or 1 DM) and 25 °C temperature.

The final value of α is expressed as a decimal point with an associated sign. For two enantiomers, the specific rotation has the same absolute value but an opposite sign.

Normally, a sodium lamp, which emits light at a particular frequency ($\lambda = 589$ nm, called line D) is used. For a measurement, a solution must be prepared in a volumetric flask, dissolving the sample with high-purity solvents (HPLC or spectroscopic grade). Normally, concentrations are 10 mg·ml^{-1} or 1 g/100 ml but can be increased if the value of the specific rotation is small. Cylindrical quartz cuvettes 10 cm long and standard volumes of 1, 5, or 10 ml are used (see Figure 5.17). These sample cells are designed with a double jacket with inlet and outlet water that can be connected to a thermostatic water bath, maintaining the temperature of the tank at the desired temperature during the measurement. The cells are usually filled with the solution using a Pasteur pipette, ensuring that no bubbles appear. It is preferable to tare with the solvent prior to measuring the sample.

The resulting measurement is expressed with its sign and one decimal place, the wavelength used for the measurement, the temperature, the concentration (in g·100 ml^{-1}), and the solvent.

Furthermore, two other magnitudes can readily be calculated from the specific rotation of a sample (with a mixture of two enantiomers) and serve to determine its purity.

- Optical purity (*op*): The optical purity of a mixture is defined as the ratio of the observed rotation of the mixture divided by the optical rotation of the pure enantiomer.

$$op = \frac{\text{(observed rotation)}}{\text{(rotation of pure enantiomer)}} \times 100\% \qquad (5.5)$$

- Enantiomeric excess (*ee*): Also expresses the relative proportion of enantiomers (+) and (−) in a mixture. Predominant *ee* is calculated as a percentage of the mixture.

$$ee = \frac{(d - l)}{(d + l)} \times 100\% = \frac{\text{(excess of one enantiomer over the other)}}{\text{(sum of the mixture)}} \times 100\% \quad (5.6)$$

5.2.8. pH meter

Many organic products in solution will make these acidic or basic, and therefore to determine their acidity or basicity is fundamental. For this, the pH meter is used, and the sample must be water soluble.

In general, this device consists of two electrodes that are joined into one and called a combined electrode. The pH meter is also often employed for conductivity measurements so that we can use it as a pH meter and/or an EC meter (electrical conductivity meter if the combined electrode also incorporates a temperature sensing probe). When the pH meter is not used, the electrode is left immersed in a solution of 0.3 M KCl (pH meter) or distilled water (EC meter).

The procedure for making a measurement is as follows:

1. Remove the electrode from the solution in which it is submerged.

2. Place a beaker (usually 150 ml) under the electrode, and with the aid of a wash bottle, wash the electrode.

3. Dry the electrode carefully using paper.

4. Dip the electrode into the desired solution, with particular care not to touch the bottom since the end of the electrode is the most sensitive part of this instrument.

5. Slightly stir the solution.

6. Remove the electrode from the solution, wash and dry (following the instructions in steps 2 and 3).

Insert electrode again into the original solution if no further action will be performed.

As with any measuring instrument, prior to use, it must be calibrated at a temperature determined using the corresponding standard solutions (see more details in the instruction manual of the meter). Figure 5.3 shows a portable pH meter with a combined electrode.

5.3. Determining spectroscopic properties

Obtaining structural information from spectroscopic data is an integral part of Organic Chemistry courses common in all universities. At the undergraduate level, the principal aim of such courses is to teach students to solve simple structural problems efficiently by using combinations of the major techniques (e.g., UV, IR, NMR, and MS).

Organic compounds are commonly isolated and purified in laboratory experiments prior to their identification, by means of either organic or spectroscopic analysis. Spectroscopic techniques require a sample preparation of the products suitable for each technique. The goal of this textbook is not to describe or explain the background of each of these spectroscopic techniques or to make a detailed description of the instruments required.

All types of radiation are characterized by a wavelength (λ), a frequency (ν), or energy (E), and the relationship between them is described in Equation 5.7:

$$E = h \cdot \nu = h \cdot \frac{c}{\lambda} \tag{5.7}$$

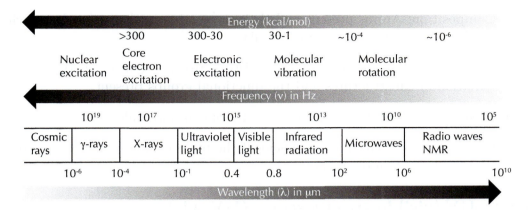

Figure 5.4: The electromagnetic spectrum and its effect on matter.

The first question to ask is what happens to the matter when it is subjected to a given radiation? Any effect depends on the energy of this radiation. In Figure 5.4 the immediate effects on matter are indicated, emphasizing from high to low energy.

When radiation is incident on a substance, not all of that substance is affected; the atom or group of atoms that absorb the radiation is called a "chromophore," and each spectroscopic technique will be different within the same molecule. In molecules, there are also atoms or groups of atoms that do not absorb radiation but that alter some of the characteristics of absorption of the chromophore, and these are termed "auxochrome" groups.

The second question is how we can use these effects on substances in order to gain information on the structure of matter and how to use that information.

In a very schematic way, the following are the most remarkable characteristics of the corresponding spectroscopic methods:

- X-ray: Overall structure of the molecule including the stereochemistry from the relative positions of atoms.

- UV/vis: Existence of chromophores and/or conjugation in the molecule from the observed absorptions.

- IR: Functional groups of the observed absorptions.

- MS [1]: Molecular formula and substructures from the observed ions.

- NMR: Functional groups, substructures, connectivity, stereochemistry, etc. from chemical shift data, peak area, and coupling constants observed.

A goal of the structural characterization is to elucidate molecule stereochemistry —i.e., the spatial arrangement of the atoms that form the molecule. This involves the successive knowledge of:

---

[1]This is not a spectroscopic technique in the sense that there is no electromagnetic radiation of the substance and no absorption of this radiation.

- Composition: Atoms present and their proportion in the molecule, which results in making a molecular formula.

- Constitution: Connections among existing atoms, resulting in the determination of substructures and the functional groups present.

- Configuration: Spatial arrangement of atoms in molecules.

- Conformation: Spatial arrangement of the molecule arising from the possibility of rotation of single bonds.

5.3.1. UV/visible spectroscopy (UV/vis)

The UV/vis spectroscopy provides information on the electronic transitions that occur in a molecule, by interaction of electromagnetic radiation with matter. This type of spectroscopy is governed by the Beer-Lambert law.

$$\text{Log}_{10}\frac{I_0}{I} = A = \epsilon \cdot l \cdot c \qquad (5.8)$$

where:

- I_0: Intensity of radiation entering the sample (incident).

- I: Intensity of radiation leaving the sample (transmitted).

- A (absorbance) $= \epsilon \cdot l \cdot c$: is the molar absorbance.

- c (concentration): Sample concentration in mol dm^{-3}.

- l (path length): Length of the cell traversing radiation cm.

- ϵ (absorptivity): Molar extinction coefficient (extinction coefficient).

The recording apparatus that allows UV/vis is called a spectrophotometer, and it measures the absorption or transmission of light through a sample. This uses radiation in the electromagnetic spectrum whose wavelength is between 100 and 800 nm, corresponding to an energy of between 286 and 36 kcal·mol^{-1}. It is based on the result of producing electronic transitions between atomic and/or molecular orbitals of the substance. Standard UV spectrometers record a range of 200–380 nm. The absorptions, in the visible region (approximately 380–800 nm), are of less importance in the solution of structural problems, because most organic compounds are colorless. The extensive region at wavelengths shorter than approximately 200 nm ("vacuum ultraviolet") also corresponds to electronic transitions, but this region is not readily used.

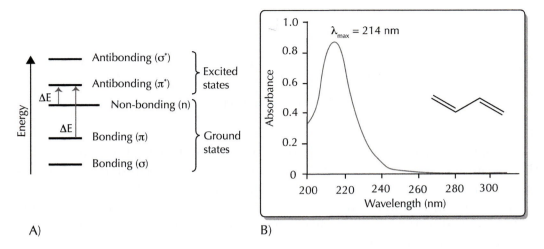

Figure 5.5: (A) Different UV/vis absorption bands of molecular orbitals. (B) Example of buta-1,3-diene UV spectrum.

Table 5.1: Some examples of chromophore groups in UV/vis.

Chromophore	Compound	λ_{max} (nm)	ϵ
C=C	Ethylene	170	15,800
	trans-2-Hexene	184	10,000
	Cyclohexene	182	7,600
	1,3-Butadiene	214	20,000
C≡C	1-Octyn	185	2,000
		222	126
C=O	Acetaldehyde	277	8 (H_2O)
		290	16 (Hexane)
C=O	Ketone	279	15
C=O	Acetic acid	204	60
C=NOH	Acetoxime	190	5,000
NO_2	Nitromethane	271	19
S=O	Cyclohexyl methyl sulfoxide	210	1,500

For analytical purposes, UV/vis absorption spectra are generally recorded in a solution of known concentration of the product. From the spectra obtained and using the Beer-Lambert law, the absorbance and molar extinction coefficient (ϵ) are determined. The values of (ϵ) vary between 10 and 10^5. UV absorption bands are characterized by the wavelength of the absorption maximum and $\log_{10}(\epsilon)$ (see Figure 5.5 (B)).

As the transitions are associated with changes of electron orbitals, they are often described in terms of the orbitals involved (e.g., $\pi \rightarrow \pi^*$, $n \rightarrow \pi^*$, etc.; see Figure 5.5 (A)), where n denotes a non-bonding orbital, the asterisk denotes an antibonding orbital, and σ, and π have the usual significance.

Figure 5.6: Examples for bathochromic and hypsochromic shifts.

A molecule may give rise to more than one band in its UV spectrum, either because it contains more than one chromophore or because more than one transition of a single chromophore is observed (see Table 5.1).

The following symbols are used for aromatic compounds: B (for benzenoid), E (for ethylenic), R (for radical-like), and K (for conjugated)[2] for the observed band type of transitions (see Table 5.2).

Table 5.2: Some examples of chromophore groups in UV/vis.

Band	Transition	$\lambda_{\max}$ (nm)	ϵ
E	$\pi \to \pi^*$	180–220	2,000–6,000
K	$\pi \to \pi^*$	220–250	10,000–30,000
B	$\pi \to \pi^*$	250–290	100–1,000
R	$n \to \pi^*$	275–330	10–100

Examples of different band values are listed in Table 5.3 for common organic compounds.

Table 5.3: Some examples of band transitions for chromophores

Compound	Band E	Band K	Band B	Band R
Benzene	184 (47,000)	-	254 (204)	-
trans-Buthylbenzene	208 (7,800)	-	257 (170)	-
Styrene	-	244 (12,000)	282 (450)	-
Acetophenone	-	240 (13,000)	278 (1,100)	319 (50)

If a structural change, such as the attachment of an auxochrome, leads to the absorption maximum being shifted to a longer wavelength, the phenomenon is termed a bathochromic shift. A shift toward shorter wavelength is called a hypsochromic shift (see Figure 5.6).

Extension of conjugation in a carbon chain is always associated with a pronounced shift toward longer wavelength and usually toward greater intensity (see Figure 5.7).

---

[2]From the German *"konjugierte."*

The stereochemistry and the presence of substituents also influence UV absorption by the diene chromophore. On the other hand, solvent polarity may affect the absorption characteristics, in particular $\lambda_{\max}$, since the polarity of a molecule usually changes when an electron is moved from one orbital to another.

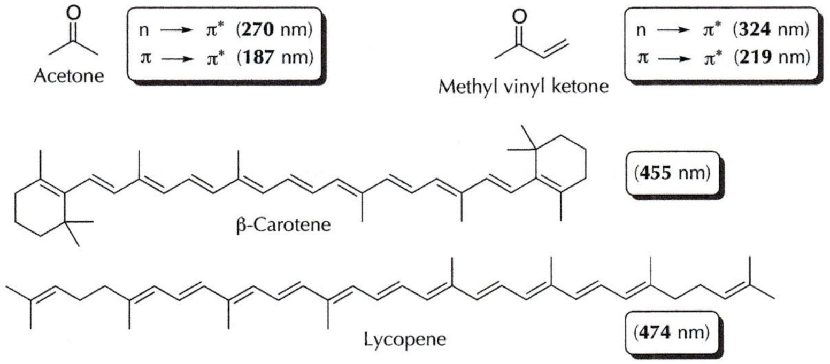

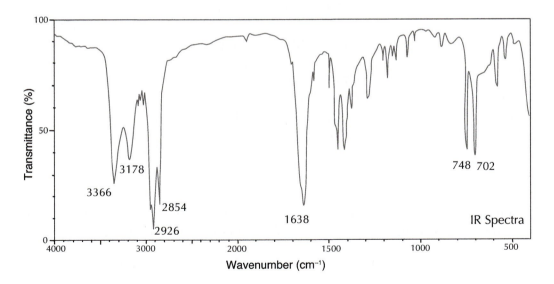

Figure 5.7: Examples of the effect of extended conjugation.

5.3.2. Infrared spectroscopy (IR)

Infrared absorption spectra are calibrated in wavelengths expressed in micrometers: $1\ \mu\mathrm{m} = 10^{-6}$ m or in frequency-related wave numbers (cm^{-1}), which are reciprocals of wavelengths: $\tilde{\nu} = 10{,}000 \cdot \lambda^{-1}$ (μ). The range accessible for standard instrumentation is usually (2.5–16 μ) or 4,000 to 666 cm^{-1}.

Figure 5.8: Example of an IR absorption spectra.

The transitions responsible for IR bands are due to molecular vibrations within the molecule. IR absorption intensities are rarely described quantitatively, except for the general classifications of s (strong), m (medium), or w (weak).

Molecules generally have a large number of bonds, and each bond may have several IR-active vibrational modes. IR spectra are complex and have many overlapping absorption bands. IR spectra are sufficiently complex that the spectrum for each compound is unique, and this makes IR spectra very useful for identifying compounds by direct comparison with spectra from authentic samples ("fingerprint region"). Very few substances are transparent over the whole of the IR range. Some groups of atoms (IR chromophores) are readily recognized from IR spectra.

Table 5.4: IR frequencies for common functional groups.

Functional group	$\tilde{\nu}$ (cm^{-1})	Functional group	$\tilde{\nu}$ (cm^{-1})
OH (H bonding)	3,100–3,200	-C≡C-	2,300–2,100
OH (non-H bonding)	3,600	-C≡N	~2,250
Ketones	1,725–1,700	-N=C=O	~2,270
Aldehydes	1,740–1,720	-N=C=S	~2,150
Aldehydes[a] & ketones[a]	1,715–1,660	C=C=C	~1,950
Cyclopentanones	1,750–1,740	NH	3,500–3,300
Cyclobutanones	1,780–1,760	N=C-	1,690–1,480
Carboxylic acids	1,725–1,700	NO$_2$	1650–1,500, 1,400–1,250
Esters	1,750-1,735	S=O	1,070–1,010
Esters[a]	1,750–1,715	Sulfones	1,350–1,300, 1,150–1,100
δ-Lactones	1,750–1,735	Sulfonamides & sulfonates	1,370–1,300, 1,180–1,140
γ-Lactones	1,780–1,760	C–F	1,400–1,000
Amides	1,690–1,630	C–Cl	780–580
-COCl	1,815–1,785	C–Br	800–560
Anhydrides	1,850–1,740[b]	C–I	600–500

[a] α,β-Unsaturated.

[b] Two bands.

When identifying the functional groups with IR spectroscopy, we will consider the IR spectrum divided into the following regions (see Figure 5.8):

- From 4,000 to 2,900 cm^{-1}: Stretching of C-H, O-H, and N-H.

- From 2,500 to 2,000 cm^{-1}: Stretching of triple bonds and accumulated double bonds.

- From 2,000 to 1,500 cm^{-1}: Stretching of C=O, C=N, and C=C.

- From 1,500 to 600 cm^{-1}: "Fingerprint region" (flexions of CH, CO, CN, CC, etc.).

According to this division, various functional groups can be identified, as indicated in Table 5.4.

5.3.3. Nuclear magnetic resonance (NMR) spectroscopy

All nuclei have charge because they contain protons and they spin. A spinning charge generates a magnetic dipole. Such nuclear magnetic dipoles are characterized by nuclear magnetic spin quantum numbers, which are designated by the letter I and can take up values equal to 0, $\frac{1}{2}$, 1, $\frac{3}{2}$, ..., etc.

If we consider a sample made up of many nuclei, the moments of these nuclei have a random distribution. However, when a magnetic field is applied, these nuclei are oriented according to the magnetic field.

In a uniform magnetic field, a nucleus of spin I may assume $2 \cdot I + 1$ orientations[3]. For nuclei with $I = \frac{1}{2}$ (e.g. ^1H, ^{13}C, ^{19}F, ^{31}P, etc.), there are just two permissible orientations[4] (see Figure 5.9).

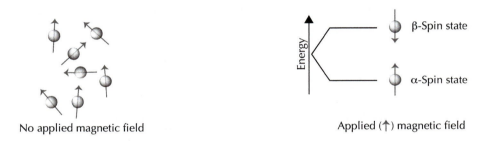

No applied magnetic field   Applied (↑) magnetic field

Figure 5.9: Orientations of nuclei when a magnetic field is applied.

These two orientations will be of unequal energy and it is possible to induce a spectroscopic transition by the absorption of electromagnetic energy of the appropriate frequency. At this time it is said to be in resonance. It is found that:

$$h \cdot \nu_o = \frac{\gamma \cdot H_o}{2 \cdot \pi} \tag{5.9}$$

where γ is a constant characteristic of the nucleus observed. This equation, is known as the Larmor equation, is the fundamental relationship in NMR spectroscopy. In NMR, the frequency of the absorbed electromagnetic radiation is not an absolute value but has a different value depending on the strength of the magnetic field applied.

---

[3]Nuclei with $I = 0$ do not interact with the applied magnetic field —e.g., $^{12}C12$, ^{16}O, etc.

[4]The α state is the most stable or lower energy, and opposed to it is the β state, less stable or with increased energy.

Table 5.5: Common IR stretching frequencies

Mass number	Atomic number	NMR signal	Examples
Even	Even	No	$^{12}C6$, $^{16}O8$
Even	Odd	Yes	$^{2}H1$, $^{10}B5$, $^{14}N7$
Odd	Even	Yes	$^{13}C6$, $^{17}O8$
Odd	Odd	Yes	$^{1}H1$, $^{11}B5$, $^{15}N7$

All the frequencies correspond to the radio frequency region of the electromagnetic spectrum, ν_o in MHz (10^6 Hz), and it depends on the strength of the field H_o applied. Therefore, for ^1H to a field $H_o = 14,000$ Gauss (1.40 Tesla), the resonant frequency would be 60 MHz, equivalent to an energy of 0.0057 cal·mol^{-1}. Not all nuclei present a resonant frequency when subjected to a magnetic field but only those nuclei with either the atomic number, the mass number, or both with an odd value. Those with the two even values do not present a resonance signal, as outlined in Table 5.5:

Table 5.6: Resonance frequencies of some nuclei in a magnetic field of 1 Tesla.

Nucleus	Natural abundance (%)	Spin	Frequency (MHz)[a]	Relative sensitivity
$^{1}H1$	99.9844	$\frac{1}{2}$	42.5759	1.000
$^{2}H1$	0.0156	1	6.53566	0.00964
$^{7}Li3$	92.57	$\frac{3}{2}$	16.547	
$^{9}Be4$	100	$\frac{3}{2}$	5.983	
$^{10}B5$	18.83	3	4.575	0.0199
$^{11}B5$	81.17	$\frac{3}{2}$	13.660	0.165
$^{13}C6$	1.108	$\frac{1}{2}$	10.705	0.159
$^{14}N7$	99.635	1	3.076	0.00101
$^{15}N7$	0.365	$\frac{1}{2}$	4.315	0.00104
$^{17}O8$	0.037	$\frac{5}{2}$	5.772	
$^{19}F9$	100	$\frac{1}{2}$	40.055	0.834
$^{31}P15$	100	$\frac{1}{2}$	17.236	0.0664

[a] Resonance frequencies at different H_o values.

Those nuclei that meet the above condition have a characteristic NMR signal when subjected to a magnetic field (H_o), so that the frequency of this signal will depend on the physical characteristics of the nuclei and sensitivity depends on the relative natural abundance of such nuclei. Resonance frequencies corre-

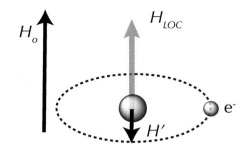

Figure 5.11: Schematic representation of the electronic shielding.

sponding to some nuclei in an applied magnetic field (H_o, 10,000 Gauss) are given in Table 5.6.

As the NMR phenomenon is not observable in the absence of an applied magnetic field, a magnet is an essential component of any NMR spectrometer. Magnets for NMR may be permanent magnets (as in many low field routine instruments), electromagnets, or in most modern instruments they are based on superconducting solenoids, cooled by liquid helium.

Figure 5.10: Time domain (left) and frequency domain (right) NMR spectra.

An intense short pulse of electromagnetic energy can be used to excite all the nuclei in an NMR sample. Following the radiofrequency pulse, the magnetism in the sample is taken as a function of time and, for a single resonance, the detected signal decays exponentially. The detected signal is called a free induction decay, or FID (see Figure 5.10, left), and this type of spectrum (known as a time domain spectrum) is converted into the more usual frequency-domain spectrum (see Figure 5.10, right) by performing a mathematical operation known as Fourier transformation (FT). Because the signal needs mathematical processing, pulsed NMR spectrometers require a computer.

The FID can be manipulated mathematically to enhance sensitivity (e.g., for routine ^{13}C NMR) at the expense of resolution or to enhance resolution (often important for ^{1}H NMR) at the expense of sensitivity.

5.3.3.1. The chemical shift (δ)

A NMR spectrum is a graph of resonance frequency vs. the intensity of absorption by the sample. All identical nuclei should resonate at the same frequency when they are in the same magnetic field. However, each nucleus is shielded by the electrons surrounding it. Like the nucleus, these electrons are also under the

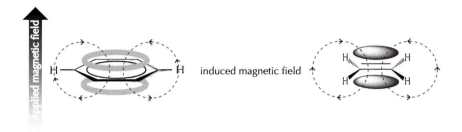

Figure 5.12: Chemical shielding for anisotropic groups.

influence of the magnetic field, H_o, and they should produce a small magnetic field, H', opposite to H_o (see Figure 5.11). Consequently, each nucleus is within an effective magnetic field, H_{LOC}, different from H_o, that can bee expressed as:

$$H_{LOC} = H_o - H' = H_o(1 - \sigma) \tag{5.10}$$

and the absorbed energy is:

$$E = h \cdot \nu_o = \mu \cdot \frac{H_o(1 - \sigma)}{l} \tag{5.11}$$

where σ is the shielding constant, which depends only on the molecular structure of the sample and not on the intensity of H_o. It is possible to calculate the maximum value of σ as a function of the electronic density around the nucleus, and its values cover the range from $\sigma = 18 \times 10^{-6}$ (Z = 1, H), 133×10^{-6} (Z = 9, F), etc. to 10^{-2} (Z = 80, Hg). Any effect that alters the electronic density around the nucleus, will alter the electronic shielding (e.g., the presence of neighboring electropositive/electronegative groups).

The chemical shielding of a nucleus may also be affected by the presence in the molecule of magnetically anisotropic groups, such as aromatic rings, double bonds, carbonyl groups, etc. For example, in an aromatic ring, the circulation of electrons forms a current loop, which gives rise to an induced magnetic field opposed to the applied magnetic field inside the ring and enhances the field outside the ring. Consequently, the nucleus is shielded or deshielded depending on its situation with respect to the anisotropic group (see Figure 5.12).

As a result of the above, each nucleus (e.g., ^{1}H) instead of a single frequency yields a range of frequencies (closely spaced) depending mainly on two factors: variation, the electron density around it and the magnetic anisotropy of the bonds of the molecule.

In this effect —i.e., the variation of the resonance frequency of a given nucleus with respect to the theoretical frequency —is termed *chemical shift* and is represented by δ, expressed for the following equation:

$$\delta = \frac{(H_o - H_{LOC})10^6}{H_o} \text{ppm} \tag{5.12}$$

$$\delta = \frac{(\nu_o - \nu_{LOC})10^6}{\nu_o} \text{(MHz) ppm} \tag{5.13}$$

The spectrum is usually calibrated in dimensionless units called "parts per million" (abbreviated to ppm), although the horizontal scale is a frequency scale, and the units are converted to ppm so that the scale has the same numbers irrespective of the strength of the magnetic field in which the measurement was made. The scale in ppm, termed the δ scale, is usually referenced to the resonance of some standard substance whose frequency is chosen as 0.0 ppm. The frequency difference between the resonance of a nucleus and the resonance of the reference compound is termed the chemical shift. Tetramethylsilane, $(CH_3)_4Si$ (abbreviated commonly as TMS) is the usual reference compound chosen for both ^1H and ^{13}C NMR.

$$\delta = \frac{(\nu - \nu_{TMS})10^6}{\nu_o}(MHz) \text{ ppm} \tag{5.14}$$

Chemical shifts can be measured in Hz but are more usually expressed in ppm. Note that for a spectrometer operating at 200 MHz, 1 ppm corresponds to 200 Hz; that is, for a spectrometer operating at 600 MHz, 1.00 ppm corresponds to exactly 600 Hz.[5]

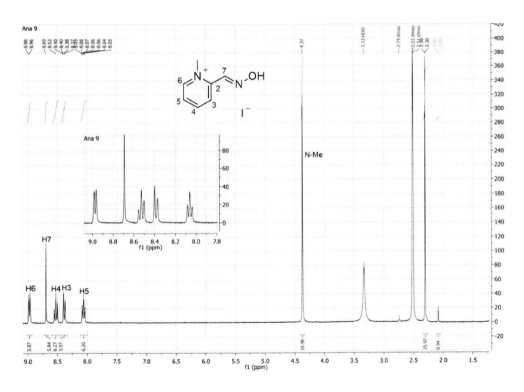

Figure 5.13: Example of ^1H NMR spectra.

---

[5]For example, a signal produced at 1.3 ppm in the first spectrometer would produce to 260 Hz relative to TMS and in the second to 780 Hz.

For the majority of organic compounds, the chemical shift range for ^{1}H covers approximately the range 0–10 ppm (from TMS) and for ^{13}C covers approximately the 0–220 ppm (from TMS) range.

The inductive effect of the methylene group adjacent to the halogen atom ($CH_3CH_2CH_2CH_2CH_2$–X) can be shown with the following data for the ^{1}H chemical shifts: X = F, Cl, Br, and I, of 4.5, 3.5, 3.4, and 3.2 ppm, respectively.

As a result of the foregoing, we can consider the proton NMR spectrum divided into a number of areas where different types of protons are located (see Figure 5.13 and Table 5.7).

Table 5.7: ^{1}H NMR chemical shifts for common functional groups.

Functional group	Proton type	Chemical shift (ppm)^a
Alkane	Cyclopropane	0.2
	R–CH_3	0.9
(acyclic)	R–CH_2–R'	1.3
(cyclic)	R–CH_2–R'	1.5
	R–CH–R'R''	1.5–2.0
Alkene	C=C–H	4.6–5.9
	C=C–CH_3	1.8
Alkyne	C≡C–H	2.0–3.0
Aromatic	Ar–H	6.0–8.5
	Ar–CH_3	2.3
Alkyl halide	CH_2–X (X=Cl, Br, I)	3.2–3.5
Alcohol	C–OH	1.0–5.0
	H–C–OH	3.4–4.0
Phenol	Ar–OH	4.0–4.7
Ether	–CH_2–O–	3.3–4.0
Aldehyde	–CHO	9.0–10.0
	–CH–CHO	2.0–2.7
Carboxylic acid	–$COOH$	10.5–13.0
Aliphatic amine	R–NH_2	1.0–3.0
Aromatic amine	Ar–NH_2	3.0–3.5
Amide	R–$CONHR'$	5.0–9.0

^{a} TMS as reference value.

Protons in groups such as alcohols (R-OH), amines (R-NH-), carboxylic acids (RCOOH), thiols (R-SH) and to a lesser extent amides (R-CO-NH-) are classified as exchangeable protons. These labile protons frequently give rise to broadened resonances in the ^{1}H NMR spectrum; and their chemical shifts are critically dependent on the solvent, concentration, and temperature; and they do not have reliably characteristic chemical shift ranges.

The general trends of ^{13}C chemical shifts somewhat parallel those in ^{1}H NMR spectra. However, ^{13}C nuclei have access to a greater variety of hybridization states (bonding geometries and electron distributions) than do ^{1}H nuclei. As a consequence, the ^{13}C chemical shift scale spans some 250 ppm, compared with the 10 ppm range commonly encountered for ^{1}H chemical shifts (see Table 5.8). While it is necessary to note that as the resonance frequency of ^{13}C in a given magnetic field is four-fold lower than that of the proton, 1 ppm in Hz is also four-fold lower than that in ^{13}C and ^{1}H (42.576 and 10.705 MHz respectively, see Table 5.6).

Table 5.8: ^{13}C NMR chemical shifts for common functional groups.

Carbon type	Chemical shift (ppm)^a	Carbon type	Chemical shift (ppm)^a
Methyl (R$-$**CH$_3$**)	8–35	C$-$Br	25–65
Methylene (R$-$**CH$_2$**$-$R$^{'}$)	15–50	C$-$Cl	35–80
Methine (R$-$**CH**$-$R$^{'}$R$^{''}$)	20–60	C$-$N	40–60
Quaternary (R$^{'}$R**C**R$^{''}$R$^{'''}$)	30–40	C$-$O	50–80
$\equiv$**C**$-$	65–80	Amide & ester	165–175
$=$**C**$-$	100–150	Carboxylic acid	175–185
Aromatic	110–170	Aldehyde	190–200
C$-$I	0–40	Ketone	205–220

^{a} TMS as reference value.

When two nuclei have identical molecular environments and hence the same chemical shift, they are termed chemically equivalent or isochronous nuclei. Assuming an achiral environment, the protons are chemically equivalent if they are interchangeable for some symmetry operation or for a rapid chemical process.

For ^{1}H NMR, the intensity of the signal (which may be determined by electronically measuring the area under individual resonance signals) is directly proportional to the number of nuclei undergoing a spin-flip, and ^{1}H NMR spectroscopy is a quantitative method.

The intensity/peak integral factor resonance in ^{13}C NMR does not correspond to a proportional increasing in the number of equivalent nuclei (as in ^{1}H). The intensity is determined by the relaxation time of the nucleus and not with the number of existing equivalent nuclei (see Figure 5.14).

5.3.3.2. Spin-spin coupling

A typical organic molecule contains more than one magnetic nucleus (e.g., more than one ^{1}H, or ^{13}H and ^{31}P, etc.). When one nucleus can sense the presence of other nuclei through the bonds of the molecule, the signals will exhibit a fine structure (splitting or multiplicity). Multiplicity arises because if an observed nucleus can sense the presence of other nuclei with magnetic moments, those nuclei could be in either the α or β state. The observed nucleus is either slightly stabilized or slightly destabilized, depending on which state the remote nuclei are in, and consequently, nuclei that sense coupled partners with an α state have

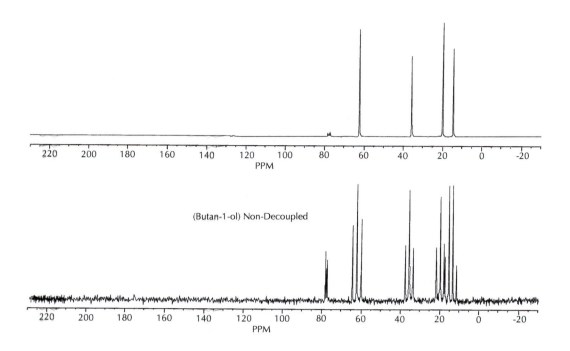

Figure 5.14: Example of ^{13}C NMR spectra.

a slightly different energy with respect to those that sense coupled partners with a β state. This would cause the resonance signal from that particular core to appear as two very close lines.

The additional fine structure caused by spin-spin coupling is not only the principal cause of difficulty in interpreting ^{1}H NMR spectra, but also provides valuable structural information when correctly interpreted. The coupling constant (related to the size of the splittings in the multiplet) is given the symbol J and is measured in Hz. By convention, a superscript before the symbol J represents the number of intervening bonds between the coupled nuclei. Labels identifying the coupled nuclei are usually indicated as subscripts after the symbol J. For example, $^2J_{ab} = 2.7$ Hz would indicate a coupling of 2.7 Hz between nuclei a and b which are separated by two intervening bonds. An important feature is that the links indicate a relationship between the magnetic moments of the nuclei and thus is a reciprocal property, i.e. if the core is coupled to the core b with a coupling constant $J_{a,b}$, the core b is coupled to the same constant.

J depends only on the number, type, and spatial arrangement of the bonds separating the two nuclei; it is a property of the molecule and is independent of the applied magnetic field. The magnitude of J, or even the mere presence of detectable interaction, constitutes valuable structural information.

Several important observations that relate to ^{1}H–^{1}H spin-spin coupling are the following: No inter-molecular spin-spin coupling is detected. Spin-spin coupling is transmitted through the bonds of a molecule and does not occur between

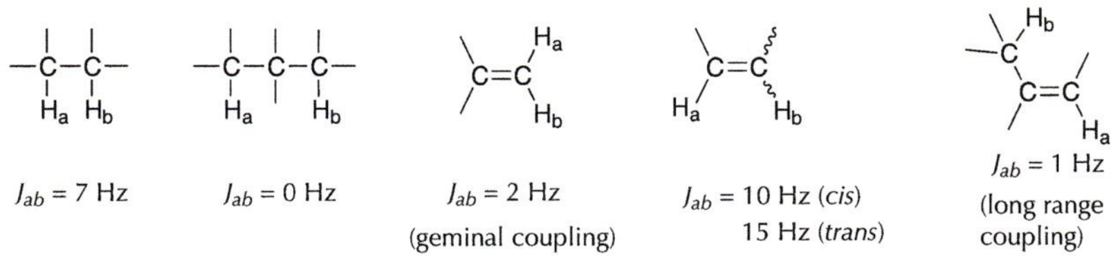

Figure 5.15: Common ^{1}H–^{1}H coupling constants.

nuclei in different molecules. The effect of coupling falls off as the number of bonds between the coupled nuclei increases (^{1}H–^{1}H coupling is generally unobservable across more than three intervening bonds). Unexpectedly large couplings across many bonds may occur if there is a particularly favorable bonding pathway —e.g., extended conjugation or a particularly favorable rigid skeleton. It also depends on the types of nuclei (their I), the type of link between the nuclei, the relative orientations of the nuclei, and the nature of the functional groups present. Typical ^{1}H–^{1}H coupling constant values are depicted in Figure 5.15.

Spin-spin coupling gives rise to multiplet splittings in ^{1}H NMR spectra. The number of peaks appearing when engaging two nuclei is given by the expression: $2 \cdot I \cdot n + 1$. Let us consider spin $\frac{1}{2}$ nuclei (^{1}H, ^{13}C), so that the number of peaks will always be $n + 1$, where n is the number of nuclei that couple to the core considered.

The NMR signal of a nucleus coupled to n equivalent hydrogens will be split into a multiplet with $(n + 1)$ lines (see Table 5.9). For simple multiplets, the spacing between the lines (in Hz) is the coupling constant. The relative intensity of the lines in multiplet will be given by the binomial coefficients of order n. These simple multiplet patterns give rise to characteristic 'fingerprints' for common fragments of organic structures.

Table 5.9: Multiplicity of the signal and relative intensities of the peaks.

No. of equivalent H causing splitting	Multiplicity	Relative peak intensities
0	Singlet	1
1	Doublet	1:1
2	Triplet	1:2:1
3	Quartet	1:3:3:1
4	Quintet	1:4:6:4:1

A spin system is defined as a group of coupled protons. Strongly and weakly coupled spins refer not to the actual magnitude of J but to the ratio of the separation of chemical shifts expressed in Hz ($\Delta\nu$) to the coupling constant J between them. For most purposes, if $(\Delta\nu) \cdot J^{-1}$ is larger than $\Delta\nu$, the spin system is considered to be weakly coupled. When this ratio is smaller than $\Delta\nu$, the spins are considered to be strongly coupled. Weakly coupled spin systems are

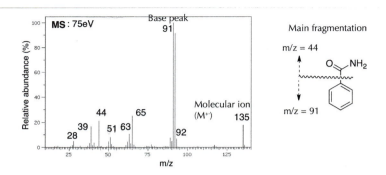

Figure 5.16: Example of a MS spectrum and main fragmentation for benzamide.

much easier to analyze than strongly coupled spin systems. Conventions used in naming spin systems. Consecutive letters of the alphabet (e.g., A, B, C, D, ...) are used to describe groups of protons that are strongly coupled. Subscripts are used to give the number of protons that are magnetically equivalent. A group of protons is magnetically equivalent when they not only have the same chemical shift (chemical equivalence) but also have identical spin-spin coupling to each individual nucleus outside the group.

The process of deriving the NMR parameters (δ and J) from a set of multiplets in a spin system is known as the analysis of the NMR spectrum. In principle, any spectrum arising from a spin system, however complicated, can be analysed. Fortunately, in a very large number of cases, multiplets can be correctly analyzed by inspection and direct measurements. These spectra are known as first order spectra and they arise from weakly coupled spin systems.

Inter-proton spin-spin coupling constants are of obvious value in obtaining structural data about a molecule —in particular, information about the connectivity of structural elements and the relative disposition of various protons.

To gain structurally useful information from NMR spectra, we must solve two separate problems. First, we need to analyze the spectrum to ascertain the NMR parameters (chemical shifts, integration of peaks, and coupling constants) for all the protons, and, second, we must interpret the values of the coupling constants in terms of established relationships between these parameters and the structure.

5.3.4. Mass spectrometry (MS)

Although this is not a spectroscopic technique (in the sense that we have seen it in this chapter, since no electromagnetic radiation absorption occurs), this technique is relevant for the characterization of organic substances and consists of the ionization of a sample and later study of the fragmentation in different ions (see Figure 5.16).

The most common method of ionization involves electron impact (EI). Collision of a molecule M with an electron e is followed by electron ejection that yields an odd-electron positively charged cation radical $[M]^{+\cdot}$ of the same mass as the initial molecule M.

$$M + e \rightarrow [M]^{+\cdot} + 2e \qquad (5.15)$$

The cation radical produced is known as the molecular ion, and its mass gives a direct measure of the molecular weight of a substance. Then it causes further fragmentation of this molecular ion. The two important types of fragmentation are:

$$[M]^{+\cdot} \rightarrow A^+ (\text{even electron cation}) + B^\cdot (\text{radical}) \qquad (5.16)$$

or

$$[M]^{+\cdot} \rightarrow C^{+\cdot} (\text{cation radical}) + D(\text{neutral molecule}) \qquad (5.17)$$

As only species bearing a positive charge will be detected, the mass spectrum will show signals due not only to $[M]^{+\cdot}$ but also to fragment ions resulting from subsequent fragmentation of $[M]^{+\cdot}$. For example, Figure 5.16 depicts the MS spectra of benzamide (m/z = 135).

Using a double-focusing mass spectrometer or a time-of-flight mass spectrometer, the mass of a molecular ion or any fragment can be determined to an accuracy of approximately 0.00001 of a mass unit (a high-resolution mass spectrum). Since the masses of the atoms of each element are known to high accuracy, molecules that may have the same mass when measured only to the nearest integer mass unit can be distinguished when the mass is measured with high precision.

For example, an ion of m/z = 43 may correspond to the following atomic combinations listed in Table 5.10.

Table 5.10: Different atomic combinations for an ion of m/z = 43.

Atomic combination	Exact mass
CHNO	43.0058
C_2H_3O	43.0184
CH_3N_2	43.0269
C_2H_5N	43.0421
C_3H_7	43.0547

There are a number of other methods for ionizing the sample in a mass spectrometer:

- Chemical ionization (CI): In CI mass spectrometry, an intermediate substance (generally methane or ammonia) is ionized by EI and the substrate is then ionized by collisions with these ions. CI is a milder ionization method than EI and leads to less fragmentation of the molecular ion.

- Fast atom bombardment (FAB): Uses collisions with neutral atoms.

- Electrospray ionization (ES): In this method, the sample is dissolved in a polar, volatile solvent and pumped through a fine metal nozzle, the tip of which is charged with a high voltage. This produces charged droplets from which the solvent rapidly evaporates to leave naked ions that pass into the mass spectrometer. ES is highly suitable for biological samples.

- Matrix-assisted laser desorption ionization (MALDI): Uses a pulse of laser light to bring about ionization.

The fragmentation pattern is a molecular fingerprint. In addition to the molecular ion peak, the mass spectrum consists of a number of peaks at lower mass number, and these result from fragmentation of the molecular ion (see Figure 5.16).

Table 5.11: Common types of ions in MS spectrometry.

Ion mass	Assignation
29	Ethyl (C_2H_5), formyl (CHO)
30	Nitrous (NO)
31	Methoxy (CH_3), hydroxymethyl (CH_2OH)
39	Cyclopropenyl (C_3H_3)
41	Alilo ($CH_2CH=CH_2$)
43	Propyl (C_3H_7), acetyl (CH_3CO)
45	Carboxyl (COOH)
46	Nitro (NO_2)
55	Butenyl (C_4H_7)
56	C_4H_8
57	*tert*-Butyl (C_4H_9), propanoyl (CH_3CH_2CO)
60	Acetic acid
65	Cyclopentadienyl (C_5H_5)
77	Phenyl (C_6H_5)
91	Benzyl (tropylium, Ph$-CH_2$)
92	Azatropilium, $C_5H_5N-CH_2$
105	Benzoyl (Ph$-CO$)
127	Iodine

The principles determining the mode of fragmentation are reasonably well understood, and it is possible to derive structural information from the fragmentation pattern in several ways. Three factors dominate the fragmentation processes:

a) Weak bonds tend to be broken most easily.

b) Stable fragments (not only ions but also the accompanying radicals and molecules) tend to be formed most readily.

c) Some fragmentation processes depend on the ability of molecules to assume cyclic transition states.

There are a number of common types of ions that are characteristic of various classes of organic compounds (see Table 5.11).

Electronic databases of the MS spectral fragmentation patterns of known molecules can be rapidly searched by computer. The pattern and intensity of fragments in the MS spectrum are characteristic of an individual compound so that the experimental mass spectrum of a compound with those in a library for positive identification, if the spectrum has been recorded previously.

5.4. Preparation of samples for spectroscopy

Organic compounds are commonly isolated and purified in laboratory experiments prior to their identification, by means of either organic or spectroscopic analysis. Spectroscopic techniques require sample preparation products suitable for each one. The following sections will describe only the procedures used to perform the these tasks.

5.4.1. UV/visible spectroscopy (UV/vis)

For analytical purposes, the spectra of UV/vis absorption is generally recorded in a solution of known concentration of the product. From the resulting spectra and, using the Beer-Lambert law, the absorbance and molar extinction coefficient (ϵ) can be determined.

Table 5.12: Characteristic absorbance of some solvents used in spectroscopy UV/vis spectroscopy, in 1 cm cuvettes.

Solvent	λ (nm)	Solvent	λ (nm)
Acetonitrile	190	1,4-Dioxane	215
Pentane	190	CH_2Cl_2	232
Hexane	195	$CHCl_3$	245
EtOH (96%)	204	CCl_4	265
H_2O	205	Toluene	285
Cyclohexane	205	Pyridine	305
Propan-2-ol	205	Ketone	330
MeOH	205	$MeNO_2$	380
Ethyl ether	215	CS_2	380

The solvents must be of high purity (HPLC grade or spectroscopic), and the concentrations of the solutions used are usually in the range 10^{-4} mol$\cdot$L^{-1}. To this end, the reagents should be weighed in precision balances, because the amounts used are small, and solutions must be prepared in volumetric flasks (see Section 4.3, p. 77). In the case of weak bands due to prohibited transitions, e.g. carbonyl groups) the concentration should be increased (10^{-2} mol·L^{-1}). For measurements, cuvettes are used.

A cuvette usually has a prism with two opposite sides sanded and two other transparent sides (see Figure 5.17). Normally, a cuvette is 1 cm wide, for easy calculation of the molar absorption coefficient, and has a cap usually made of Teflon. To take measurements in the visible range, cuvettes are manufactured in Pyrex, since this material does not absorb above 300 nm, and for measurements in the UV range, quartz cuvettes are used. If the absorbance is large (>1), dilute the original solution using a volumetric flask. The absorption of the solvent itself should be considered, which should not interfere with the chromophore in the sample. Table 5.12 lists the absorbance of some typical solvents for reference.

From the UV/vis spectra, information can normally be obtained and the maximum absorption intensity is expressed indicating the solvent. For example, nitromethane, dissolved in hexane, exhibits a characteristic absorption band with λ_{max} values (hexane) and $\epsilon = 203 = 4400$. UV/vis spectrometry is commonly used in the quantitative determination of solutions of transition metal ions and organic compounds with high conjugation.

5.4.2. Infrared spectroscopy (IR)

Very few substances are transparent over the whole of the IR range: NaCl, KCl, NaBr, and KBr are most common (see Table 5.13). The cells used for determining IR spectra in solution typically have NaCl windows, and liquids can be examined as films on NaCl plates. Solution spectra are generally measured in $CHCl_3$ or CCl_4, but this leads to a loss of information at longer wavelengths where the solvent absorbs. Organic solids may also be examined as mulls (fine suspensions) in heavy oils. The oils absorb IR radiation but only in well-defined regions of the IR spectrum. Solids may also be examined as dispersions in compressed KBr or KCl discs.

The vibration characteristics of the bonds within a molecule produce the IR spectra. IR spectra of the substances can be recorded in the three states of aggregation (gas, liquid, and solid) as well as in solution. The choice of the appropriate method depends on the nature and physical properties of the sample.

Gas samples do not require special preparation (apart from their purification). As they usually present low absorbance values, they are injected into a long cell (10 cm) provided with a stopcock, the ends of which are composed of NaCl plates that are transparent at IR (see Figure 5.17).

In solution, the samples are prepared by diluting the compound in $CHCl_3$ (alcohol free) or CCl_4, introducing the sample in a special NaCl cuvette of 1 cm length, because this material is transparent to the radiation in this region. The most common devices have what is called a double beam. This device generates a beam of IR light that is divided in two. One passes through the sample, and the other through a reference, which is usually the solvent in which the sample is dissolved or the substance with which the sample is mixed. The two signals are compared, and then the data are recorded so that the effect of the solvent is offset.

Table 5.13: Materials used in manufacturing IR cells

Material	Transmission limit (cm^{-1})	Refractive index at 4,000 cm^{-1}	Solubility (water) at 20 °C ($g \cdot 100 \, ml^{-1}$)
Irtran-2	800	2.3	Insoluble
NaCl	650	1.5	36
KCl	500	1.5	35
KBr	400	1.5	53
AgCl	400	2.0	Insoluble
CsI	150	1.7	80

If the sample is liquid, it is usually prepared by placing a drop of liquid (with a water content less than 2%) between two polished pads of high purity salt (NaCl, KBr, or CaF_2), which does not absorb in the IR (see Table 5.14). We must be careful that as the NaCl is soluble in water, the sample and washing agents must be anhydrous.

Table 5.14: Possible interference regions (sample-solvent), in the IR spectrum.

Solvent	Absorption zone (cm^{-1})
Toluene	3,100–2,800 & 750–650
Hexane	3,000–2,800 & 1,500–1,400
Et_2O	3,000–2,700 & 1,200–1,050
CCl_4	2,850–700
CH_3CN	2,300–2,200 & 1,600–1,300
CS_2	2,200–2,100 & 1,600–1,400
CH_2Cl_2	1,300–1,200 & 820–650
$CHCl_3$	1,250–1,175 & 820–650
DMSO	1,100–900

In solid state, two standard procedures are used:

- As a suspension in Nujol (paraffin oil): grinding approximately 1 mg of solid and a drop of Nujol, in an agate mortar and placing the paste obtained between two NaCl pellets. The pellet is placed in a special holder held by screws (care is needed not to over-tighten the screws as the pellet could be easily broken).

- As KBr, the solid is mixed with about 10-fold the proportion of KBr in an agate mortar and then, using an IR pellet die set, compressed in a mechanical die press, resulting in the formation a translucent pellet that we place in a special holder.

It should be borne in mind that according to the sample preparation method used, the resulting spectra may vary slightly (see Figure 5.17).

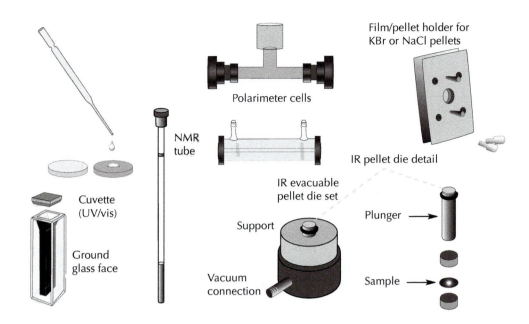

Figure 5.17: Different sample cells/containers for spectroscopy.

5.4.3. Nuclear magnetic resonance (NMR) spectroscopy

NMR spectra are registered mainly in solution, and the solvent of choice should be selected to fit different characteristics: the solvent should dissolve the sample and present no signals that interfere with the sample. It is recommended that the solvent contain deuterium (^{2}H), which cannot be bonded strongly with the sample molecules. There are many solvents that fulfill such characteristics, such as deuterochloroform ($CDCl_3$), DMF, etc.

In Organic Chemistry, ^{1}H and ^{13}C NMR are frequently used but also, depending on the availability of NMR equipment, spectra of other nuclei such as ^{15}N, ^{19}F, ^{31}P, etc. can be recorded. NMR spectra are usually recorded in solution (see Table 5.15). The solvent chosen should be deuterated, typically $CDCl_3$, with a degree of deuteration >99%, although depending on the polarity of the sample, there are many choices: DMSO-d_6, C_6D_6, CD_3OD, etc. ^{1}H spectra also have protons signals coming from water and from other solvents (not deuterated), and ^{13}C spectra also have the carbons signals from other solvents.

The solvent choice will depend, among other factors, on the solubility of the sample of that solvent and on overlapping signals (solvent + sample) given by the spectrum. As a reference signal, TMS or residual solvent signal can be used. The solvent affects the chemical shift of the sample signals. The amount of sample depends on the spectrum chosen to perform, for a spectrum of ^{1}H NMR of 1–5 mg in 0.75 ml of deuterated solvent are enough, whereas ^{13}C usually requires 10–50 mg. The spectrum is typically conducted at r.t. but, in special cases, may be performed at high or low temperatures. Samples are placed in a special glass tube about 5 mm thick (NMR tubes). It is not recommended to use metal spatulas (when handling solids), since traces of paramagnetic metals can affect the resolution of the spectrum.

Table 5.15: Solvents commonly used in NMR spectroscopy.

Solvent	Formula	m.p. (°C)	b.p. (°C)	δ_H	$\delta_H(H_2O)$	δ_C (multiplicity)
Carbon tetrachloride	CCl_4	−23	77	-	-	96.0 (singlet)
Chloroform	$CDCl_3$	−64	61	7.24	1.5	77.0 (triplet)
Methanol	CD_3OD	−98	64	3.35	4.9	49.3 (septet)
Acetone	CD_3COCD_3	−95	56	2.04	2.8	29.3 (septet)
Benzene	C_6D_6	6	80	7.27	0.4	128.0 (triplet)
Dimethyl sulfoxide	CD_3SOCD_3	19	189	2.49	3.3	39.7 (septet)
Water	D_2O	0	100	4.65	4.8	-

Chapter 6

Functional-Group Analysis

6.1. Introduction

Qualitative organic analysis has remarkable relevance in the teaching of Organic Chemistry. Unlike inorganic analysis, it can identify every element of a substance, since for organic compounds the analysis tests are designed to identify organic functional groups in molecules. Most organic compounds are formed by a few elements and are always the same (C, H, O, halogens, etc.). Thus, identifying organic compounds in the same way as for inorganic compounds is not useful because the information gained in this way is not relevant for the structural elucidation of a substance.

Although at first glance it may seem extremely difficult to identify an organic compound, such identification can be approached through a systematic and well-defined protocol, and the tactics employed consist essentially of a successive and progressive elimination of structural possibilities.

Usually the procedure begins with the identification the organoleptic properties of the unknown substance and continues with a series of chemical tests in order to determine the components (other than C or H). When they are clear enough, the procedure will continue to identify the functional groups present in the molecule, generally using a solubility test. Once they are determined, the next step is to prepare derivatives, which will enable the clear identification of the substance by comparing them with those outlined in the literature.

6.2. Qualitative elemental analysis

With an unknown organic substance, the identification protocol should start by performing qualitative (and quantitative, if possible) elemental analysis of the constituent elements, thus determining the nature of the elements present in the molecule and the proportion in which they occur. Quantitative analysis usually involves the combustion of given weight of the substance, determining chromatographically (or gravimetrically) the amount and nature of the gases formed in the combustion.

For the qualitative analysis, the most common method (and/or simplest) is called the Lassaigne test or sodium fusion, which is the mineralization of the organic substance, transforming the elements present into a group of easily identifiable anions:

$$(C, H, O, S, N, X, ...) \longrightarrow NaCN + Na_2S + NaX + etc. ...$$

Procedure 6.2.1:

In a Pyrex test tube a small piece of metallic sodium (medium-pea sized) is placed and heated by flame until the sodium is melted. Then about 200 mg of the unknown substance is added carefully and repeatedly and the tube is heated to cherry red. Then, it is cooled and 2 ml of EtOH is added to the tube. When the bubbling stops, about 10 ml of deionized water is added. The mixture is heated to boiling and filtered. The solution obtained is divided into several aliquots in which the inorganic ions are identified.

Nitrogen and sulfur detection

Procedure 6.2.2:

To 1 ml of the former filtrate, solid sodium bicarbonate is added until saturation. One ml of a 1% of *p*-nitrobenzaldehyde solution in DMSO is added to one or two drops of this solution. The appearance of a purple color indicates the presence of nitrogen. A green color indicates the presence of sulfur. If both elements are present, only purple coloration appears, and the presence of sulfur must be confirmed by the lead acetate test.

Sulfur detection

Procedure 6.2.3:

About 2 ml of alkaline solution is acidified with acetic acid and a few drops of lead acetate solution are added. The formation of a black precipitate of lead sulfide indicates the presence of sulfur in the unknown substance.

$$Pb(CH_3COO)_2 \ + \ Na_2S \ \longrightarrow \ PbS \downarrow \ + \ 2\,CH_3COONa$$
(Lead sulfide)

Procedure 6.2.4:

About 2 ml of the former alkaline solution is acidified with HCl and gently heated. On the side of the test tube is placed a small piece of paper impregnated with lead acetate. The appearance of a blackish brown spot of lead sulfide will indicate the presence of sulfur.

Detection of nitrogen   This procedure requires the prior removal of sulfur.

Procedure 6.2.5:

About 3 ml of the alkaline filtrate is placed in a test tube, and some ferrous sulfate is added with a spatula; the test tube is gently boiled for a few seconds and the solution is allowed to cool. Then, it is acidified with diluted HCl (or diluted H_2SO_4). The formation of an intense blue color (due to Prussian blue solution or a suspension), indicates the presence of nitrogen.

$$6\,NaCN \ + \ FeSO_4 \ \longrightarrow \ Na_4[Fe(CN)_6] \ + \ Na_2SO_4$$
(Sodium ferrocyanide)

$$Na_4[Fe(CN)_6] \ + \ 2\,Fe_2(SO_4)_3 \ \longrightarrow \ Fe_4[Fe(CN)_6]_3 \ + \ Na_2SO_4$$
Ferric ferrocyanide
(Prussian blue)

Halogen detection (Beilstein test)

Procedure 6.2.6:

Heat a red copper wire in the flame of a Bunsen burner until no color is observed. Cool the wire, submerge it in the alkaline solution, and return it to the flame. The appearance of a green color indicates the presence of halogens.

Halogen detection (silver halide test)

Procedure 6.2.7:

Acidify 3 ml of the alkaline solution with concentrated HNO_3. If the sample contained sulfur and/or nitrogen, boil the solution to remove sulfhydric and hydrocyanic acids that may have formed. Then, add 4–5 drops of silver nitrate solution. The formation of a white or yellow precipitate that darkens rapidly in light indicates the presence of halogens. If the white precipitate is decanted and dissolved in ammonia, then, it is chloride. If it is pale yellow, and it is hardly soluble then it is bromide. If it remains completely insoluble in ammonium hydroxide and shows an intense yellow color, it is iodide.

6.3. Organoleptic and physical properties

After the qualitative elemental analysis and the determination of the elements that constitute the molecule, conduct a very simple examination of the substance: state, color, odor, crystalline form (if it is solid), ignition test (if it burns completely or a solid waste remains, flame type, etc.), and any other property that may give additional clues as to the identity of that substance.

Next, the main physical constants are determined: m.p. for solid and b.p. for a liquid. If values of these measurements are in a wide range, it is not a pure compound. This will give an idea of the degree of purity of the substances and will indicate the need for their purification, either by recrystallization (after completing the solubility tests) or by simple or vacuum distillation. Other parameters may be the determination of the refractive index, density, or optical rotation (if they are optically active). Finally, if possible, the spectroscopic data, mainly IR and NMR, would be useful.

6.4. Classification by solubility (solubility tests)

Given the dissimilar solubility of organic compounds in different solvents, these can be classified into several groups. The phrase "like dissolves like" leads to the classification of organic molecules into various types of more common functional groups. Thus, non-ionic compounds are not appreciably dissolved in water unless the molecules are ionized in aqueous solution or can be associated with water molecules through hydrogen bonding. Those without hydrogen, and capable of forming hydrogen bonds, rapidly dissolve in organic solvents such as ether and other related solvents. However, it should be borne in mind that as we move up in a homologous series, solubility, as well as other physical properties, tends to approach the hydrocarbons of the same number of carbon atoms. Compounds having a basic character are dissolved in acid media and those that have acidic groups in basic media. Through these simple procedures, types and functional groups can be divided and selected. However, it should be noted that the solubility tests are not infallible and some exceptions are known. Therefore, they will never be fully conclusive and it will be necessary to perform other types of tests for the definitive identification of a compound.

> **Procedure 6.4.1:**
>
> Place 0.2 ml (0.1 g or if solid) of the unknown substance in a test tube. Add water, stirring in portions of 3 ml (see Figure 6.1). If the product is soluble, repeat the process in another test tube but with ethyl ether as a solvent.

If the compound is insoluble in water, its solubility is tested in a 5% NaOH solution. All compounds soluble in NaOH must be solubility tested with a 5% sodium bicarbonate solution. These solubility tests are performed as indicated for the test with water (0.2 ml or 0.1 g of substance and 3 ml of solution). If the compound is insoluble in an NaOH solution, its solubility in 5% HCl is tested.

Finally, if the substance is insoluble in water, NaOH, and HCl, and if it contains no nitrogen or sulfur, solubility is tested in concentrated H_2SO_4.

Then major types of compounds classified according to their different solubility are listed (see Figure 6.1).[1]

6.5. Functional-group analysis

Organic Chemistry has a series of reactions which are characteristic of functional groups, and therefore it is possible to characterize almost any functional groups presents in the molecules. Sometimes, various functional groups can give the same reaction, so that it will require some other characteristic reaction to be sure of their nature, or some polyfunctional compounds may even give different reactions to each function separately. Therefore, reactions listed below should not be considered infallible/excluded for identifying functional groups. For example, the iodine test to give charge transfer complexes by almost all functional groups. When used in alkene characterization, it is assumed, after testing solubility, that the unknown molecule is a hydrocarbon and it is necessary to ascertain whether it has a saturated or unsaturated structure.

For characterization, generally, a derivative is prepared, and the m.p. is determined and compared with bibliographic data. Although the IR bands are

---

[1]This figure is for informative use only and includes relatively common substances in practices of the Organic Analysis laboratory, and the functional groups are the most common in Organic Chemistry.

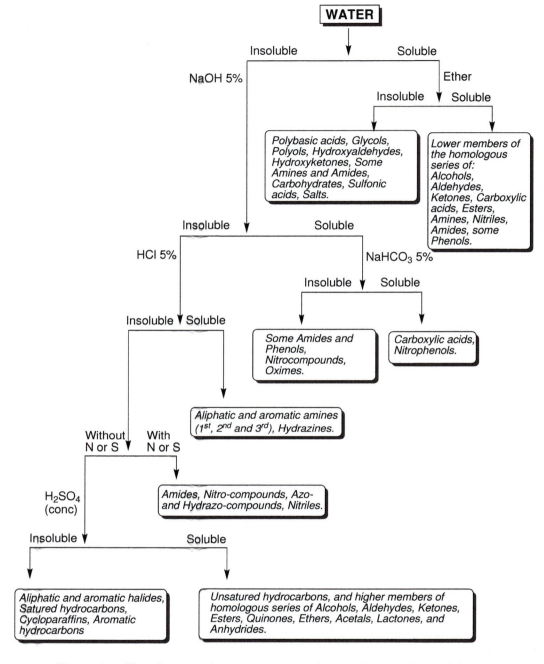

Figure 6.1: Classification of organic compounds according to their solubility.

described, this is not a requisite in order to identify the functional group but also for speed or confirmation purposes.

6.5.1. Alkanes

Because of the very low reactivity of alkanes, there are only a few specific reactions for such molecules. Usually, they are identified by indirect procedures or for exclusion, once a group of test reactions has been made. For characterization, it is usually valuable to determine b.p. and density.

The main features related to these compounds are:

- The IR spectrum is exceptionally easy. The most intense bands appear around 2,900, 1,460, and 1,370 cm^{-1}. (If methyl groups are not present, the last peak is absent).

- They yield negative tests in the formation of charge transfer complex with iodine. (In case of a positive test, a small iodine crystal on a few drops of the compound gives a characteristic violet solution).

- They are insoluble in cold concentrated H_2SO_4.

For characterization, generally b.p., density, and refraction index are determined.

6.5.2. Alkenes

The most remarkable properties are based on the unsaturated bonds. The main features include:

- The IR spectrum of alkenes is much more complex than in saturated hydrocarbons. $C = C$ absorption appears at 1,680–1,620 cm^{-1} region.

- The iodine charge transfer test involves the addition of several drops of the compound to a small iodine crystal. A brown solution constitutes a positive test.

- They are soluble, at least partially, in concentrated H_2SO_4.

Bromine in CCl$_4$ test

Procedure 6.5.1:

Dissolve 2 drops (approximately 50 mg) of the product in 1 ml of CCl_4, and add dropwise a solution of bromine in CCl_4 (2%) until the bromine color persists. If more than 2 drops must be added so that the color (brownish) persists, the test is considered positive.

$$\overset{\diagdown}{\diagup}C = C\overset{\diagup}{\diagdown} \; + \; Br_2 \; \longrightarrow \; -\overset{|}{\underset{|}{C}}-\overset{Br}{\underset{Br}{\overset{|}{C}}}-$$

Alkynes are able to give positive results with this test. Check the same test with a substance that does not add bromine. In case of a substitution reaction, some bubbling is observed because of the formation of hydrobromic acid gas.

Baeyer test (Baeyer permanganate)

Procedure 6.5.2:

Dissolve a drop (25 mg) of the compound in 2 ml of water or acetone in a test tube. Dropwise, with vigorous stirring, add an aqueous solution of 1% potassium permanganate. The result is positive if it is necessary to add additional drops of permanganate solution to maintain the violet color. If positive, the initially violet solution turns brown.

6.5.3. Alkynes

Alkynes show some common reactions with alkanes as they are soluble in concentrated H_2SO_4, and they can add bromine in CCl_4 solution, and are oxidized by a cold aqueous permanganate solution.

In the case of terminal alkynes, the IR spectrum shows a sharp characteristic band in the $3,310$–$3,200$ cm^{-1} range. The stretching band corresponding to the triple bond appears with an average intensity of $2,140$–$2,100$ cm^{-1}. If not terminal the triple-bond band appears in the $2,250$–$2,150$ cm^{-1} range.

Alkaline mercuric iodide test   This test is for the terminal alkyne detection.

Procedure 6.5.3:

Prepare the reagent dissolving 6.6 g of mercuric chloride in a solution of 16.3 g of potassium iodide in 16.3 ml of water, and then add 125 ml of 10% NaOH. Cool the mixtures and dissolve the alkyne to 20-fold its volume of EtOH. Add this solution dropwise to 2 ml of the above reagent. The appearance of a white or off-white precipitate is indicative of terminal alkyne. Stir for 2 to 3 min, filter out and wash with 50% EtOH to get a mercuric derivative, and handle very carefully when dry because many of these derivatives are explosive when dry.

6.5.4. Aromatic hydrocarbons

Aromatic hydrocarbons burn with a characteristic dark flame, and they give negative results in tests with bromine in CCl_4 and with alkaline permanganate. Most alkylated aromatic hydrocarbons are liquid, but those with more than one aromatic ring (fused or not) are solid.

- Solubility in H_2SO_4.

  Simple aromatic hydrocarbons are insoluble in H_2SO_4 but soluble fuming H_2SO_4. When two or more alkyl substituents are sulfonated they easily dissolve in concentrated H_2SO_4. Condensed aromatic hydrocarbons, such as naphthalene and anthracene, react slowly with concentrated H_2SO_4.

- IR spectrum absorption bands have ring features.

Nitration test   Nitration of aromatic rings take places generally by increasing the temperature by treating with a cold mixture of concentrated H_2SO_4/HNO_3. This reaction can be used for a preparatory level.

Procedure 6.5.4:

Add 500 mg of product to about 3 ml of concentrated H_2SO_4 and then an equal volume of concentrated HNO_3 dropwise, stirring the mixture after each addition. The reaction is heated 5–10 min in a water bath at 60 °C. Remove the tube from the bath and shake frequently. Cool it and pour over ice water. Filter the precipitate under vacuum, and wash it with water. The product can be purified by recrystallization from aqueous EtOH.

Aroylbenzoic acid test

Procedure 6.5.5:

In a flask with reflux condenser, place 400 mg of phthalic anhydride, 10 ml of carbon disulfide, 800 mg of anhydrous aluminum chloride, and 400 mg of the aromatic hydrocarbon. Heat the mixture for 30 min in a water bath, and then cool it with a water bath. Slowly decant the carbon disulfide layer. Add cooling 5 ml of concentrated HCl and then 5 ml of water. Shake the mixture vigorously until solid. Collect the solid by filtration and wash it with cold water. If oil forms, if not crystallized, extract it with diluted ammonium hydroxide and then neutralize with HCl (conc.).

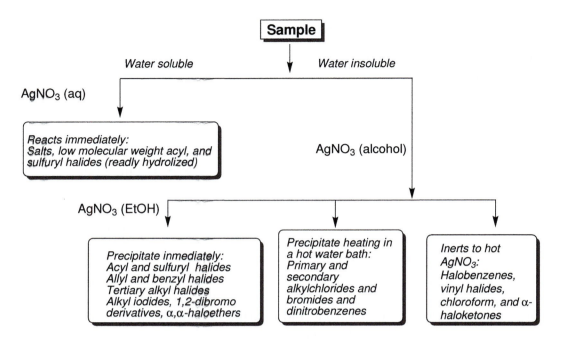

Figure 6.2: Classification of halogenated compounds according to their reaction with $AgNO_3$.

6.5.5. Alkyl halides

The presence or absence of halogen in organic compounds can be determined by the Beilstein test, consisting of the formation of copper halides, which give a green or blue flame, or by analysis of the filtrate left after fusion with sodium.

Different types of halogenated compounds are classified by the following two tests:

Silver nitrate test

> ### Procedure 6.5.6:
>
> Add a drop of halogenated compound to 2 ml of 2% ethanolic silver nitrate. If after 5 min no reaction is observed, heat the solution for several minutes until boiling; note the color of the precipitate and add 2 drops of 5% HNO_3 solution.

Some organic acids give insoluble silver salts, while silver halides are insoluble in diluted HNO_3. Generally silver salts of organic acids are soluble. For water-soluble organic compounds, the test is performed with aqueous silver nitrate.

Depending on the results, halogenated products can be classified as shown in Figure 6.2.

Sodium iodide in acetone test

Two drops (or 100 mg) of the product are dissolved in a small amount of acetone for 1 to 2 ml of reagent (15 g of sodium iodide in 100 ml of acetone) by letting the solution stand for 3 or 4 min at r.t. If any change occurs, heat the test tube to 50 °C for 6 min. Cool to r.t. and record the observed changes.

Depending on the results, we can classify halides as follows:

- If a precipitation occurs at room temperature in 3 min: primary bromides, acyl halides, allyl halides, α-halocarbon compounds, CCl_3.

- If a precipitation occurs when heated to 50 °C for 6 min: primary and secondary chlorides, bromides, and secondary and tertiary bromoform.

- If no reaction is detected at 50 °C when heated for 6 min: vinyl and aryl halides, geminal polychlorinated compounds.

- If precipitate and iodine (brown) are formed: neighborhood halides, sulfonyl halides, triphenylmethyl halides.

Alkyl halide characterization: picrates of "S"-alkylthiouronium These are a group of solid derivatives easily obtained from halides of primary and secondary alkyls (tertiary do not react) with Procedure 6.5.8.

Procedure 6.5.8:

Alkyl halide (1 mmol) and thiourea (150 mg) in 5 ml of ethylene glycol are dissolved in a flask equipped with a condenser. The mixture is heated for 30 min. Primary alkyl iodides require heating to a temperature of approximately 65 °C, while the other halides require temperatures about 120 °C. Then add 1 ml of saturated picric acid in EtOH and heat for 15 min. Cool the reaction mixture and add 5 ml of cold water. Let stand in an ice bath and filter the precipitate under vacuum.

Aryl halide characterization The usual procedure for their characterization is a nitration reaction, as indicated in the case of simple aromatic hydrocarbons.

6.5.6. Alcohols

As evidence for the existence of an alcohol, the IR spectrum analysis is normally used. The absence of the carbonyl band and the appearance of a band of strong intensity in the 3,600–3,400 cm^{-1} region[2] indicates the presence of an alcohol. For their identification, and classification the following tests are used:

Reaction with metallic sodium

Procedure 6.5.9:

A small piece of metallic sodium (medium-pea sized) is added to 3 ml of alcohol, and the result is observed. The test is considered positive if hydrogen gas is released. A few ml of EtOH is added to destroy excess sodium, and the gas release must cease before being poured into the appropriate recycling container.

Lucas test (ZnCl$_2$/HCl (conc.))   Tertiary alcohols react easily with ZnCl$_2$/HCl (conc.) to give alkyl chlorides that are insoluble in water, while secondary alcohols react slowly, and primary ones remain practically inert. The test is not valid for aryl alcohols or those that are insoluble in water.

$$R{-}OH \ + \ HCl \ (conc.) \ + \ ZnCl_2 \ \longrightarrow \ R{-}Cl \ + \ H_2O$$

Procedure 6.5.10:

The reagent is prepared by dissolving 16 g of anhydrous zinc chloride in 10 ml of concentrated HCl, cooling the mixture to avoid loss of HCl. Seal the flask reagent, because the ZnCl$_2$ is strongly hygroscopic and loses effectiveness. Add 3 or 4 drops of alcohol to 2 ml of Lucas' reagent in a small test tube. Close the tube and shake vigorously for 15 s, and let stand at r.t.

Depending on the results, the following classification can be made:

• The tertiary alcohols, allylic and benzylic, react almost immediately, separating two layers (aqueous and the alkyl chloride).

---

[2]This band may also be indicative of amines and phenols, which can be distinguished by the different acidic and basic character; amines are basic and phenols are considerably more acidic than alcohols and can be dissolved in NaOH 5%.

- The secondary alcohols after a few minutes produce an opalescent solution due to the formation of insoluble chloride.

- The primary alcohols with less than six carbons dissolve and with more than six carbons do not dissolve and the aqueous phase remains clear.

If there is doubt concerning whether the alcohol is secondary or tertiary, the test is repeated with HCl (conc.). Tertiary alcohols continue to react quickly, producing insoluble alkyl chloride, while secondary alcohols do not react.

Oxidation test ($K_2Cr_2O_7/H_2SO_4$) Primary and secondary alcohols react rapidly with chromic acid to give a greenish suspension due to the formation of Cr(III), whereas tertiary alcohols do not give this reaction. The presence of other easily oxidizable functions such as aldehydes or phenols can interfere, since they also react with this reagent.

Procedure 6.5.11:

 Prepare the reagent by dissolving 5 g of potassium dichromate in 50 ml of water and adding 5 ml of concentrated H_2SO_4. Add drops of alcohol to 2 ml of the reagent and stir the mixture. Then score the result.

6.5.6.1. Characterization of alcohols: obtaining esters

Many esters have been used as an aid in alcohol characterization, comprising a wide variety of phthalic acid esters, xanthates, benzoates and acetates, the latter two being particularly suitable for the characterization of glycols and polyhydroxy compounds.

Test with acetyl chloride (CH_3COCl) Vigorously react with acetyl chloride to form an ester and release HCl, which can be detected with indicator paper.

$$R-OH \ + \ CH_3-COCl \ \longrightarrow \ CH_3-COOR \ + \ HCl \ (gas)$$

Noticeboard 6.5.1

⚡ DANGER!

Because of the danger of both the acetyl chloride and the benzoyl chloride, such tests must be performed in a fume hood while using gloves.

Procedure 6.5.12:

After placing 0.5 ml of alcohol in a dry test tube, carefully add dropwise 0.3 ml of acetyl chloride[a], observing the reaction that occurs by heat release. Pour the solution onto 10 ml of ice/water and smell the sample (esters from simple alcohols typically have fruity odors).

[a]Read the Danger Noticeboard 6.5.1.

Reaction with benzoyl chloride (Schotten-Baumann reaction)

Procedure 6.5.13:

In a test tube, 2 ml of alcohol is dissolved in 8 ml of NaOH (10%). Afterward, a total of 2 ml of benzoyl chloride[a] is added in 0.5 ml portions, shaking the test tube for 1 min after each addition. The reaction mixture is poured into 10 ml of cold water and the solid ester formed is collected by filtration. It can be recrystallized from EtOH/H_2O.

[a]Read the Danger Noticeboard 6.5.1.

Preparation of 3,5-dinitro- and p-*nitro-benzoates*   If benzoate formation does not usually give a solid product or it is difficult to crystallize, the formation of 3,5-dinitro and *p*-nitrobenzoate derivatives can be performed.

Procedure 6.5.14:

In a round-bottom flask, place 100 mg of alcohol, 100 mg of *p*-nitrobenzoyl chloride or 3,5-dinitrobenzoyl chloride, and 2 ml pyridine, together with a boiling chip. Place a reflux condenser and heat for 1 h. Let stand overnight. Cool the reaction mixture and add 5 ml of water and 2 to 3 drops of H_2SO_4. Shake vigorously and collect the crystals by vacuum filtration. Add crystals to 5 ml of NaOH 2% to remove corresponding nitrobenzoic acid (3,5-dinitro-: m.p. = 202 °C; *p*-nitro-: m.p. = 241 °C). Filter, wash several times with cold water, and recrystallize from EtOH or EtOH/H_2O.

6.5.6.2. Characterization of alcohols: obtaining urethanes.

Alcohols (must be anhydrous) react with isocyanates to produce urethanes.

Procedure 6.5.15:

Place 0.2 ml of alcohol and 0.2 ml of phenyl isocyanate in a dry tube. If no reaction occurs immediately, heat in a water bath at 60–70 °C for 5 min. Cool in an ice bath. Filter and recrystallize from petroleum ether or CCl_4.

6.5.7. Glycols

The physical properties of these compounds result directly from the formation of hydrogen bonds and are manifested in relatively high b.p., high viscosity, and their water solubility. In terms of chemical properties, they are typical of other alcohols and are specific because of the presence of two hydroxyl groups on adjacent carbon atoms. Among these specific reactions the following two can be highlighted:

Borax test

Procedure 6.5.16:

Prepare 100 ml of borax (sodium tetraborate) 10% with enough phenolphthalein to produce a pink color. Add one or two drops of glycol to 0.5 ml of this reagent, observing the disappearance of the color. The color reappears on heating and dims as the solution cools.

Periodic acid oxidation (H_5IO_6)

Procedure 6.5.17:

In a test tube containing 2 ml of reagent, [a] add 1 drop of HNO_3 (conc.) and shake vigorously. Add a drop of compound to test and stir for about 15 s. Add 1 or 2 drops of $AgNO_3$ solution 3%. The instantaneous formation of a white precipitate of silver iodate means a positive test result.

---

[a]The reagent is prepared by dissolving 0.5 g of periodic acid (H_5IO_6) in 100 ml of deionized water.

6.5.8. Phenols

Phenols are compounds having intermediate acidity between those of carboxylic acids and alcohols. These latter can be distinguished by their solubility in NaOH 5% (alcohols do not dissolve if they are not water soluble) and are not solubilized in sodium bicarbonate solution 5% (acids will be soluble) except that phenol has highly electron-withdrawing substituents such as a nitro group.

Ferric chloride test (FeCl₃)   Most phenols give strongly stained solutions with this reagent (blue, green, purple, etc.). If the color is light yellow, the same as $FeCl_3$, the reaction is considered negative. Some phenols do not give the indicated color, such as with the hydroquinone, since it is oxidized to quinone, and the reagent gives no staining. Acids, except phenolic ones, do not react, although some give yellowish solutions or precipitate.

It is necessary to indicate that, as a precaution, the absence of color is not a negative test for the presence of phenol, and a positive test should be considered with caution because other functions also give color changes: aliphatic as well as aromatic carboxylic acids, enols, oximes, hydroxamic acids, and sulfinic acids.

Procedure 6.5.18:

Add several drops of $FeCl_3$ solution 2.5% (compare the result of staining with a blank test) to 1 ml of a dilute aqueous solution of compound. The color resulting in the test may not be permanent, and therefore record the results immediately after mixing.

If the starting compound is not very soluble, it is necessary to perform the above test with alcoholic solutions instead of aqueous ones. Also, it is usually sufficient to carry out the same test using a drop of each reagent on a piece of paper filter and noting the color of the area of intersection of the two spots.

Bromine water test

Procedure 6.5.19:

Add to an aqueous 1% solution of phenol a saturated solution of bromine water (5 ml of bromine per 100 ml water) dropwise until the bromine color persists. A positive result is given by the appearance of a precipitate caused by the formation of bromophenol and strongly acidic reaction mixture.

6.5.8.1. Characterization of phenols

In principle, the usual procedure is the same as alcohols —i.e., esterification as indicated above. However, it is also possible to prepare an ether-type compound such as aryloxyacetic acids.

Procedure 6.5.20:

Dissolve 200 mg of phenol in 1 ml of 6 N NaOH (24 g in 100 ml of water) in a small test tube. An additional small amount of water can be added to dissolve the sodium phenoxide, if necessary. Add to this solution 0.5 ml of an aqueous solution of chloroacetic acid (50%). Pour the mixture into a flask fitted with a reflux, and heat in a water bath at 90–100°C for 1 h. Cool the solution, and add 2 ml of diluted HCl. The mixture is extracted several times with small portions of ethyl ether. The ether extract is washed with water and extracted with a solution of sodium carbonate (5 %). Acidify the sodium carbonate solution extracted with HCl to precipitate the aryloxyacetic acid. After being filtered, it can be recrystallized from hot water.

6.5.9. Aldehydes and ketones

Aldehydes and ketones have common reactions due to the presence of the carbonyl group in both cases, but also show different reactions because aldehydes are oxidized to acids by the action of mild oxidizing agents.

The presence of the carbonyl group is signalled by the appearance of an absorption IR band (typically the strongest one in the spectrum) in the region of $1,650–1,850 \, cm^{-1}$. They differ from the esters by the absence of two strong bands in the region of $1,050–1,250 \, cm^{-1}$ that appear in these.

Bisulfite combination   All aldehydes and most ketones yield this reaction except for sterically hindered ones.

Procedure 6.5.21:

Vigorously shake the mixture containing the aldehyde or ketone with a saturated solution of sodium hydrogen sulfite (bisulfite). It will raise the temperature because the addition reaction is exothermic. Collect the crystalline material, wash with EtOH and ethyl ether, and let dry. The addition compound decomposes with a solution of 10% sodium carbonate or diluted HCl. The appearance of a precipitate demonstrates the presence of a carbonyl group.

Formation of 2,4-dinitrophenylhydrazone   Aldehydes and ketones are identified by the formation of 2,4-dinitrophenylhydrazones by a reaction with 2,4-phenylhydrazine, which renders a precipitate. A yellow crystalline product indicates a saturated carbonyl compound, while an orange precipitate indicates the presence of an α,β-unsaturated system, and a red compound denotes the presence of a ketone or aromatic aldehyde.

Procedure 6.5.22:

 To a test tube containing 1 ml of reagent, [a] add a drop of carbonyl compound dissolved in EtOH if liquid or about 50 mg if solid. Stir the mixture vigorously. If a solid is not formed immediately, let stand for 15 min. The formation of a precipitate between yellow and red is considered a positive result. Determine the m.p., considering the possibility that it could be 2,4-dinitrophenylhydrazine (m.p. = 198 °C).

---

[a]Dissolve 3 g of 2,4-dinitrophenylhydrazine in 15 ml of concentrated H_2SO_4. Add this solution with stirring in another 20 ml of water and 70 ml of EtOH. Mix both solutions and filter.

6.5.9.1. Aldehydes: differences from ketones

Tollen's reagent   The reagent is an ammoniacal solution, $Ag(NH_3)OH$, to be prepared when required for use.

$$AgNO_3 \;+\; NH_4OH \;\longrightarrow\; Ag(NH_3)OH$$

$$RCHO \;+\; 2\,Ag(NH_3)OH \;\longrightarrow\; \begin{array}{l} RCOOH \;+\; 2\,NH_3 \;+\; 2\,AgO \;+\; H_2O \\ \qquad\qquad \text{(mirror)} \end{array}$$

Procedure 6.5.23:

Prepare two solutions:

- Solution A: Dissolve 3 g of silver nitrate in 30 ml of water.
- Solution B: Prepare a solution of NaOH (10%).

When the reagent is required, mix 1 ml of each of the two solutions (A and B) in a clean test tube and add dropwise a diluted ammonia solution until complete dissolution of silver oxide. Add a few drops of a diluted solution of the compound to the above mixture. In a positive test, silver is deposited as a mirror in the test tube wall either cold or after heating in a hot water bath. Use dilute HNO_3 to clean the test tube in order to remove the silver deposited.

Ketones do not give this reaction, except for hydroxyketones and 1,2-diketones that are reducing and some nitrogen compounds such as hydrazines, hydroxylamines, and aminophenols, which are easily distinguished from them.

Fehling's solution The reagent is prepared upon use by mixing Fehling A (cupric solution) with Fehling B (alkali solution of potassium sodium tartrate) in equal parts. It forms a complex with cupric ion, which is reduced by the same compounds that reduced the Tollen's reagent. If the reaction is positive a red Cu_2O precipitate is formed.

$$R-CHO \; + \; 2\,Cu^{++} \; + \; 4\,HO^- \; \longrightarrow \; R-COOH \; + \; Cu_2O \; + \; 2\,H_2O$$

Procedure 6.5.24:

Prepare two solutions.

- Solution A: Dissolve 34.64 g of copper sulfate in 500 ml of water.
- Solution B: Dissolve 17.6 g of potassium sodium tartrate and 7.7 g of NaOH in 50 ml of water.

When required for use, 3 ml of each solution is mixed. Add a few drops of the liquid compound to this mixture and heat the solution for 2 min in a water bath. The appearance of a red precipitate indicates that the test is positive.

6.5.9.2. Ketones: differences from aldehydes

The most commonly used is the iodoform reaction. It is applied to identify methyl ketones. It should be stressed that, in addition to this reaction and to EtOH and acetaldehyde, other substances that yield methyl ketones by oxidation also give a positive iodoform test.

Iodoform test for methyl ketones This test is based on the breakage of the carbonyl compound by methyl carbonyl bond and subsequent oxidation to carboxylic acid.

$$I_2 \; + \; NaOH \; \longrightarrow \; HIO_3 \; + \; NaI$$

$$R-CO-CH_3 \; + \; 3\,I_2 \; \longrightarrow \; CHI_3 \; + \; R-COONa \; + \; NaI \; + \; H_2O$$

Procedure 6.5.25:

Dissolve 5 or 6 drops of the compound (approximately 100 mg) in 2 ml of water. If necessary add dioxane.[a] To dissolve the sample, add 1 ml of 10% NaOH and reagent[b] dropwise and stir until the dark iodine color appears. Let stand for a few minutes. If no precipitate forms, heat in a bath at 60 °C and, if the color disappears on heating, add more reactive until the color remains. Then, add a few drops of NaOH solution, dilute with 4 ml of water, and let stand for 15 min. A positive result is indicated by the appearance of a yellow precipitate (m.p. = 119–121 °C) with a characteristic medicinal odor.

[a]Read the Danger Noticeboard 6.5.2.

[b]Prepare the reagent by dissolving 20 g of potassium iodide and 10 g iodine in 100 ml of water.

Noticeboard 6.5.2

⚠ DANGER!

The test must be performed in a fume hood, using gloves.

6.5.9.3. Aldehyde and ketone characterization

Phenylhydrazones

Procedure 6.5.26:

In a test tube, place 100 mg of carbonyl compound, 4 ml of MeOH, and 4 drops of phenylhydrazine. Boil the mixture for 1 min, add a drop of glacial acetic acid, and again boil for 3 min. Add cold water dropwise until a permanent turbidity appears. Cool, collect, and wash the crystal with 1 ml of water containing acetic acid. Recrystallize from hot MeOH, adding water drop by drop until reaching turbidity.

Nitrophenylhydrazones

Procedure 6.5.27:

Place 100 mg of nitrophenylhydrazine in a test tube or Erlenmeyer flask with 10 ml of MeOH. Add 5 drops of concentrated HCl and heat if necessary for complete dissolution. Dissolve approximately 100 mg of compound in 1 ml of MeOH and add to the reagent. Heat the mixture in a steam bath for 2 min and let stand for 20–30 min. To ensure crystallization, it is advisable to add water until reaching persistent turbidity.

Semicarbazones

Procedure 6.5.28:

In a test tube, place 100 mg of semicarbazide hydrochloride, 150 mg of sodium acetate, 1 ml water, and 1 ml of EtOH. Add 100 mg of carbonyl compound. If the mixture is cloudy, add more EtOH to make it clear. Shake the mixture for a few minutes and let stand. The reaction can be accelerated by heating in a steam bath for 10 min and then cooling in an ice bath. Filter the crystals and wash with cold water. They are usually recrystallized from MeOH or EtOH, sometimes with the aid of water.

Oximes   These can be produced in a similar way as semicarbazones with heating but also as described in Procedure 6.5.29.

Procedure 6.5.29:

Reflux a mixture of 100 mg of aldehyde or ketone, 100 mg of hydroxylamine hydrochloride, 2 ml of EtOH, and 0.5 ml of pyridine in a water bath for 15–60 min. Remove the solvent under reduced pressure. Add several milliliters of cold water and blend thoroughly. Collect the oxime, and recrystallize from EtOH or EtOH/H_2O.

6.5.10. Quinones

Quinones are colored crystalline compounds (mostly yellow) with a pungent odor. The most characteristic band in the IR spectrum is the carbonyl band near $1,670 \text{ cm}^{-1}$.

Iodide test   Most quinones release iodine from acidified solutions of potassium iodide.

Oximes and semicarbazones   Quinones usually form oximes and semicarbazones that often do not have the usual structure.

Reduction dihydroxybenzenes (diphenols)

Procedure 6.5.30:

Suspend the quinone in diluted HCl and add a pinch of zinc dust. When the solution has become colorless, neutralize with sodium bicarbonate and extract the dihydroxybenzene with ether. Evaporate the solvent and identify it.[a]

---

[a]Pyrocatechol (*o*-dihydroxybenzene) and hydroquinone (*p*-dihydroxybenzene) reduce Tollens' reagent in cold and Fehling's in hot. Resorcinol (*m*-dihydroxybenzene) reduces both in hot. Pyrogallol (1,2,3-trihydroxybenzene) also reduces the silver salts to metallic silver.

Quinoxaline formation   This is a test for *o*-quinones.

Procedure 6.5.31:

Dissolve the *o*-quinone alcohol or glacial acetic acid and add an equivalent amount of *o*-phenylenediamine in alcohol. Heat the mixture in a steam bath for 15–20 min, cool, and dilute with water to crystallize. Filter and Recrystallize from EtOH/H$_2$O.

6.5.11. Carbohydrates

Mono- and disaccharides are colorless solids or viscous liquids that readily dissolve in water. The IR spectrum exhibits strong absorption of hydroxyl and usually does not have carbonyl absorption. Sugar "reductors" are oxidized by Tollens' reagent and Fehling's solution.

Molisch test This is a test for pentoses and hexoses.

Procedure 6.5.32:

Place 5 mg of substance in 0.5 ml of water. Add 2 drops of a solution of α-naphthol in 10% EtOH. With a dropper, place 1 ml of concentrated H_2SO_4 onto a slide so that by tilting the tube wall the H_2SO_4, being more dense, is deposited on the bottom without mixing with water. If a carbohydrate is present, a red ring is formed at the interface of the two liquids. Let the solution stand a few minutes, and dilute with 5 ml of water, whereupon a dark purple precipitate will appear.

6.5.12. Carboxylic acids

To distinguish a water-insoluble, carboxylic acid from other acidic compounds, solubility tests can be used. These are the only soluble components in a solution of $NaOH/NaHCO_3$ (except some nitrophenols). The soluble components in water release CO_2 from the bicarbonate solution. Other compounds that also give CO_2 are the amine salts, the sulfonic acid, and derivatives that are easily hydrolyzed, giving acidic products.

They have a very wide and characteristic OH band in the IR spectrum and a carbonyl band near $1{,}700\ \mathrm{cm}^{-1}$.

Carboxylic acids can be characterized by the formation of the following derivatives:

Amide formation

Procedure 6.5.33:

In a 25 ml flask equipped with reflux and drying tube, place 100 mg of the acid (or its sodium salt) and 4 ml of thionyl chloride. The mixture is boiled gently for 30 min and then the mixture is poured over 15 ml of ice-cold concentrated ammonia. Collect the precipitated amide by filtration. If, instead of the amide, anilide or *p*-toluide is to be prepared, the excess of thionyl chloride (b.p. = 78 °C) is distilled. The acid chloride is dissolved in 5 ml of benzene, and a solution of 2 g of the amine is added to 15 ml of benzene. The reaction mixture is shaken with 5 ml of diluted HCl (to remove excess aniline) and decanted. The benzene layer is washed with 5 ml of water, the solvent is removed, and anilide is recrystallized from water or $EtOH/H_2O$.

Phenacyl esters

Procedure 6.5.34:

In a flask fitted with a reflux condenser place 150 mg of acid and 1 ml of water. Add a drop of phenolphthalein and neutralize with 10% NaOH until pink. Add 1 or 2 drops of diluted HCl to clear the color. Add an alcoholic solution of phenacyl bromide (150 mg in approximately 6 ml of alcohol). Boil the solution at reflux for at least 2 h; cool, add 1 ml of water, and scrape with a rod to induce crystallization. Filter the ester solid, washing with 5% sodium carbonate solution and then several times with cold water. In the case of the sodium salt of the acid, dissolve 150 mg of it in water, put a drop of phenolphthalein and adjust the pH as in the previous case.

6.5.13. Esters

Most esters are liquids or solids of low melting point, many with characteristic odors of flowers and fruits. In its IR spectrum, strong carbonyl bands 1,780–1,720 cm^{-1} appear, accompanied by two strong absorptions in the C–O region of 1,300–1,050 cm^{-1}. The main method for characterizing an ester involves identifying the alcohol and acid that comprise it.

Hydrolysis of esters   Esters can be saponified by boiling with aqueous or alcoholic alkali solutions, with concentrated HCl or 40% H_2SO_4.

Procedure 6.5.35:

In a small flask, place 200 mg to 1 g of ester, add 2 to 10 ml of 25% aqueous NaOH, and stir magnetically. Mount a condenser and reflux for 30 min (if the ester boils below 110 °C) or for 2 h (if it boils above that temperature). Cool and acidify with diluted acid (recommended to use phosphoric acid). Recover the acid by filtration or extraction.

6.5.13.1. Conversion into amides

By heating with concentrated ammonia solutions, many esters are transformed into amides. Likewise, by heating for several hours with hydrazine, hydrates become hydrazides.

6.5.14. Ethers

Like hydrocarbons, ethers are not very reactive but can be distinguished from those for the charge transfer test with iodine and H_2SO_4 solubility (not diaryl ethers). When ethers are allowed to stand exposed to air and light, they tend to form highly explosive peroxides.

Ferrox paper test

Procedure 6.5.36:

Such paper is prepared separately by dissolving 1 g of ferric chloride and 1 g of potassium thiocyanate in 10 ml portions of MeOH. Both solutions are mixed and filtered. Immerse in the solution two or three times strips of filter paper and dry after each immersion. The strips should be stored in the dark. Ethers color the ferrox paper to red due to the ferric thiocyanate complex released.

6.5.15. Amines

Most simple amines are easily recognized by their solubility in diluted mineral acids. Water-soluble amines can be determined by their basic reaction by litmus or other indicators.

Primary amines, both aliphatic and aromatic, present in the IR spectrum as a weak but recognizable doublet in the region of $3,500-3,300$ cm^{-1} and a strong absorption in the region of $1,640-1,560$ cm^{-1}. Secondary amines have an isolated band in the region of $3,450-3,310$ cm^{-1}. Tertiary amines do not exhibit useful absorption features.

Cupric ion test

Procedure 6.5.37:

Add 10 mg or a small drop of the compound to 0.5 ml of a 10% solution of cupric sulfate. The appearance of a color or a precipitate (blue or blue-green) is indicative of the presence of an amine.

Hinsberg test Suspend the amine in a solution of NaOH. By adding benzene-sulfonyl chloride:

- Primary amines form sulfonamides that remain dissolved in the strongly alkaline solution. To acidify, white solid sulfonamide is precipitated.

- The secondary amines form sulfonamides that are not in solution but as white solids precipitated directly from the alkaline reaction mixture insoluble in water, dilute acids, and alkalis.

- Tertiary amines do not react, remain undissolved, and are dissolved when acidified.

Procedure 6.5.38:

In a test tube, place 0.1 ml (100 mg) of amine, 200 mg of *p*-toluenesulfonyl chloride, and 5 ml of 10% NaOH. Cap the tube and stir for 5 min. Remove the cap and heat for 1 min. If no reaction occurs, the substance is probably a tertiary amine. If a precipitate appears, dilute with 5 ml of water and shake. If it does not dissolve, it is probably a secondary amine. If dissolved, acidify with dilute HCl. If it appears again, the precipitate will be a primary amine.

Nitrous acid test (2-naphthol test for aromatic amines)

Procedure 6.5.39:

Dissolve 100 mg of amine in 3 ml of 2N HCl (or 5%), and cool in an ice bath. Add 1 ml of 10% solution of sodium nitrite in water. The release of nitrogen gas indicates that it is an aliphatic primary amine. The release of nitrogen gas after gentle heating indicates that it is an aromatic primary amine.

If it is an aromatic primary amine, the corresponding diazonium salt can be trapped with a solution of 100 mg of 2-naphthol in 2 ml of 10% NaOH. Added dropwise to the solution previously indicated gives a red precipitate of the corresponding azo dye. The colorant may also be orange.

If an insoluble yellow compound (oil or solid) is formed on treatment with nitrite, it is a secondary amine.

If no reaction is observed, it is an aliphatic tertiary amine.

If it is an aromatic tertiary amine, the *p*-nitro derivative can be formed (if the position is free), recognized as a yellow compound, water soluble, and that by the action of alkali gives a greenish precipitate.

6.5.15.1. Characterization of amines

Amines can be characterized by the formation of the following derivatives:

Sulfonamide formation

Procedure 6.5.40:

Reflux for 5–10 min a mixture of 150 mg of sulfonyl chloride (benzene or *p*-toluenesulfonyl) and 200 mg of the amine in 4 ml of dry benzene. The mixture is allowed to cool. By filtration, remove the precipitate of amine hydrochloride. Benzene from the filtrate is evaporated to obtain the crude sulfonamide, which can be recrystallized from EtOH.

Benzamide formation

Procedure 6.5.41:

Suspend approximately 150 mg of the amine in 1 ml of 10% NaOH, and add 0.5 ml of benzoyl chloride dropwise while stirring vigorously and then cool. After 10 min, carefully neutralize to pH = 8. Collect the benzamide, wash with water, and recrystallize from EtOH/H_2O. An alternative method involves heating the amine and benzoyl chloride and dissolving in 2 ml of pyridine at reflux for 30 min. Then pour into ice water, and collect the derivative by vacuum filtration.

Acetamide formation

Procedure 6.5.42:

Dissolve approximately 200 mg of water-insoluble amine in 10 ml of 5% HCl. Add 5% NaOH with a burette until incipient turbidity. Remove cloudiness with a few drops of 5% HCl. Add some ice and 1 ml of acetic anhydride. Stir the mixture, and add 1 g of hydrated sodium acetate dissolved in 2 ml of water. Cool in an ice-water bath, and collect the solid. Recrystallize from EtOH/H_2O.

Picrate formation   These are used mainly as tertiary amine derivatives.

> **Procedure 6.5.43:**
>
> Dissolve approximately 100 mg of amine in 5 ml of EtOH. Add 5 ml of saturated picric acid in EtOH. Heat the solution to a boil and let cool slowly. Recrystallize the formed yellow crystals from MeOH or EtOH.

6.5.16. Amides and related compounds

Almost all are colorless crystalline solids. Important in their classification is the observation of the IR spectrum in which two bands are observed in the region of the carbonyl group (amide I and II) in addition to the N–H stretching group near $3,400 \text{ cm}^{-1}$. As its classification is not always easy, the hydrolysis reaction can be used.

Sodium hydroxide test

> **Procedure 6.5.44:**
>
> Heat to boiling a mixture of 200 mg primary amide with 5 ml of 10% NaOH. Ammonia released can be recognized by its odor and/or by indicator paper.

Anilide conversion

> **Procedure 6.5.45:**
>
> Heat the primary amides together with the releasing of ammonia and yielding the corresponding anilide.

Mercury oxide test   Some amides react with oxides of mercury to form mercuric salts.

6.5.17. Nitrocompounds

Most nitroalkanes are colorless liquids, which become yellow when stored. They give off a distinctive odor, and are both insoluble and denser in water. Most aromatics are yellow crystalline solids. Aromatics present in the IR spectrum are seen in two strong bands at 1560 and 1350 cm^{-1}. Generally, their chemical analysis are not used for its identification, but nitrocompounds can be reduced to a wide range of nitrogen products.

Zinc test

> **Procedure 6.5.46:**
>
> Dissolve 500 mg of product in 10 ml of 50% EtOH and add 500 mg of ammonium chloride and 500 mg of zinc powder. Stir the mixture, and heat to boiling. Let cool for 5 min, filter, conduct the test using Tollens' reagent in the filter.

The unknown product is usually reduced to hydrazine, hydroxylamine, or amino phenol compounds, which are easily oxidized with Tollens' reagent.

6.5.18. Sulfur compounds: thiols

There are many compounds that contain sulfur due to its different oxidation states —for example: thiols, sulfides, disulfides, sulfoxides, sulfones, sulfenic acids, sulfinic acids and sulfonic acids. However, their study is beyond the scope of this chapter, and therefore thiols are treated here only as an example.

Thiols have an odor reminiscent of hydrogen sulfide (SH_2). They are usually insoluble in water but soluble in aqueous alkali to form soluble salts. A yellow precipitate is produced when a few drops of the thiol are added to a solution of lead acetate in EtOH.

6.6. Identification of an unknown substance

In summary, in order to systematize the recognition of an unknown substance, the following guidelines can be followed:

1. Physical examination of the product (organoleptic properties).

2. Determination of physical constants.

3. Qualitative elemental analysis.

4. Study of the IR spectrum.

5. Classification by solubility.

6. Examination of reaction characteristics of the functional groups.

7. Study the table data of the possible product.

8. Preparation of solid derivatives.

9. Product identification.

It is advisable to fill out the a form sheet for each of the unknown products (see Figure 6.3).

FUNCTIONAL ANALYSIS SAMPLE No.:

1–2. ORGANOLEPTIC PROPERTIES AND PHYSICAL PROPERTIES:

a) State: _____ Melting point: _____
b) Color: _____ Boiling point: _____
c) Odor: _____

3. Elemental analysis: Contains: _____

4. Infrared spectrum:
Absorption (cm⁻¹) Assignment
_____ _____
_____ _____

5. SOLUBILITY:

Soluble in: _____
Insoluble in: _____
Group: _____

6. ORGANIC FUNCTIONAL ANALYSIS: State reagent, result, and deductions.

a) _____
b) _____

7. POTENTIAL PRODUCTS (indicate the reason):

_____

8. FORMATION OF DERIVATIVES:

a) _____
b) _____

9. PRODUCT IDENTIFICATION:

Figure 6.3: Form sheet for functional group analysis of a sample.

Chapter 7

Basic Operation Experiments

The experiments included in this chapter have been chosen with relatively simple processes that serve as a pretext for the development of such techniques, so that the student acquires the necessary skills to independently perform an experiment in the laboratory. The processes will be of increasing complexity, both theoretically and practically. In all cases, further information is available in the corresponding sections of Chapters 3 and 4, where the techniques and equipment necessary for each of the basic operations are described in detail.

In all cases, an estimate of the difficulty of each experiment (indicated by an icon of one, two, or three Erlenmeyer flasks) and the approximate time needed for each experiment are given. This classification, sorted by difficulty and time estimation, holds for all experiments described in the book.

Below, a series of experiments are described, including the following techniques:

- Recrystallization of organic compounds in water and in organic solvents.

- Magnetic stirring of a reaction and centrifugation.

- Extraction and decantation of a mixture of two immiscible liquids.

- Reflux, simple distillation, fractional distillation, and steam distillation.

- Removal of solvents under reduced pressure (rotary evaporator).

- Solid-liquid extraction (Sohxlet).

- Thin-layer chromatography (TLC) and column chromatography (CC).

7.1. Recrystallization and solubility tests: from water and from organic solvents

Estimated time	Difficulty	Basic lab operations
2 h		✔ Gravity filtration (see p. 100). ✔ Vacuum filtration (see p. 103). ✔ Recrystallization from water (see p. 105). ✔ Recrystallization from organic solvent (see p. 105). ✔ Reflux (see p. 89).

7.1.1. Goal

To study the recrystallization and isolation purification technique for organic compounds. Determination of acidic, basic, or neutral character of various organic compounds by solubility tests.

Naphthalene   Acetanilide   Benzoic acid

7.1.2. Background

For dissolving a solute in a solvent, the solute-solvent attractive interactions must be stronger than the intermolecular attractions solute-solute and solvent-solvent. The more similar these attractions become, the more easily the solute-solvent attractive interactions will be established. In general, it can be said that "like dissolves like."

- Non-polar compounds are dissolved in apolar solvents because solute-solvent intermolecular forces of similar intensity are set to those present in the undissolved solute (van der Waals interactions).

- Polar compounds are dissolved in polar solvents because solute-solvent intermolecular forces are of similar intensity to those present in the undissolved solute.

- A non-polar compound is not soluble in a polar solvent such as water, because non-polar compounds can form hydrogen-bonding interactions between solute and solvent molecules.

- Molecules that can establish links by hydrogen bonding with water (hydrophilic) or having ionizable groups (such as strong acids, acetic acid, MeOH, NaCl, etc.) will be soluble in water.

A key factor to consider is whether the solvent is aqueous or organic (see Noticeboard 7.1.1 on p. 211).

7.1.3. Procedure

A) **Recrystallization of benzoic acid from water**: Weigh 1 g of benzoic acid, and transfer to a beaker. Add approximately 20 ml of deionized water and heat the mixture to boiling. Gradually add portions of hot water, stirring until completely dissolved. Remove the beaker from the heat source, filter with a conical funnel (using a filter paper rolled into a cone), and cool the filtrate to r.t. for approximately 20 min. Separate the crystals obtained by Büchner filtration. Dry the solid and weigh when dry.

B) **Recrystallization of acetanilide from water**: Weigh 1 g of acetanilide, and transfer to a beaker. Add 20 ml of water and heat gently. The appearance of an oil is observed in the solution, which dissolves with the addition of several portions of hot water. If a colored solution results, add a small amount of activated carbon (approximately 0.3 g). Filter the hot solution with a conical funnel (using a filter paper rolled into a cone). Cool the mixture for approximately 30 min and filter off the crystals by Büchner filtration (see Figure 7.1).

C) **Recrystallization of naphthalene from EtOH**: Transfer 2.5 g of naphthalene to a 50 ml round-bottom flask and add 12 ml of EtOH and a magnetic bar. Attach a reflux condenser and heat the mixture to boiling with magnetic stirring. With a Pasteur pipette, at the upper reflux condenser part add EtOH in 1 ml portions, maintaining reflux until complete dissolution of the solid (see Noticeboard 7.1.1 on p. 211). Filter while warm with a conical funnel (using a filter paper rolled into a cone). Collect the filtrate in a 50 ml Erlenmeyer, and cover it with a suitable plug. Allow to cool to about r.t. for 30 min and filter by Büchner filtration. Dry and weigh the resulting solid.

D) **Other examples of recrystallization from organic solvents**: Recrystallization of 2.5 g of benzoic acid from 15 ml of MeOH (washing the crystals with a mixture of MeOH/water, 50%); recrystallization of 1 g of acetanilide from 20 ml toluene. Filter into a hot Hirsch funnel (do not allow the solute to solidify in the funnel).

E) **Solubility tests (acid, basic, or neutral character of an organic compound)**: The amounts of compounds used for solubility testing are approximate. Use 2–3 drops if a liquid or approximately 10 mg if a solid. If it is a fine solid powder, a small amount of solid is spread on a watch glass with the back of a spatula or a mortar. It is not necessary to weigh the solid;

use only enough to cover the tip of a small spatula. Add the appropriate amount of unknown problem substance (solid or liquid) to a small test tube and proceed with the F) and G) solubility tests.

F) **Solubility in water**: Add approximately 6 drops of water to the test tube containing the substance. Shake the tube. If the compound is soluble, a homogeneous solution will form with water, while an insoluble liquid will remain as a separate phase. A water-soluble organic compound (solid or liquid) may be a low-molecular-weight polar compound (up to five carbon atoms). If the compound does not dissolve completely with 6 drops of water, add more water to 1 ml and observe whether the compound is solubilized. Check the pH of the water to determine whether the substance is partly or fully soluble in water and whether the compound has changed the pH of water.

- If the pH paper turns red: Acidic compound soluble in water.

- If the pH paper turns blue: Basic compound soluble in water.

- If the paper does not change pH color: Neutral compound soluble in water or water-insoluble compound.

G) **Acid-base solubility test**:

- Solubility in NaOH (5%): Add approximately 1 ml of NaOH 5% in small portions, approximately 6 drops each, in the test tube containing the test compound. Vigorously shake the test tube after adding each portion of solvent. The formation of a homogeneous solution, a color change or gas or heat release will indicate solubility. If soluble, then the test compound acts as an organic acid. The most common acidic organic compounds are carboxylic acids and phenols. Carboxylic acids are generally considered stronger than phenols, but both react with NaOH.

- Solubility in $NaHCO_3$ (5%): Add approximately 1 ml $NaHCO_3$ 5% solution in small portions, approximately 6 drops each, in the test tube containing the test compound. Vigorously shake the test tube after adding each portion of solvent. The formation of a homogeneous solution, a color change or gas or heat release will indicate solubility. If soluble, then the test compound acts as a strong organic acid. If not, then it is a weak organic acid, if dissolved in NaOH. Typically, only carboxylic acids react with $NaHCO_3$. Phenols are weak organic acids, unless they present nitro groups (nitrophenol is more acidic than phenol).

- Solubility in HCl (5%): Add about 1 ml of HCl 5%, in small portions, approximately 6 drops to the test tube containing the compound. Vigorously shake the test tube after adding each portion of solvent. The formation of a homogeneous solution, a colour change or gas or heat release will indicate solubility. If the compound is soluble in HCl, then it is an organic base. Amines are the most common organic bases. If the compound is insoluble in all previous solutions, then the test compound is not an acidic or basic organic compound.

Table 7.1: Physico-chemical properties of the reagents used.

Compound	M_w	M.p. (°C)	B.p. (°C)	Density (g·ml^{-1})	Danger[a] (GHS)
Benzoic acid	122.12	125	249	1.08	
Acetanilide	135.16	113–115	304	-	
Naphthalene	128.17	79.5–81.0	218	-	
EtOH	46.07	−114.1	78.5	0.790	
MeOH	32.04	−98	64.7	0.791	
Hexane	86.18	−95	69	0.659	
CH$_2$Cl$_2$	84.93	−97	40.0	1.33	
Toluene	92.14	−93	110.6	0.867	
NaOH	40.00	318	1,390	2.130	
NaHCO$_3$	84.01	300	-	2.160	Non-hazardous
HCl	36.46	−30	>100	1.200	

[a] For brevity, only GHS icons are indicated. The information offered in the Material Safety Data Sheet (MSDS) should be consulted.

Noticeboard 7.1.1

 DANGER!

The use of organic solvents (EtOH in an experiment) always requires the mixture to be heated, setting up a reflux. Caution must be taken when filtering a hot solution. If not handled in this way, flammable vapors can be released into the atmosphere, and nearby flames or heat sources would pose a serious risk of fire and explosion.

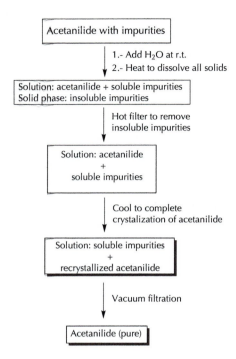

Figure 7.1: General scheme for isolation and purification of acetanilide.

7.2. Stirring and decanting: preparing biodiesel from cooking oil

Estimated time	Difficulty	Basic lab operations
2 h	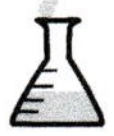	✔ Liquid-liquid extraction (see p. 111).

7.2.1. Goal

To learn techniques for magnetic stirring and separation of two immiscible liquids by decantation.

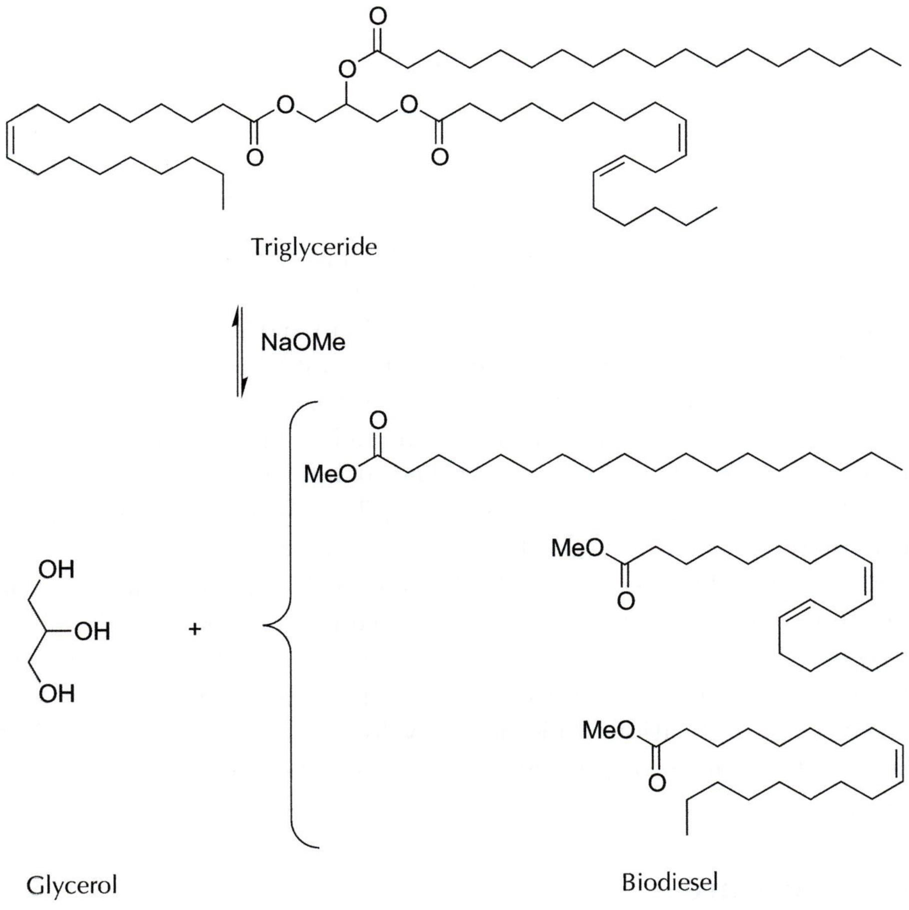
Triglyceride
NaOMe
MeO
MeO
MeO
OH
OH
OH
+
Glycerol
Biodiesel

7.2.2. Background

Biofuels are an alternative to fossil fuels as an energy source. Vegetable oils could in theory be used as fuel, but because of their high viscosity, they cannot be used directly. It is necessary to transform them into other less viscous materials that can be used in vehicle engines. This transformation is performed by means of a transesterification reaction.

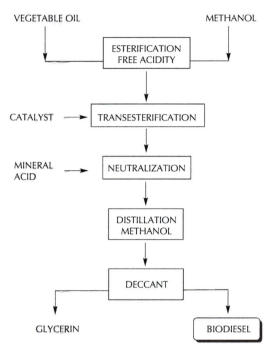

Vegetable oils are compounds belonging to the family of triglycerides. By treatment with a base and an alcohol (MeOH or EtOH), an oil becomes a glycerin and methyl or ethyl ester. The transesterification reaction is a catalytic process composed by two reaction steps. In the first, the NaOH reacts with MeOH in an acid-base sodium methoxide-forming process, a very strong base, and water. In the second step, sodium methoxide acts as a nucleophile attacking the carbonyl carbon of vegetable oil, yielding a tetrahedral intermediate. In the final step the methyl ester is separated from glycerol. When the experiment is performed with raw material such as vegetable oil, the reaction proceeds smoothly and provides a good yield of biodiesel. If the raw oil is a cooked one, the process is more complex because of the presence of free fatty acids from the hydrolysis of triglycerides. In the basic transesterification process, the free fatty acids are deprotonated, leading to the formation of carboxylates that do not react by base-catalyzed esterification. This not only lowers the reaction yield but also complicates the process of separation as long-chain carboxylates give rise to emulsions.

7.2.3. Procedure

In a 100 ml Erlenmeyer equipped with a clamp, stir 40 ml of MeOH and 1.3 g of solid NaOH at r.t. to make a solution. Conversely, add 130 ml of vegetable oil[1]

---

[1]Sunflower, soybean, or olive oils can be used as feedstock.

to a 250 ml round-bottom flask. Heat the flask with the oil to 50 °C using a water bath with magnetic stirring (water bath, see p. 83). Add sodium methoxide solution to the oil and stir the mixture for 1 h, maintaining the indicated temperature. Transfer the flask contents to a separatory funnel. Allow the mixture to stand until the two layers separate. The less dense layer (upper) corresponds to biodiesel, the denser (bottom) to the glycerol. Most of the decanting process occurs in the first hour; however, very often the process takes several hours to completely separate the two layers.[2] Use the separatory funnel stopcock to separate the two layers. If the glycerol does not pour freely, the biodiesel can be removed at the top of the separatory funnel with a pipette. The resulting volume is measured with a graduated cylinder and weighed in the same cylinder to determine density (see ref. [6]).

Table 7.2: Physico-chemical properties of the reagents used.

Compound	M_w	M.p. (°C)	B.p. (°C)	Density (g·ml^{-1})	Danger[a] (GHS)
NaOH	40.00	318	1,390	2.130	
MeOH	32.04	−98	64.7	0.791	
MeONa	54.02	-	-	0.970	
Glycerol	92.09	20	182	-	Non-hazardous

[a] For brevity, only GHS icons are indicated. The information offered in the Material Safety Data Sheet (MSDS) should be consulted.

---

[2]The reaction can be left overnight to ensure that the process is completed.

7.3. Simple distillation: isolation of alcohol from wine

Estimated time	Difficulty	Basic lab operations
2 h		✔ Simple distillation (see p. 120).

7.3.1. Goal

To determine the percentage of EtOH in wine "alcohol by volume" (ABV) by distillation and subsequent measurement of the density of the distillate.

| CH₃OH | CH₃CH₂OH | Glucose | Fructose | Tannin (galic acid) |

CH$_3$OH — Methanol   CH$_3$CH$_2$OH — Ethanol   Glucose   Fructose   Tannin (galic acid)

7.3.2. Background

One of the characteristics to be specified in the documentation of alcoholic beverages —for example, wines —is the percentage (%) of alcohol by volume of EtOH, called alcohol by volume (ABV). In a sample of alcohol and water, this can be determined directly by measuring the density using a hydrometer or alcoholmeter. Wine, however, is a complex mixture containing water, EtOH, sugars, organic acids, pigments, and other ingredients; therefore, the ABV cannot be directly determined from the density of the wine, but instead a simple distillation process becomes necessary. The volatile components found in considerable quantity are water and EtOH, having a b.p. of 100.0 and 78.3 °C, respectively. Both can form azeotrope boiling at 78.2 °C and have a composition of 96% of mass of EtOH (97% in v/v). In red wines, the ABV of EtOH is expressed in percentage volume and is somewhat higher than 10%. In red wine distillation, any fraction containing alcohol at 100% cannot be produced because the more volatile component is the azeotrope. In this experiment, fractions will not be produced; on the other hand, all EtOH contained in the sample will be distilled, with the goal of determining the ABV content of the corresponding red wine.

7.3.3. Procedure

Using a graduated cylinder, measure 100 ml of wine and transfer it to a 250 ml round-bottom flask. Wash the graduated cylinder with 50 ml of deionized water to remove any remaining liquid. Add water to the flask contents.[3] Add a boiling chip and mount a simple distillation in the flask containing the sample of wine and begin heating with a hot plate, maintaining a homogeneous boiling.[4] Collect the distillate in a graduated cylinder to reach a volume approximately 90 ml. Allow the contents of the graduated cylinder to cool to r.t. and add deionized water to reach 100 ml (the same volume as the starting sample). Using an alcoholometer, determine the alcohol content of the wine (see ref. [5]).

Table 7.3: Physico-chemical properties of the reagents used.

Compound	M_w	M.p. (°C)	B.p. (°C)	Density (g·ml^{-1})	Danger[a] (GHS)
Red wine					
EtOH	46.07	−114.1	78.5	0.790	

[a] For brevity, only GHS icons are indicated. The information offered in the Material Safety Data Sheet (MSDS) should be consulted.

---

[3]The addition of this extra water will ensure that the flask does not run dry during the distillation process.

[4]Do not forget to open the water coolant.

7.4. Extraction and "rotary evaporator": obtaining cloves and cinnamon oils

Estimated time	Difficulty	Basic lab operations
2 h		✔ Gravity filtration (see p. 100). ✔ Rotary evaporator (see p. 117). ✔ Reflux (see p. 89).

7.4.1. Goal

To extract essential oils from cinnamon and/or cloves by hot extraction using an organic solvent under reflux and then removing the solvent under reduced pressure (rotary evaporator).

Cinnamaldehyde

Eugenol

7.4.2. Background

Many plants are commonly used for their pleasant fragrances or flavors caused by the presence of certain molecules that provide them with their organoleptic properties. The use of plants to give food flavor and aroma is a practice known since antiquity. Cinnamon, cloves, garlic, mint, and vanilla are some common examples of this experiment as rooted in different cultures and are also fundamental elements of the human diet that enrich food and make it attractive. On the other hand, the use of fragrances and essences for healing or for ornamental purposes has also been known for thousands of years. Essential oil is a concentrated substance, usually volatile, extracted from a plant. Essential oils are generally located in the seeds or flowers but may exist in other plant parts such as stems, leaves, or roots. The term "essential" comes from the belief that the oil represented the *essence* of the odor or taste of a particular plant species. Essential oils are used in almost all types of food production, such as the manufacture of candies, pastries, pickles, drinks, and perfume. They can be isolated from a plant by various processes or a combination of processes, such as mechanical pressing, grinding, maceration, solvent extraction, or distillation.

7.4.3. Procedure

Weigh 5.0 g of spice (cinnamon or clove), and finely grind using a mortar. Transfer the solid to a 100 ml round-bottom flask, add 35 ml of ethyl acetate, and use a stirbar (or boiling chip). Fit a condenser, reflux the mixture for 30 min. Cool the flask contents to r.t. and filter by gravity into a dry Erlenmeyer with a fluted paper filter. Wash the round-bottom flask with 10 ml of ethyl acetate, and pour the solvent onto the solid remaining in the fluted paper filter to ensure that all the extract from the species is obtained. Then dry the ethyl acetate solution over anhydrous sodium sulfate, and filter by gravity into a 100 ml tared round-bottom flask. Remove the solvent under reduced pressure (rotary evaporator), weigh the contents, and determine the percentage of essential oil from the starting sample. For cinnamon, the main essential oil component will be cinnamaldehyde, while for cloves it will be eugenol (see ref. [7]).

Table 7.4: Physico-chemical properties of the reagents used.

Compound	M_w	M.p. (°C)	B.p. (°C)	Density (g·ml^{-1})	Danger[a] (GHS)
Cinnamon or clove					
Cinnamaldehyde	132.16	−7.5	248	1.050	!
Eugenol	164.20	−7.5	254	1.067	✠ !
Ethyl acetate	88.11	−84	77.1	0.902	🔥 !
Na_2SO_4	142.04	884	-	2.630	Non-hazardous

[a] For brevity, only GHS icons are indicated. The information offered in the Material Safety Data Sheet (MSDS) should be consulted.

7.5. Liquid-liquid extraction (with centrifuge): isolation of caffeine from soluble coffee

Estimated time	Difficulty	Basic lab operations
2 h		✔ Gravity filtration (see p. 100). ✔ Vacuum filtration (see p. 103). ✔ Liquid-liquid extraction (see p. 111). ✔ Centrifugation (see p. 104). ✔ Recrystallization from organic solvent (see p. 105). ✔ Rotary evaporator (see p. 117).

7.5.1. Goal

To get caffeine from a commercial soluble coffee by extraction with an organic solvent and centrifugal separation from the resulting aqueous phase and the organic phase.

Caffeine

7.5.2. Background

Caffeine is an alkaloid in the xanthine family. When pure, it is an odorless white solid of m.p. 234–236 °C partially soluble in water (100 mM). As is known, caffeine has a stimulating effect and is found in plants such as coffee or tea. Instant coffee or soluble coffee is a widespread product on the market that is made from infusion by the lyophilization process.[5] According to the manufacturers, instant coffee contains between 55 and 62 mg of caffeine per cup prepared with a teaspoon of the product. In this practice, caffeine is obtained from an extract from the commercial product.

---

[5]In this process, the product is frozen and subsequently a vacuum is drawn to remove water by sublimation.

7.5.3. Procedure

Place 2 g of sodium carbonate and 2 g of instant soluble coffee in a 25 ml beaker.[6] Add hot deionized water (9 ml), stir the mixture, and heat until boiling. Then cool the mixture to r.t. Pour the contents of the beaker into a 15 ml centrifuge tube, and add 2 ml of CH_2Cl_2; place a stopper on the test tube, and shake vigorously for 1 min. Then centrifuge for 90 s. Separate the yellowish solution of CH_2Cl_2 from the aqueous phase using a Pasteur pipette. Repeat the operation twice.[7] Combine the organic extracts and dry with anhydrous sodium sulfate. Remove the solid by gravity filtration. Wash the desiccant with 3 ml of solvent in the filter paper, and mix together with the rest of the CH_2Cl_2. Transfer the solution to a tared flask, remove the solvent using a rotary evaporator, weigh the flask, and calculate the caffeine obtained. Caffeine can be purified by recrystallization from EtOH (see ref. [8]).

Table 7.5: Physico-chemical properties of the reagents used.

Compound	M_w	M.p. (°C)	B.p. (°C)	Density (g·ml^{-1})	Danger[a] (GHS)
Soluble coffee					
Na_2CO_3	105.99	851	-	2.532	!
CH_2Cl_2	84.93	−97	40.0	1.33	
Na_2SO_4	142.04	884	-	2.630	Non-hazardous
EtOH	46.07	−114.1	78.5	0.790	
Caffeine	194.19	234–236.5	-	1.230	!

[a] For brevity, only GHS icons are indicated. The information offered in the Material Safety Data Sheet (MSDS) should be consulted.

Spectroscopic data (caffeine):

- IR (KBr): 1702 (s), 1662 (s), 1551 (m), 1487 (m), 1241 (m), 746 (s) cm^{-1}.

- ^{1}H-NMR (CDCl$_3$): $\delta_H = 7.51$ (s, 1H), 4.00 (s, 3H), 3.58 (s, 3H), and 3.41(s, 3H) ppm.

- ^{13}C-NMR (CDCl$_3$): $\delta_C = 155.32$, 151.66, 148.67, 141.57, 107.51, 33.57, 29.70, and 27.88 ppm.

- MS: m/z = 194.19 (M$^+$)(PB), 109.

---

[6]This ensures that none of the hydrogens is protonated, which will facilitate dissolution in an organic solvent.

[7]Alternatively, the two phases can be separated by decanting in a separatory funnel, but this is less effective.

7.6. Liquid-liquid extraction (with separatory funnel): separating components of a mixture

Estimated time	Difficulty	Basic lab operations
4 h		✔ Gravity filtration (see p. 100). ✔ Vacuum filtration (see p. 103). ✔ Liquid-liquid extraction with reaction (see p. 111). ✔ Recrystallization from water (see p. 105). ✔ Rotary evaporator (see p. 117).

7.6.1. Goal

To learn how different organic compounds can be isolated according to their acid-base properties, which change their solubility in the organic and aqueous solvents, using liquid-liquid extraction.

Naphthalene   *p*-Chloroaniline   Phenol   Benzoic acid

7.6.2. Background

Liquid-liquid extraction is one of the most common operations in the Organic Chemistry laboratory, since many reactions involve the use of this technique for the isolation of the products. Carboxylic acids and phenols can react with bases such as sodium bicarbonate, sodium carbonate, or sodium hydroxide, yielding a proton and forming the corresponding water-soluble anions. Furthermore, amines yield water-soluble ions generated by protonation, such as ammonium salts. This experiment involves the separation of the components of a mixture consisting of naphthalene, *p*-chloroaniline, phenol, and benzoic acid (5 g/L for each component), which will be dissolved in an organic solvent such as CH_2Cl_2, depending on the acidic or basic character of the components of the mixture.

> Noticeboard 7.6.1
>
> **WARNING!**
>
> Never throw a layer away until the end of the experiment or before being absolutely certain that the layer is no longer needed.

7.6.3. Procedure

A) Two-components example (mixture an acidic and a neutral compound):
Given 30 ml of CH_2Cl_2 solution of benzoic acid and naphthalene, the amount of each component in the 30 ml is 4 g. Pour the solution into a separatory funnel and treat with a solution of 1.0 M sodium bicarbonate (2 × 20 ml). Mix the contents thoroughly by shaking vigorously for about 2–3 min with periodic venting. Allow the layers to separate completely, and then drain off the organic layer (bottom) into a 100 ml Erlenmeyer flask. Afterward, drain off the aqueous layer (top) into another 100 ml Erlenmeyer. The two components are now separated into two layers. To isolate benzoic acid from the aqueous solution, neutralize the sodium salt adding dropwise 3 M HCl to the aqueous solution, checking that the mixture is acidic (pH indicator paper). Then vacuum filter the white solid using a Büchner and wash the solid with portions of 10 ml of cold H_2O. Weigh the solid, dry, calculate the yield, and determine the m.p. Naphthalene can be isolated from the organic layer. Dry the solution with anhydrous sodium sulfate for a few minutes. Remove the desiccant by gravity filtration, transfer the filtrate in a tared round-bottom flask, and evaporate the solvent to dryness on the rotary evaporator, to give naphthalene as a solid. Calculate the yield and the m.p.

B) Three-components example (mixture of an acidic, a basic, and a neutral compound):
Take 30 ml of an unknown solution containing an organic solvent such as CH_2Cl_2 and three different compounds dissolved with neutral, basic, and acidic character such as naphthalene, *p*-chloroaniline, and benzoic acid, respectively. The amount of each in the 30 ml of solution is approximately 4 ml (or 4 g, if they are solid). Place 30 ml of the unknown solution in a separatory funnel of 250 ml. Add 20 ml of 5% HCl and shake vigorously. Let stand and then separate the two layers (label the aqueous layer as solution A and the organic layer as solution B). Repeat the process with a new quantity of acid and finally collect the aqueous extracts (top layer) in a beaker (solution A). Then make these aqueous extracts (solution A) basic by adding small NaOH portions (10%) and cooling the mixture to r.t. if necessary (an oily substance will appear). Extract this basic solution with CH_2Cl_2 (2 × 10 ml); after collecting both extracts (organic layer, bottom), dry with anhydrous Na_2SO_4 for a few minutes. Filter and then remove the

solvent with the rotary evaporator, which in this case yields the basic component of the mixture, *p*-chloroaniline. Discard the remaining aqueous layer. In a separatory funnel, extract the organic layer (solution B) with 5% NaOH solution (2 × 20 ml), separating the two layers. Dry the corresponding organic layer (bottom) with anhydrous Na_2SO_4 for a few minutes. Remove the desiccant by gravity filtration. Transfer the filtrate to a tared round-bottom flask, and evaporate the solvent to dryness on the rotary evaporator, to give naphthalene as a solid. Calculate the yield and the m.p. Acidify the aqueous extracts (from solution B) by adding small portions of HCl (conc.). Cool the mixture and vacuum filter the crystalline product (benzoic acid) in a Büchner.[8] Wash the benzoic acid crystals with 20 ml of cold water and dry the solid.

C) Four-components example (mixture of a basic, a neutral, and two acidic compounds):

Take 30 ml of CH_2Cl_2 solution containing four different dissolved compounds with neutral, basic, and acid character, such as naphthalene, *p*-chloroaniline, benzoic acid, and phenol, respectively. The amount of each in the 30 ml of solution is approximately 4 ml (or 4 g, if solid). Place the solution in a separatory funnel, and extract with 5% HCl (2 × 20 ml). This extraction removes *p*-chloroaniline from the organic phase as the water-soluble *p*-chloroanilinium chloride salt (label flask A), leaving the other three components in the organic solution. Aqueous extracts of HCl (flask A) are made basic by the addition of small 10% NaOH portions, cooling the mixture to r.t. if necessary. Extract this basic solution with CH_2Cl_2 (2 × 10 ml). Collect both extracts, dry with anhydrous Na_2SO_4, filter, and remove the solvent with a rotary evaporator, the product, in this case, being the basic component of the mixture, the *p*-chloroaniline. Afterward, extract the organic layer using a 5% aqueous solution of $NaHCO_3$ (2 × 20 ml) (label flask B). This removes the benzoic acid from the organic solution as sodium benzoate but leaves the less acidic phenol and naphthalene in the organic layer. Acidify the basic aqueous extracts (from flask B) by adding small portions of HCl (conc.). Cool the mixture (flask C) and vacuum filter the crystalline product (benzoic acid) in a Büchner.[9] The remaining organic layer (from flask B) is then extracted with 5% NaOH. Phenol can be extracted from the aqueous layer (flask D) as the water-soluble sodium phenolate salt, leaving the neutral naphthalene as the sole compound in the remaining organic layer (flask E). Each component can now be isolated. Acidify the basic aqueous

---

[8]Check that filtering on the addition of a small amount of HCl does not cause the appearance of more crystalline substance. If it appears, add slightly more acid and filter again on the Büchner itself.

[9]Check that filtering on adding a small amount of HCl does not cause the appearance of more crystalline substance. If so, add slightly more acid and filter again in Büchner itself. Wash the crystals of benzoic acid with 20 ml of cold water and dry, leaving a stream of air through the Büchner.

extracts (flask D) by adding small portions of HCl (conc.) until slightly acidic. Extract this acidic solution with CH_2Cl_2 (2 × 10 ml); after both extracts are collected, dry with anhydrous Na_2SO_4. Filter and then remove the solvent with a rotary evaporator, yielding, in this case, phenol as a liquid compound. Finally, pour the remaining organic layer (flask E) through the separatory funnel into an Erlenmeyer, and dry with anhydrous Na_2SO_4 for a few minutes. Gravity filter the solution and transfer the filtrate to a rotary evaporator. Naphthalene will appear after solvent removal.

Table 7.6: Physico-chemical properties of the reagents used.

Compound	M_w	M.p. (°C)	B.p. (°C)	Density ($g \cdot ml^{-1}$)	Danger[a] (GHS)
Naphthalene	128.17	79.5–81.0	218	-	☠ ! 🌿
p-Chloroaniline	127.57	72.5	232	1.140	☠ ☠ 🌿
Benzoic acid	122.12	125	249	1.08	🧪 !
Phenol	94.11	40–42	182	1.07	☠ ☠ 🧪
NaOH	40.00	318	1,390	2.130	🧪
HCl	36.46	−30	>100	1.200	🧪 !
CH_2Cl_2	84.93	−97	40.0	1.33	☠
Na_2SO_4	142.04	884	-	2.630	Non-hazardous

[a] For brevity, only GHS icons are indicated. The information offered in the Material Safety Data Sheet (MSDS) should be consulted.

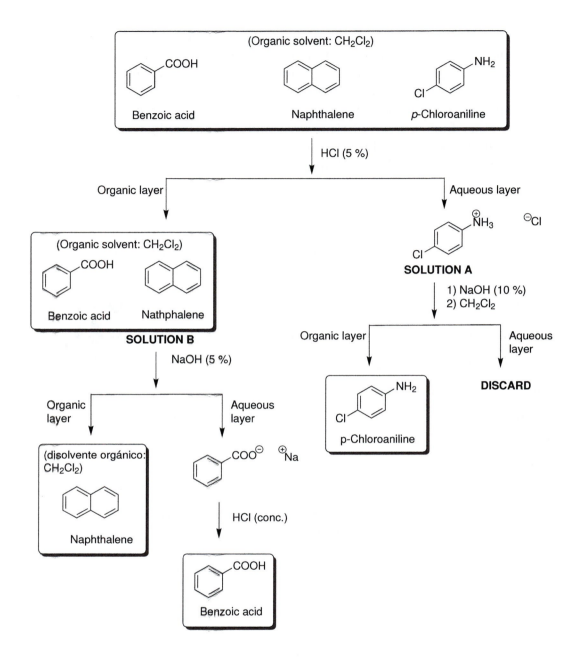

7.7. Reflux, simple, and fractional distillation: ethyl acetate preparation

Estimated time	Difficulty	Basic lab operations
4 h		✔ Vacuum filtration (see p. 103). ✔ Liquid-liquid extraction (see p. 111). ✔ Liquid-liquid extraction with reaction (see p. 111). ✔ Recrystallization from organic solvent (see p. 105). ✔ Simple distillation (see p. 120). ✔ Fractional distillation (see p. 122). ✔ Reflux (see p. 89).

7.7.1. Goal

To learn some basic techniques in the Organic Chemistry laboratory, including:

- Set up a reflux as a general heating technique of a reaction in order to raise the boiling temperature of a given solvent.

- Simple distillation as a process for purifying liquids.

- The use of fractional distillation to purify mixtures of liquids with similar b.p.

For this goal, a simple ester such as ethyl acetate will be synthesized.

7.7.2. Background

The carboxylic acid esters may be formed by a reaction between a carboxylic acid and an alcohol in the presence of a protic acid as a catalyst, such as HCl or H_2SO_4 (so-called Fischer esterification) or Lewis acid catalyst such as boron trifluoride,[10] or by reacting an acid derivative such as acid chloride or anhydride with an alcohol. Esters are versatile compounds in Organic Chemistry and are widely used because they are easily converted into a variety of other functional groups.

---

[10]Usually forming a complex with diethyl ether.

7.7.3. Procedure

Place 40 ml of glacial acetic acid and 45 ml of EtOH (96%) into a 250 ml round-bottom flask and cool in an ice-water bath. Then, add 5 ml of concentrated H_2SO_4 dropwise with magnetic stirring. Next set up the flask with a reflux condenser and a drying tube, and reflux the reaction for 1 h. After the reflux is finished, allow the flask content to cool to r.t.

Arrange an assembly for simple distillation to separate the ethyl acetate formed together with the unreacted EtOH and some trailing of the acetic acid. Distill until no more liquid is yielded, with H_2SO_4 and unreacted acetic acid remaining in the flask.

Introduce the distillate into a separatory funnel and shake with an Na_2CO_3 solution (10%), taking care to open the valve immediately upon mixing the two liquids in order to release the overpressure, since CO_2 gas is produced in the neutralization of acetic acid.[11] Continue the ethyl acetate washing until no acid reaction is evident.[12]

Separate the organic layer (top) and wash with a solution of 20 g of $CaCl_2$ in 25 ml of water.[13] Once again, decant into two layers, discarding the aqueous (bottom) layer. Transfer the organic layer to a dry Erlenmeyer flask and add anhydrous Na_2SO_4 to remove any water it may contain. Stir the mixture until it reaches a clear solution.

When the ethyl acetate is dry, decant the solid ($CaCl_2$) and purify the resulting liquid by fractional distillation. Using a graduated cylinder, measure the distillate obtained between 75–78 °C and calculate the amount by weight, bearing in mind that the density is $0.9 \ \text{g·ml}^{-1}$ (see ref. [12]).

---

[11]The organic layer is on top of the separatory funnel and the ethyl acetate is less dense than water.

[12]Check the pH of the aqueous solution with indicator paper. This should be a sign that there is acid in the organic layer.

[13]Calcium forms a water-soluble complex with EtOH.

Table 7.7: Physico-chemical properties of the reagents used.

Compound	M_w	M.p. (°C)	B.p. (°C)	Density (g·ml^{-1})	Danger[a] (GHS)
H_2SO_4 96–98%	98.08	3	-	1.80–1.84	
Glacial acetic acid	60.05	16.2	118	1.049	
Ethyl acetate	88.11	−84	77.1	0.902	
EtOH	46.07	−114.1	78.5	0.790	
Na_2SO_4	142.04	884	-	2.630	Non-hazardous
$CaCl_2$	110.98	782	>1,600	2.15	
$Na_2CO_3 \cdot 10\,H_2O$	286.14	32–34	-	1.44	

[a] For brevity, only GHS icons are indicated. The information offered in the Material Safety Data Sheet (MSDS) should be consulted.

Spectroscopic data (ethyl acetate):

- IR: 1742 (s), 1242 (s) cm^{-1}.
- ^{1}H-NMR: δ_H = 4.1 (c, 2H, J=7.2 Hz, CH_2), 2.0 (s, 3H, Ac), 1.3 (t, 3H, J=7.2 Hz, CH_3-CH_2) ppm.
- ^{13}C-NMR: δ_C = 171.0, 60.3, 20.9, and 14.1 ppm.
- MS: m/z = 88 (M$^+$), 43 (BP).

7.8. Steam distillation and vacuum distillation: isolation of limonene from citrus peel

Estimated time	Difficulty	Basic lab operations
4 h	🧪🧪🧪	✔ Gravity filtration (see p. 100). ✔ Rotary evaporator (see p. 117). ✔ Vacuum distillation (see p. 125). ✔ Steam distillation (see p. 127).

7.8.1. Goal

To learn steam-distillation and vacuum-distillation techniques to produce an essential oil from a plant raw material (orange peel).

(R)-(+)-Limonene   *(S)*-(−)-Limonene

7.8.2. Background

Limonene, a compound of the terpene family, is present in the essential oil of citrus peel. The limonene structure has a chiral center, and thus it is found in nature as two enantiomers the (R)- and (S)-limonene. Isomer (R)- has the characteristic smell of oranges, while the (S)- smells like lemons. In oranges, essential oil comprises 95% of (R)-limonene, whereas lemon peel contains mostly (S)-limonene. Limonene is widely used as a fragrance in cosmetics, a flavoring in the food industry, and even as a biodegradable domestic degreaser in industrial cleaning. Also, it is used as starting material for producing p-cymene by catalyzed dehydrogenation and as an insecticide against ants, aphids, and other pests, (R)-limonene being especially effective.

7.8.3. Procedure

Peel two oranges, striving to have the minimum amount of the white part of the peel. After peeling, chop the skin into small pieces of approximately 1 cm^2. Weigh the sample and place the pieces in a 250 ml round-bottom flask with a ground-glass joint, and add 100 ml of deionized water.

Place the flask on a hot plate, insert a stir bar and perform the steam distillation. Among the various configurations of steam distillation, the simplest one (with a Claisen adapter, addition funnel, distillation head, thermometer, condenser, extension, and collector) should be chosen.

Connect the hot plate, regulating the temperature so that the dropping rate of the distillate is approximately 20 drops/min.[14] Add water from the addition funnel so that the volume of liquid in the flask is kept constant, to prevent the sugars contained in the orange peel from caramelizing in the heat. Collect the distillate (approximately 100 ml), which will appear as an emulsion.

After the distillation, the content of the collector is transferred to a separatory funnel. Add CH_2Cl_2 (2×25 ml) and decant. Combine the organic phases and dry with anhydrous Na_2SO_4. Remove the desiccant by gravity filtration. Pour the CH_2Cl_2 onto a round-bottom flask, previously dried and tared, and remove the solvent under reduced pressure in a rotary evaporator. Weigh the flask and determine the percentage of essential oil of orange peel.

Limonene thus distilled can be purified by vacuum distillation.[15] Temperature must be carefully controlled with efficient magnetic stirring to prevent splashes. The distilling flask must be cooled with an ice bath. Apply vacuum with a water pump (approximately 20 mm Hg). The distillation temperature for limonene is 75–80 °C (at a pressure of 3 to 5 Torr, is 35–38 °C; see ref. [11]).

Table 7.8: Physico-chemical properties of the reagents used.

Compound	M_w	M.p. (°C)	B.p. (°C)	Density (g·ml^{-1})	Danger^a (GHS)
Orange peel					
CH_2Cl_2	84.93	−97	40.0	1.33	☣
Na_2SO_4	142.04	884	-	2.630	Non-hazardous
(R)-$(+)$-Limonene	136.23	-	176-177	-	🔥 ! 🌿

^{a} For brevity, only GHS icons are indicated. The information offered in the Material Safety Data Sheet (MSDS) should be consulted.

Spectroscopic data ((+)-limonene):

- IR (liquid): 2963, 2924, 2873 (s), 1658, 1645 (w), 1448 (m), 1377 (m) cm^{-1}.

- ^{1}H-NMR (DMSO-d$_6$): δ_H = 5.40 (m, 1H), 4.70 (s, 2H), 2.08 (m, 1H), 2.05(m, 1H) 1.97 (m, 1H), 1.89 (m, 1H), 1.79 (m, 1H), 1.73 (s, 3H) 1.65 (s, 3H) and 1.47 (m, 1H) ppm.

- ^{13}C-NMR (DMSO-d$_6$): δ_C = 150.1, 133.7, 120.8, 108.5, 41.2, 30.9, 30.7, 28.1, 23.5, and 20.8 ppm.

- MS: m/z = 136.2 (M$^+$), 121, 107, 93, 68 (BP), 67.

---

[14]If necessary, cover the flask with aluminum foil and/or different assembly parts to avoid heat losses, which prevent proper distillation.

[15]To have an adequate amount, samples can be obtained from various practices for a manageable sample volume.

7.9. Solid-liquid extraction (Soxhlet): sunflower oil from seeds

Estimated time	Difficulty	Basic lab operations
4 h		✔ Liquid-liquid extraction (see p. 111). ✔ Solid-liquid extraction (Soxhlet) (see p. 116). ✔ Simple distillation (see p. 120). ✔ Steam distillation (see p. 127).

7.9.1. Goal

To learn how to set up Soxhlet extraction to obtain oil from sunflower seeds.

Fatty acids present in the sunflower oil

Myristic acid

Arachidonic acid

Linoleic acid

Oleic acid

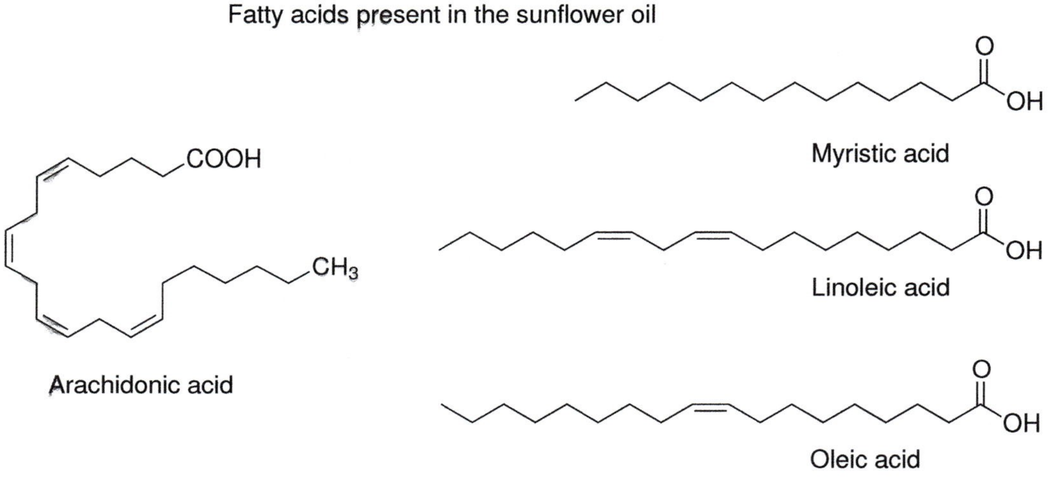

7.9.2. Background

Vegetable oils are mainly triglycerides from seeds or fruits. They are a key component of the human diet and are also used for other purposes, including the production of biofuels (biodiesel), lubricants, and cosmetics such as soaps and creams. They are often produced by mechanical or chemical processes or a combination of the two. One of the most common procedures to produce vegetable oils is extraction from the oil-rich parts of the plant using organic solvents such as hexane. In the case of sunflower seed, the oil content can be as high as 40% of the dry weight of seeds. In this experiment, oil from sunflower seeds is obtained by means of continuous extraction (Soxhlet equipment) with hexane.

7.9.3. Procedure

Weigh 20 g of sunflower seeds. Using a mortar, grind seeds to form a material as homogeneous as possible. Prepare a piece of cartridge filter paper, and place the crushed seeds inside. Then fill the Soxhlet flask with 250 ml of hexane and add a stir bar. Place the cartridge with the crushed seeds in the extractor. Set up the system and keep it running for two or three extraction cycles, overall approximately 30 min.

Disassemble the Soxhlet and add 120 ml of warm water to the flask. Then, connect the flask to a distillation adapter (three-way adapter) with a thermometer, water-jacketed condenser, distilling adapter, and receiving flask. Continue with the steam distillation (internal source of vapor) until only distilled water remains to ensure that the hexane has been completely removed from the oil.

Allow the flask content and the receiving content to cool to r.t. and then pour independently into a separatory funnel for separation. The distillate will be composed of a mixture of hexane (upper phase) and water, and separation of the two phases allows recovery of the organic solvent (hexane) for reuse. Then the residue remaining after distillation will consist of a mixture of water and oil (upper phase). Once the oil is separated, measure the volume in a graduated cylinder, previously tared, weigh, and calculate the yield and the approximate density of the oil (see ref. [10]).

Table 7.9: Physico-chemical properties of the reagents used.

Compound	M_w	M.p. (°C)	B.p. (°C)	Density (g·ml^{-1})	Danger[a] (GHS)
Sunflower seeds					
Oleic acid	282.46	13–14	194–195	0.890	!
Hexane	86.18	−95	69	0.659	🔥☠!🐟

[a] For brevity, only GHS icons are indicated. The information offered in the Material Safety Data Sheet (MSDS) should be consulted.

7.10. TLC and CC chromatography: separation of the pigments from spinach leaf[16]

Estimated time	Difficulty	Basic lab operations
3 h	 🧪🧪	✔ Gravity filtration (see p. 100). ✔ Chromatography (see p. 135).

7.10.1. Goal

To extract photosynthetic pigments and separate them by CC and TLC techniques.

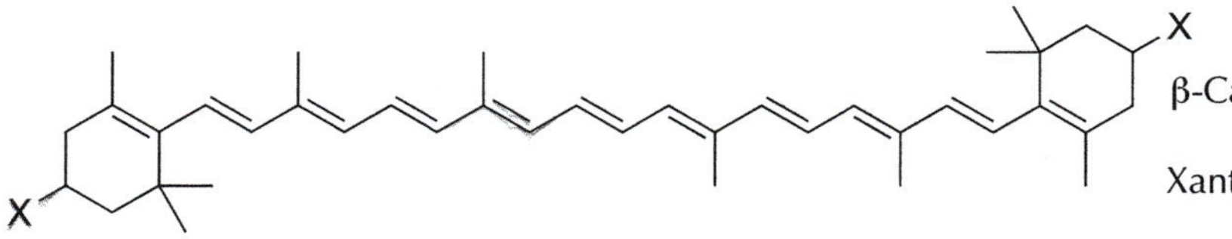

β-Carotene (X = H)

Xanthophyll (X = OH)

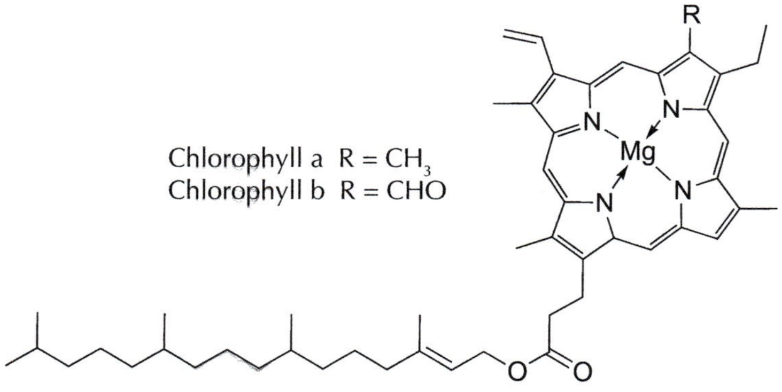

Chlorophyll a R = CH$_3$
Chlorophyll b R = CHO

7.10.2. Background

Spinach leaves contain a number of natural products such as carotenes, chlorophylls, xanthophylls, soluble vitamins, etc. Spinach leaves present β-carotene and chlorophyll, these being primarily responsible for the leaf color, together with minor amounts of xanthophyll components. While β-carotene is yellow, chlorophyll is green and has two main types: "a" and "b." The a type has a methyl group, whereas b has a formyl group attached to the porphyrin ring. Mg^{++} ion loss in chlorophyll leads to the formation of pheophytin (a and b). Moreover,

---

[16]This experiment can be carried out with other greens if spinach is not available. Alternatively, the student can use other substances such as ink.

xanthophylls are a family of compounds, derived from carotenes with oxygen atoms in the structure. Spinach leaves contain less β-carotene than chlorophyll.

In this experiment, the pigments from spinach leaves are extracted and separated by means of thin-layer (TLC) and column chromatography (CC). Chromatography is a technique for separating a mixture of substances based on the different affinities for these products in the stationary phase (silica gel) and a mobile phase (eluent).

7.10.3. Procedure

A) Sample preparation: Weigh approximately 2 g of spinach leaves. Wash the leaves with water, remove the veins, cut up with scissors, and put the pieces in a mortar together with 22 ml of acetone, 3 ml of hexane, and a small amount of calcium carbonate $CaCO_3$ using a spatula (to prevent degradation of photosynthetic pigments). Grind the mixture until the leaves become discolored and the solvent takes on a deep-green color. Transfer only the resulting liquid to a separatory funnel (prevent the transfer of solids), and add 20 ml of hexane and 20 ml of aqueous solution of NaCl 10%. Stir the mixture and decant. Discard the aqueous layer (bottom) and the organic layer (top) and wash with 5 ml of deionized water. Decant again (with additional 20 ml of NaCl solution) and transfer the organic layer to a 50 ml Erlenmeyer flask. Dry the solution with anhydrous Na_2SO_4 and remove the desiccant by gravity filtration. Divide the sample into two parts. Take 1 ml of solution and reserve in a test tube for the analysis by TLC, and use the remaining sample for CC separation.

B) TLC analysis of spinach pigments:

- Prepare a squared piece of filter paper slightly smaller in height and width than the chromatography tank, and place it against the inside part of the wall, ensuring a complete vision of the TLC-foil throughout the experiment.

- Prepare a hexane/acetone solution (7:3).

- Place the eluent inside the tray (to about 5 mm high), and seal it; this allows the vapors of the solvent to saturate the entire tray in order to impregnate the filter paper.

- With a pencil (use only graphite pencils) and a ruler, draw a horizontal line about 5 mm from the base of a TLC-foil (Al TLC 5 × 7.5 cm silica gel 60F).[17]

- With the help of a glass capillary, add a drop on the line at the TLC-foil.

---

[17]The line should be above the level of solvent when the TLC-foil is placed in the tray.

- The TLC-foil will quickly enter the tray; keep it upright and slightly inclined so that from the outside the solvent level in the TLC-foil can be seen at all times.

- When the solvent level is positioned a few millimeters from the top of the TLC-foil, remove the TLC-foil from the tray and allow it to air dry (colored spots appear along the TLC-foil).

- Quickly cover the tray to keep its atmosphere saturated with solvent.

- Using a graphite pencil, mark the level reached by the solvent.

- Colored spots appear along the TLC spots, carotenoids moving more quickly than chlorophyll.

- Determine the value of R_f.

Table 7.10: R_f values and color of different pigments in the spinach leaf.

Pigment	Color	R_f
Carotene	Yellow-orange	0.93
Pheophytin a	Gray	0.55
Pheophytin b	Gray (not visible)	0.47–0.54
Chlorophyll a	Blue-green	0.46
Chlorophyll b	Green	0.42
Xanthophylls	Yellow	0.41
Xanthophylls	Yellow	0.31
Xanthophylls	Yellow	0.17

C) Chromatographic separation of pigments by CC:

- Put cotton or glass wool in the bottom of the column.

- Pour 10 ml of eluent (hexane/acetone, 7:3) into the column.

- Weigh 20 g of silica gel on a 40 ml Erlenmeyer flask, and add approximately 50 ml of eluent, stirring with a glass rod or spatula, to form a slurry.

- Fill the column with the slurry, opening the column stopcock until the liquid is at the right edge of the stationary phase (it is very important for the column never to empty, completely or partially, and not to form air bubbles), collecting the solvent in a container for reuse.

- Sand can be added to the silica gel to form a layer of 1 or 2 cm thick on the edge of the stationary phase.

- Pipette 10 ml of the sample prepared from spinach leaves, and add them to the column.

- Add eluent by opening the column key and collecting it in an Erlenmeyer.

- When a band of yellow discoloration (β-carotene) appears, collect the liquid in test tubes.

- Change the test tube when the tube begins to fill or when the content changes color (see ref. [9]).

Table 7.11: Physico-chemical properties of the reagents used.

Compound	M_w	M.p. (°C)	B.p. (°C)	Density (g·ml^{-1})	Danger[a] (GHS)
Acetone	58.08	−94	56	0.791	
Hexane	86.18	−95	69	0.659	
CaCO$_3$	100.09	800	-	2.93	Non-hazardous
Chlorophyll b	907.47	183–185	-	-	Non-hazardous
Chlorophyll a	893.49	150–153	-	-	Non-hazardous
β-Carotene	536.87	176–184	-	1.000	Non-hazardous

[a] For brevity, only GHS icons are indicated. The information offered in the Material Safety Data Sheet (MSDS) should be consulted.

7.11. References

1. T. E. Sample and L. F. Hatch, *3-Sulfolene: a butadiene source for a Diels-Alder synthesis: an undergraduate laboratory experiment*, Journal of Chemical Education **45** (1968), no. 1, 55, DOI 10.1021/ed045p55.

2. L. C. McKenzie, L. M. Huffman, J. E. Hutchison, C. E. Rogers, T. E. Goodwin, and G. O. Spessard, *Greener solutions for the organic chemistry teaching lab: exploring the advantages of alternative reaction media*, Journal of Chemical Education **86** (2009), no. 4, 488, DOI 10.1021/ed086p488.

3. A. I. Vogel and B. S. Furniss, *Vogel's Textbook of Practical Organic Chemistry*, Longman, 1989.

4. R. E. Buckles, *The use of the Perkin reaction in organic laboratory classes*, Journal of Chemical Education **27** (1950), no. 4, 210–211, DOI 10.1021/ed027p210.

5. M. A. Amerine and C. S. Ough, *Wine and must analysis*, John Wiley & Sons, 1974.

6. D. Bladt, S. Murray, B. Gitch, H. Trout, and C. Liberko, *Acid-catalyzed preparation of biodiesel from waste vegetable oil: an experiment for the undergraduate organic chemistry laboratory*, Journal of Chemical Education **88** (2011), no. 2, 201–203, DOI 10.1021/ed9000427.

7. M. S. Ntamila and A. Hassanali, *Isolation of oil of clove and separation of eugenol and acetyl eugenol. An instructive experiment for beginning chemistry undergraduates*, Journal of Chemical Education **53** (1976), no. 4, 263, DOI 10.1021/ed053p263.

8. D. L. Pavia, *Coffee, tea, or cocoa. A trio of experiments including the isolation of theobromine from cocoa*, Journal of Chemical Education **50** (1973), no. 11, 791, DOI 10.1021/ed050p791.

9. M. H. Anwar, *Separation of plant pigments by thin layer chromatography*, Journal of Chemical Education **40** (1963), no. 1, 29, DOI 10.1021/ed040p29.

10. M. D. Luque de Castro and L. E. García-Ayuso, *Soxhlet extraction of solid materials: an outdated technique with a promising innovative future*, Analytica Chimica Acta **369** (1998), no. 1-2, 1–10, DOI 10.1016/S0003-2670(98)00233-5.

11. J. H. Beatty, *Limonene —a natural insecticide*, Journal of Chemical Education **63** (1986), no. 9, 768, DOI 10.1021/ed063p768.

12. E. Fischer and A. Speier, *Darstellung der ester*, Berichte der Deutschen Chemischen Gesellschaft **28** (1895), no. 3, 3252–3258, DOI 10.1002/cber.189502803176.

Chapter 8

Organic Synthesis Experiments

In this chapter, twenty related experiments concerning the reactivity of functional groups that are usually studied in a basic course of Organic Chemistry are described.

Most of the experiments described correspond to changes that take place in one step, except the last three experiments, which are devoted to the synthesis of a compound in several steps and to the study of the concept of protecting functional groups in a synthetic pathway. In some cases, a difunctional compound is also used to develop the concept of chemoselectivity against different reactives but within the same family of compounds.

The practices were chosen in order to have the widest possible range of substrates (aliphatic or aromatic), reactive types (substitution, elimination, addition, oxidation-reduction, etc.), and according to reaction conditions (reflux, room temperature, cold conditions, etc.) using different methods of isolation and purification of organic compounds. Consideration of them all together or a few among them would make it possible to apply the main basic operations of laboratory experiments from low to medium complexity.

8.1. Alkene oxidation: preparation of adipic acid from cyclohexene

Estimated time	Difficulty	Basic lab operations
3 h	⚗	✔ Vacuum filtration (see p. 103). ✔ Recrystallization from water (see p. 105).

8.1.1. Goal

To perform the oxidative cleavage of a double bond of a cycloalkene.

8.1.2. Background

Potassium permanganate ($KMnO_4$) is an oxidant frequently used in Organic Chemistry. In this experiment, this reagent is used to perform the oxidative cleavage of the double bond of cyclohexene to produce adipic acid, a dicarboxylic acid.

8.1.3. Procedure

Prepare a solution of 6.3 g of potassium permanganate with 100 ml of water, and place in a 250 ml Erlenmeyer flask. The solid is slowly dissolved, so it is advisable to stir until it is completely dissolved. Separately, in another 250 ml Erlenmeyer, prepare a second solution of 1.24 ml (1.24 g) of cyclohexene in 10 ml of acetone. Magnetically stir this solution and then add the permanganate solution in 10 ml portions for 10 min. After the addition, heat the reaction mixture in a water bath at 55–60 °C for 30 min. Afterward, remove the flask from the water bath and add 1 g of sodium bisulfite to the brownish mixture. After 5 min, cool the Erlenmeyer in an ice bath.

Filter the reaction mixture under vacuum. Transfer the filtrate to a 500 ml beaker. To the solution, add HCl to reach pH = 2.0 (check with a pH test paper). Concentrate the solution by heating it on a hot plate (add boiling chips to avoid projections) to a volume of around 10 ml. When cooled (using an ice bath) a white precipitate is formed. Filtrate the white solid under vacuum and dry under air stream. Recrystallize the adipic acid from water (m.p. = 152 °C, estimated yield 60%).

Table 8.1: Physico-chemical properties of the reagents used.

Compound	M_w	M.p. (°C)	B.p. (°C)	Density (g·ml^{-1})	Danger[a] (GHS)
KMnO$_4$	158.04	240	-	2.71	
Cyclohexene	82.14	−104	83	0.779	
Acetone	58.08	−94	56	0.791	
HCl	36.46	−30	>100	1.200	
NaHSO$_3$ (40%)	104.06	-	-	1.48	
Adipic acid	146.14	151–154	337.5	1.36	

[a] For brevity, only GHS icons are indicated. The information offered in the Material Safety Data Sheet (MSDS) should be consulted.

8.2. Dehydration of alcohols: synthesis of cyclohexene from cyclohexanol

Estimated time	Difficulty	Basic lab operations
4 h	🔺🔺🔺	✔ Gravity filtration (see p. 100). ✔ Washing in a separatory funnel (see p. 111). ✔ Simple distillation (see p. 120). ✔ Reflux (see p. 89).

8.2.1. Goal

To produce an alkene (cyclohexene) by dehydration reaction of an acid-catalyzed alcohol (cyclohexanol).

Cyclohexanol H_3PO_4 (85%)   Cyclohexene

8.2.2. Background

A secondary alcohol such as cyclohexanol undergoes dehydration through an $E1$ mechanism, where the key intermediate is the cyclohexyl cation, which can react either by an elimination or a substitution reaction. For good yield of the alkene, it is necessary to suppress the competitive substitution side reaction. In this experiment, the elimination reaction is favored by using strong acids (with anions that are relatively weak nucleophiles), high reaction temperatures, and distillation of cyclohexene as it forms.

The dehydration reaction of an alcohol catalyzed by acids is a reversible unimolecular elimination reaction following Saytzev's rule. As a dehydrating agent, any strong acid (sulfuric acid, phosphoric acid, or oxalic acid) can be used. Traditionally, this reaction has been performed by the sulfuric acid method, but here the phosphoric acid method is described, since it has certain advantages with respect to the sulfuric acid method, such as the non-carbonization of reagents and the reduced risk for the experimenter.

8.2.3. Procedure

Place 20 g (21 ml) of cyclohexanol and 5 ml of phosphoric acid (85%) in a 100 ml round-bottom flask. Prepare a simple distillation assembly. Keep the 50 ml collector flask immersed in an ice-water bath, because cyclohexene is highly volatile and flammable. The distillation must be carried out slowly. Stop the distillation when approximately 8 ml appears in the distillation flask. Before disposing of the distillation residue in the waste container, pour into a beaker with ice-water. Transfer the distillate to a separatory funnel and add 10 ml of water, followed by 10 ml of saturated NaCl (brine) solution.[1] Separate the organic layer together with cyclohexene from the separatory funnel. Dry in an Erlenmeyer with 2 g of CaCl$_2$ anhydrous, separating the desiccant from the cyclohexene by decantation[2]. Quickly collect the cyclohexene in a dry 50 ml round-bottom flask and add 2 or 3 pieces of boiling chip. Perform a simple distillation and collect the fraction[3] after distilling at 78–83 °C. The collecting flask should be immersed in an ice-water bath. The estimated yield is 75% (see ref. [3], p. 491).

Noticeboard 8.2.1

 DANGER!

Perform the distillation in a fume hood.

Table 8.2: Physico-chemical properties of the reagents used.

Compound	M$_w$	M.p. (°C)	B.p. (°C)	Density (g·ml^{-1})	Danger[a] (GHS)
Cyclohexene	82.14	−104	83	0.779	
CaCl$_2$	110.98	782	>1,600	2.15	!
Cyclohexanol	100.16	20–22	160–161	0.968	!
Phosphoric acid	98.00	40	158	1.685	
NaCl	58.44	801	1,413	2.165	Non-hazardous

[a] For brevity, only GHS icons are indicated. The information offered in the Material Safety Data Sheet (MSDS) should be consulted.

---

[1] It is recommended to do these washes as fast as possible.

[2] Due to the volatility of cyclohexene, filtration is not recommended.

[3] Toward the end of the distillation, the residue becomes yellow to dark brown and the distillation temperature approaches 83 °C.

8.3. Oxidation of alcohols: 4-chlorobenzoic acid from 4-chlorobenzyl alcohol

Estimated time	Difficulty	Basic lab operations
2 h	⚗	✔ Gravity filtration (see p. 100). ✔ Vacuum filtration (see p. 103). ✔ Recrystallization from organic solvent (see p. 105). ✔ Reflux (see p. 89).

8.3.1. Goal

To obtain a carboxylic acid by oxidation of a primary alcohol, using hypochlorite as the oxidant.

4-Chlorobenzyl alcohol   4-Chlorobenzoic acid

8.3.2. Background

An alternative method to the use of chromium(VI), which is carcinogenic, for the alcohol oxidation is the use of a green oxidizing agent such as hypochlorite. One of the most common sources of hypochlorite is bleach; bleach is actually an aqueous solution of sodium hypochlorite. In the case of primary alcohols, hypochlorite oxidation yields carboxylic acids if an excess is used, since the reaction proceeds via an aldehyde, which is afterward oxidized to acid.

8.3.3. Procedure

Dissolve in an Erlenmeyer 0.5 g of 4-chlorobenzyl alcohol and 5 ml of acetonitrile. Furthermore, weigh 2.4 g of commercial calcium hypochlorite (65%) and transfer to a round-bottom flask. Add 20 ml of water and stir with a stir bar. Still stirring, add 2 ml of glacial acetic acid dropwise. Attach the reflux condenser and place in a water bath. With the aid of a Pasteur pipette, add the alcohol solution through the condenser. Once the addition is complete, heat the reaction for 1 h in a water bath (see p. 83) while stirring vigorously, and do not let the bath temperature exceed 50 °C. Next, cool to r.t.

Transfer the reaction crude to a separatory funnel, add 10 ml of water, wash the flask with 10 ml of diethyl ether, pour over the funnel, and extract. Then, extract again with diethyl ether (2×10 ml). Put the ether extracts together and treat with two portions of 10 ml of saturated aqueous sodium bicarbonate solution. Collect the aqueous layer and acidify with HCl 5% to pH = 3.0 to give a precipitate that is filtered under vacuum. Recrystallize the product from MeOH.

Table 8.3: Physico-chemical properties of the reagents used.

Compound	M_w	M.p. (°C)	B.p. (°C)	Density (g·ml^{-1})	Danger[a] (GHS)
Glacial acetic acid	60.05	16.2	118	1.049	
4-Chlorobenzyl alcohol	142.58	68–71	234	1.200	Non-hazardous
Acetonitrile	41.05	−48	81–82	-	
Calcium hypochlorite	142.98	100	-	2.350	
Diethyl ether	74.12	−116	34.6	0.71	
NaHCO$_3$	84.01	300	-	2.160	Non-hazardous
HCl	36.46	−30	>100	1.200	
MeOH	32.04	−98	64.7	0.791	

[a] For brevity, only GHS icons are indicated. The information offered in the Material Safety Data Sheet (MSDS) should be consulted.

8.4. Nucleophilic substitution reactions: synthesis of 1-bromobutane

Estimated time	Difficulty	Basic lab operations
4 h	(flask icons)	✔ Vacuum filtration (see p. 103). ✔ Liquid-liquid extraction (see p. 111). ✔ Simple distillation (see p. 120). ✔ Reflux (see p. 89).

8.4.1. Goal

To produce butyl bromide from butyl alcohol.

$$\text{Butan-1-ol (}\sim\!\!\sim\!\!\text{OH)} + NaBr \xrightarrow{\frac{1}{2} H_2SO_4} \text{1-Bromobutane (}\sim\!\!\sim\!\!\text{Br)} + \frac{1}{2} Na_2SO_4 + H_2O$$

8.4.2. Background

Reacting a primary alcohol with a hydrogen halide produces a primary alkyl halide. The reaction follows an $S_N 2$ mechanism, as the dehydration step is competitive. The reaction requires a strong acid to protonate the hydroxyl group. Aqueous HBr, HI or HBr gas can be used. In this experiment, the HBr is generated in situ by reacting NaBr with H_2SO_4.

$$NaBr + H_2SO_4 \longrightarrow HBr + NaHSO_4$$

In this transformation, some side reactions occur: The butan-1-ol can react with the HSO_4^- ions present in the solution to yield a hydrogen sulfate ester ($R-OSO_3H$). This inorganic ester can in turn trigger an elimination reaction that leads to but-1-en (a gas reflux is lost during processing or during the reaction) or replacement with butan-1-ol to yield di-*n*-butyl ether (which must be removed during the processing of the reaction).

8.4.3. Procedure

In a 250 ml round-bottom flask, place 30 g of sodium bromide and 30 ml of water. Shake the flask until most of the salt has been dissolved. Add 18.5 g of butan-1-ol, and cool the flask to 5–10 °C in an ice bath. Slowly add 25 ml of concentrated H_2SO_4.

Place the flask with a reflux condenser and a trap adapted for gases containing NaOH solution, and reflux the mixture for 30 min with magnetic stirring.

Noticeboard 8.4.1

 DANGER!

During the refluxing, HBr vapors are are released, which are corrosive and toxic (10–20 times more toxic than CO_2 and as toxic as Cl_2). The reaction should be performed in a fume hood and the reflux system connected to a trap for HBr.[a] Make sure that in the trap, the funnel is not submerged, since liquid may enter in the reaction flask.

[a]Mounting the bubbler for HBr shown in Figure 4.32, p. 134. The liquid in the glass is aqueous 5% NaOH (caution: caustic).

During the reflux the reaction mixture forms two layers. Cool the reaction flask (ice bath can be used) to a temperature at which it can be manipulated. Add a magnetic bar and prepare the assembly for a simple distillation. Distill until the temperature of the distillation mixture reaches 110–115 °C. The distillate consists of two phases (1-bromobutane and water), which are most apparent at the beginning of the distillation. At the end of the distillation, 1-bromobutane should no longer be visible in the drops of distillate.[4] The distillation residue (strong acid!) must be collected by pouring onto ice and diluting with water (pour to the acid waste container).

Transfer the distillate to a 125 ml separatory funnel, and add approximately 25 ml of water. Shake the funnel and allow the two phases to separate. Collect the lower layer of the 1-bromobutane in an Erlenmeyer flask.[5] Discard the top layer. Add 25 ml of concentrated H_2SO_4 and cool the 1-bromobutane with ice.

Swirl the flask to mix the contents. If the mixture starts to heat, cool the flask in an ice bath. Pass the mixture to a separatory funnel. The concentrated H_2SO_4 is denser (d = 1.84) than 1-bromobutane (d = 1.28), and therefore 1-bromobutane now forms the upper layer. Gently shake the separatory funnel to avoid the formation of an emulsion and let stand for 5–10 min.[6]

---

[4]To check that there is no codistillation of oil with the water at the end of the distillation, collect a few milliliters of distillate in a test tube. Shake the test tube, and if there is oil, droplets will be visible.

[5]Be very careful and pay attention to the identification of the phases in the funnel during this experiment. It is wise to keep all layers in labeled flasks until the end of the experiment to avoid accidentally pulling the wrong layer.

[6]The formation of emulsions is frequent in extraction and will require a rest period for the two phases to separate.

Noticeboard 8.4.2

 DANGER!

Extreme care is needed when handling concentrated H_2SO_4 in the funnel. The funnel should be frequently turned on to release the excess pressure. An oversight, even a leaky stopcock, can lead to the H_2SO_4, spilled on clothing or the work area. Any spillage onto skin or clothing should be washed immediately with water.

Collect lower layer (caution: strong acid) and discard; pour over ice and dilute with water; eliminate in the acid waste container. Wash the 1-bromobutane remaining in the funnel with 25 ml of water to remove the residual H_2SO_4. As 1-bromobutane is denser than the aqueous solution, 1-bromobutane will now form the lower layer. Shake the funnel, collecting this lower layer into a clean flask. Remove the aqueous phase remaining in the funnel and replace 1-bromobutane in the funnel. Extract the 1-bromobutane with 25 ml of a solution of 10% NaOH. In this extraction, as in the above, 1-bromobutane is in the lower layer in the separatory funnel.

Collect 1-bromobutane in another clean flask, add 2 g of $CaCl_2$ anhydrous, stopper the flask tightly and let the mixture stand to clarify the liquid (best overnight).[7]

Decant the clean 1-bromobutane to a 50 ml round-bottom flask using a disposable dropper to transfer remaining liquid. Be very careful not to transfer any solid calcium chloride. Add a magnetic bar and distill 1-bromobutane. Collect the fraction at 98–103 °C (if the distillate is cloudy it means it has moisture and should be dried and redistilled). Estimated yield will be 50%.

If the experiment cannot be completed in one session, the breakpoint is after the reflux period: while the 1-bromobutane is being dried with anhydrous $CaCl_2$ (see ref. [3], p. 561).

---

[7]Since the 1-bromobutane is quite volatile, the drying vessel should have a glass stopper or septum.

Table 8.4: Physico-chemical properties of the reagents used.

Compound	M_w	M.p. (°C)	B.p. (°C)	Density (g·ml^{-1})	Danger[a] (GHS)
CaCl$_2$	110.98	782	>1,600	2.15	❗
Butan-1-ol	74.12	−90	116–118	0.810	🔥 ⚠ ❗
H$_2$SO$_4$ 96–98%	98.08	3	-	1.80–1.84	⚠
NaBr	102.89	755	1,393	3.20	Non-hazardous
NaOH	40.00	318	1,390	2.130	⚠
1-Bromobutane	137.02	−112	100–104	1.276	🔥 ❗ 🌿

[a] For brevity, only GHS icons are indicated. The information offered in the Material Safety Data Sheet (MSDS) should be consulted.

8.5. Synthesis of an ether: preparation of β-naphthyl methyl ether

Estimated time	Difficulty	Basic lab operations
4 h		✔ Gravity filtration (see p. 100). ✔ Vacuum filtration (see p. 103). ✔ Recrystallization from organic solvent (see p. 105). ✔ Reflux (see p. 89).

8.5.1. Goal

To produce an asymmetric ether industrial interest, from naphthalene-2-ol (β-naphthol) using two synthetic methods.

8.5.2. Background

2-Methoxynaphthalene (β-naphthol methyl ether) is a white solid with m.p. 73 °C and with the trade name of Nerolin. It used in perfumery for its floral scent, as stabilizer in special powders, and as intermediate for the synthesis of non-steroidal anti-inflammatory drugs (NSAIDs). It can be produced by two synthetic methods: using the Williamson reaction or by the treatment of MeOH and β-naphthol with H_2SO_4 (hot). Williamson synthesis is a process that allows the preparation of a wide range of symmetric and asymmetric ethers by an S_N2 mechanism. The second procedure used in this experiment is based on the use of the alkyloxonium ion formed by the reaction of MeOH with H_2SO_4. This alkyloxonium ion at a later step is attacked by a molecule of β-naphthol to yield the corresponding ether (β-naphthol methyl ether).

8.5.3. Procedure

A) To produce β-naphthol methyl ether by Williamson synthesis:
   In a 100 ml round-bottom flask with a stir bar, mix 2.88 g (20 mmol) of β-naphthol (2-hydroxynaphthalene), 1.46 g (26 mmol) of KOH, and 20 ml of MeOH. Put in a stopper and stir the mixture at r.t. until dissolution of the

solid. Using a syringe, add 1.4 ml (23 mmol) of methyl iodide (CH$_3$I). Next, after setting up a reflux condenser and drying tube, heat for 1 h. Cool the mixture to r.t. and add water until a precipitate appears. Filter the solid under vacuum, and recrystallize from hot EtOH (estimated yield 75%; m.p. 73 °; see ref. [3], p. 987).

Noticeboard 8.5.1

 DANGER!

Do the experiment in a fume hood because of the danger of methyl iodide (CH$_3$I).

B) To produce β-naphthyl methyl ether in acid medium:

In a 100 ml round-bottom flask, place 5 g of β-naphthol, 25 ml of MeOH, and 5 ml H$_2$SO$_4$. Reflux the mixture for 1 h. Cool the crude reaction, and pour into 100 ml of ice water. The ether precipitate is collected by vacuum filtration in a Büchner. Wash the precipitate twice in a Büchner with ice water (once with 20 ml of NaOH (10%) and again with ice water). Recrystallize the product obtained from EtOH, using activated carbon if necessary (estimated yield 70%; see ref. [3], p. 987).

Table 8.5: Physico-chemical properties of the reagents used.

Compound	M_w	M.p. (°C)	B.p. (°C)	Density (g·ml^{-1})	Danger[a] (GHS)
β-Naphthol	144.17	120–122	285–286	1.280	
MeOH	32.04	−98	64.7	0.791	
EtOH	46.07	−114.1	78.5	0.790	
H$_2$SO$_4$ 96–98%	98.08	3	-	1.80–1.84	
CH$_3$I	141.94	−64	41–43	2.280	
KOH	56.11	361	1,320	2.044	
NaOH	40.00	318	1,390	2.130	

[a] For brevity, only GHS icons are indicated. The information offered in the Material Safety Data Sheet (MSDS) should be consulted.

8.6.   Aromatic electrophilic substitution of an ester: methyl benzoate nitration

Estimated time	Difficulty	Basic lab operations
1 h		✔ Gravity filtration (see p. 100). ✔ Vacuum filtration (see p. 103). ✔ Recrystallization from organic solvent (see p. 105). ✔ Reflux (see p. 89).

8.6.1.   Goal

To learn to perform an electrophilic aromatic substitution reaction such as the nitration of methyl benzoate.

COOCH$_3$

$\xrightarrow{\text{HNO}_3 \text{ (conc.)}}$
H$_2$SO$_4$ (conc.)

COOCH$_3$
NO$_2$

Methyl benzoate   Methyl 3-nitrobenzoate

8.6.2.   Background

An example of an electrophilic aromatic substitution (S$_E$Ar) reaction is the nitration of an aromatic ring, using as the electrophile nitronium ion, NO$_2^+$. The methoxycarbonyl group of benzoate deactivates the ring with orientation at the meta position.

8.6.3.   Procedure

Place 1.8 ml (2 g) of methyl benzoate in a 25 ml flask, and add 4 ml H$_2$SO$_4$ (conc.) while stirring. Cool the mixture in an ice bath.

> ### Noticeboard 8.6.1
>
> ### ⚡ DANGER!
>
> Nitrous fumes. The whole process of nitration must be done in a fume hood.

Place HNO$_3$ in another flask and add H$_2$SO$_4$ while stirring and cool in ice. Using a Pasteur pipette, add dropwise a solution of HNO$_3$ to methyl benzoate solution while stirring, keeping the temperature within a range of 0–10 °C, with the help of an ice bath. The addition requires about 30 min. Then let the solution stand for another 10 min at r.t. Pour the solution over ice to form a precipitate. Vacuum filter 3-nitrobenzoate, and wash well with water in a Büchner. Recrystallize from EtOH to produce an almost colorless solid (estimated yield 55%; see ref. [3], p. 1071).

Table 8.6: Physico-chemical properties of the reagents used.

Compound	M_w	M.p. (°C)	B.p. (°C)	Density (g·ml^{-1})	Danger[a] (GHS)
Methyl benzoate	136.15	−12	198–199	1.094	❗
H$_2$SO$_4$ 96–98%	98.08	3	-	1.80–1.84	☣
HNO$_3$ 68–70%	63.01	-	120.5	1.413	☣ 🔥
Methyl 3-nitrobenzoate	181.15	78–80	279	-	Non-hazardous
EtOH	46.07	−114.1	78.5	0.790	🔥

[a] For brevity, only GHS icons are indicated. The information offered in the Material Safety Data Sheet (MSDS) should be consulted.

8.7. Electrophilic aromatic substitution (S_EAr): preparation of a synthetic detergent

Estimated time	Difficulty	Basic lab operations
2 h	![flask icon]	✔ Liquid-liquid extraction (see p. 111).

8.7.1. Goal

To prepare a synthetic liquid and solid detergents through sulfonation of an aromatic ring.

Dodecylbenzene → Dodecylbenzene sulfonic acid → Sodium dodecylbenzenesulfonate

Dodecylbenzene-sulfonic acid + Triethanol amine → Triethanolamine dodecylbenzenesulfonate

8.7.2. Background

Detergents are surfactants —that is, they greatly reduce the surface tension of water when used in very low concentrations. Detergents have a hydrophilic polar head, such as the salts of carboxylic or sulfonic acids, allowing the water solubility. Besides, the carbon chain confers to this part of the molecule hydrophobicity, which helps to clean fats and oils from clothing. To facilitate the biodegradation of these detergents, it is preferable to use linear instead of branched hydrocarbon chains. Soaps have traditionally been prepared by a reaction of saponification of vegetable oils (esters of glycerol), whereby the sodium salts of fatty acids are produced.

In this experiment, a synthetic liquid detergent will be prepared in the form of ammonium salt of triethanolamine, through a reaction of sulfonation of dodecylbenzene. Subsequently, the resulting dodecylbenzenesulfonic acid will be transformed into a solid detergent by treatment with 40% NaOH.

8.7.3. Procedure

Place 10 g (11.4 ml) of dodecylbenzene in a 100 ml Erlenmeyer flask. During 30 min of constant stirring, add dropwise 20 g (10 ml) of fuming H$_2$SO$_4$ (caution in handling) to the dodecylbenzene, maintaining the temperature between 40 and 45 °C with an ice-water bath. After the addition, heat the reaction mixture with a water bath at 50 °C for 5 min (see p. 83), while stirring continuously. Cool the mixture to r.t. and place in a separatory funnel until separation into two layers. Once separated, the resulting 2/3 of dodecylbenzenesulfonic acid is carefully neutralized, dropwise adding a solution of 40% NaOH. Neutralize the remaining part with triethanolamine, following the same procedure for preparing the liquid detergent (see refs. [5,6]).

Table 8.7: Physico-chemical properties of the reagents used.

Compound	M$_w$	M.p. (°C)	B.p. (°C)	Density (g·ml^{-1})	Danger[a] (GHS)
Dodecylbenzene	246.43	3	331	0.856	Non-hazardous
NaOH	40.00	318	1,390	2.130	⚠
H$_2$SO$_4$ 96–98%	98.08	3	-	1.80–1.84	⚠
Triethanolamine	149.19	17.9–21	190–193	1.124	Non-hazardous
Dodecylbenzenesulfonic acid	326.49	-	82	1.06	⚠ !
Triethanolamine dodecylbenzenesulfonate	475.68	-	264	1.200	!

[a] For brevity, only GHS icons are indicated. The information offered in the Material Safety Data Sheet (MSDS) should be consulted.

8.8. Synthesis of azo dyes: methyl orange

Estimated time	Difficulty	Basic lab operations
3 h	⚗️⚗️	✔ Vacuum filtration (see p. 103).

8.8.1. Goal

To produce azo dyes —i.e., methyl orange —by the coupling reaction of a diazonium salt.

8.8.2. Background

Each amine type yields a different product on reacting with nitrous acid (HNO_2), an unstable reagent formed in situ in the presence of the amine by the action of a mineral acid to sodium nitrite. When a primary aromatic amine, dissolved in a cold aqueous mineral acid, is treated with sodium nitrite, a diazonium salt is formed:

$$ArNH_2 \ + \ NaNO_2 \ + \ 2HX \xrightarrow{\text{cool}} Ar{-}\overset{\oplus}{N{\equiv}N} \ + \ X^{\ominus} \ + \ NaX \ + \ H_2O$$

Since these decompose slowly even at the temperature of an ice bath, their solutions are used immediately after preparation. The large number of reactions of diazonium salts can be grouped into two types.

- Substitution or replacement: In which nitrogen is lost as N_2, another group remaining in its place at the aromatic ring.

$$ArN_2^{\oplus} \quad + \quad Z: \quad \longrightarrow \quad ArZ \quad + \quad N_2$$

- Coupling: Where nitrogen remains in the molecule. The coupling of diazonium salts with aromatic amines and phenols generates azo-compounds, which are key intermediates in the dye industry.

$$ArN_2^{\oplus} \quad + \quad \langle\!\langle\bigcirc\rangle\!\rangle\!-\!G \quad \longrightarrow \quad Ar\!-\!N\!=\!N\!-\!\langle\!\langle\bigcirc\rangle\!\rangle\!-\!G$$

A diazo dye is produced by the following steps:

- Diazotization of an aromatic ring containing a primary amino group.

- Preparing a solution of an amino-aromatic compound in a diluted acid or a phenolic substance in diluted alkali.

- Mixing the two above-mentioned solutions with the subsequent formation of the corresponding diazo dye, in a reaction termed a "coupling reaction." For this reaction to occur, the solution should be alkaline or slightly acidic.

In this experiment, for the production of methyl orange, the diazotization of sulfanilic acid is the first step. Dimethylaniline is dissolved in diluted HCl, and finally the two solutions are mixed in order for the coupling reaction to occur.

8.8.3. Procedure

Independently, prepare the following solutions:

- 5 g of sulfanilic acid (4-aminobenzenesulfonic acid) and 2 g of sodium carbonate in 100 ml water.

- 2 g of sodium nitrite in 15 ml water.

- 4 ml of concentrated HCl in 25 ml water.

- 3 ml of dimethylaniline, 15 ml of water and 3 ml of concentrated HCl.

Cool the solutions (with an external ice bath) and only when they are cool, proceed as follows:
Into the dissolution of sulfanilic acid and carbonate, pour a solution of sodium nitrite. Maintain the mixture in an ice bath and, while stirring, slowly add the HCl solution. After the addition is finished, keep the whole mixture cold and slowly add the dimethylaniline solution. The reaction changes in color to reddish. Once the process ends, add 40 ml of NaOH 10%, whereupon the reaction mixture turns orange because of the formation of the sodium salt of dye. Add alkali until the solution is slightly alkaline (test the pH with an indicator paper). Decant the reaction crude into a 500 ml beaker, and then add 30 g of salt and heat the contents of the beaker to 50–60 °C. Once cooled to r.t., filter under vacuum the resulting pasty solid. Pass a stream of air through the Büchner for 10 min. Once

dry,[8] weigh the solid, and calculate the yield (estimated yield 70%; see ref. [3], p. 951).

Table 8.8: Physico-chemical properties of the reagents used.

Compound	M_w	M.p. (°C)	B.p. (°C)	Density (g·ml^{-1})	Danger[a] (GHS)
Sulfanilic acid	173.19	>300	-	-	!
NaNO$_2$	69.00	271	320	2.164	☠☠🌱
HCl	36.46	−30	>100	1.200	🧪 !
N,N-Dimethylaniline	121.18	1.8	192.5	0.958	☠🌱
Na$_2$CO$_3$·H$_2$O	124.00	851	-	2.25	!
NaOH	40.00	318	1,390	2.130	🧪
NaCl	58.44	801	1,413	2.165	Non-hazardous
Methyl orange	337.33	>300	-	-	☠

[a] For brevity, only GHS icons are indicated. The information offered in the Material Safety Data Sheet (MSDS) should be consulted.

---

[8]For this product, it is especially recommended to use a vacuum desiccator.

8.9. Auto-oxidation–reduction of aromatic aldehydes: cannizzaro reaction

Estimated time	Difficulty	Basic lab operations
3 h		✔ Gravity filtration (see p. 100). ✔ Vacuum filtration (see p. 103). ✔ Liquid-liquid extraction (see p. 111). ✔ Liquid-liquid extraction with reaction (see p. 111). ✔ Washing in a separatory funnel (see p. 111). ✔ Recrystallization from water (see p. 105). ✔ Rotary evaporator (see p. 117). ✔ Vacuum distillation (see p. 125).

8.9.1. Goal

The purpose of this practice is to produce an alcohol and carboxylic acid in one step using the Cannizzaro reaction of benzaldehyde.

8.9.2. Background

The dilute base added to aldehydes or ketones leads to the formation of β-hydroxyaldehyde or β-hydroxyketones by aldol condensation. When the aldehyde or ketone has no hydrogens in the α position, aldol condensation does not take place but rather undergoes auto-oxidation–reduction (disproportionation) in the presence of strong bases to give an equimolar mixture of the alcohol and the corresponding acid (the latter in salt form). For example, benzaldehyde produces benzyl alcohol and sodium benzoate in the presence of sodium hydroxide.

Intermediate ion
(loses $H^{\ominus}$ easily)

Formaldehyde also yields the Cannizzaro reaction, like aliphatic aldehydes (such as trimethylacetaldehyde) containing no α hydrogens, and aromatic derivatives such as benzaldehyde.

8.9.3. Procedure

In a 250 ml Erlenmeyer flask, place 10 ml (10.4 g, 98 mmol) of benzaldehyde and 9 g of NaOH (dissolved in 5 ml of water). Cover securely with a stopper (septum), and shake vigorously for 20 min by hand, until a solid product and a colorless liquid simultaneously appear. After the reaction, add enough water to dissolve the solid formed (sodium benzoate), but avoiding excess water. Pass the solution through a separatory funnel, wash the flask with a few milliliters of CH_2Cl_2 to drag any solid residues left in the flask, and extract the solution with CH_2Cl_2 (3 × 15 ml). Save both phases, collect the organic extracts, and wash with $NaHSO_3$ 10% (2 × 20 ml) and with Na_2CO_3 10% (1 × 20 ml). Dry over Na_2SO_4, filter, and evaporate in a rotary evaporator. A subsequent vacuum distillation can be made to purify the benzyl alcohol (colorless liquid slightly soluble in water). Acidify the aqueous alkaline layer from the first extraction with HCl in order to precipitate benzoic acid as a white mass. Cool the mixture, filter under vacuum, and wash with water. Benzoic acid is a white crystalline solid, slightly soluble in water (0.2 g/100 g at 20 °C, 2.2 g/100 g at 75 °C). Recrystallize from water to achieve purity (estimated yield of benzoic acid 80%; see ref. [3], p. 1029).

Table 8.9: Physico-chemical properties of the reagents used.

Compound	M_w	M.p. (°C)	B.p. (°C)	Density (g·ml^{-1})	Danger[a] (GHS)
Benzaldehyde	106.12	−26	178–179	1.044	❗
Benzoic acid	122.12	125	249	1.08	☣❗
NaOH	40.00	318	1,390	2.130	☣
NaHSO$_3$ (40%)	104.06	-	-	1.48	☣❗
Na$_2$CO$_3$ · H$_2$O	124.00	851	-	2.25	❗
HCl	36.46	−30	>100	1.200	☣❗
CH$_2$Cl$_2$	84.93	−97	40.0	1.33	☣
Na$_2$SO$_4$	142.04	884	-	2.630	Non-hazardous
Benzyl alcohol	108.140	−15	203–205	1.040	❗

[a] For brevity, only GHS icons are indicated. The information offered in the Material Safety Data Sheet (MSDS) should be consulted.

8.10. Synthesis of α, β-unsaturated ketones: Claisen-Schmidt reaction

Estimated time	Difficulty	Basic lab operations
2 h	🔬	✔ Gravity filtration (see p. 100). ✔ Vacuum filtration (see p. 103). ✔ Recrystallization from organic solvent (see p. 105). ✔ Rotary evaporator (see p. 117). ✔ Reflux (see p. 89).

8.10.1. Goal

The purpose of this experiment is to make an aldol condensation between a non-enolizable aldehyde and enolizable ketone.

8.10.2. Background

The Claisen-Schmidt reaction is an aldol condensation type, consisting of the synthesis of α, β-unsaturated ketones by condensing an aromatic aldehyde with a ketone. As the aromatic aldehyde possesses no hydrogens in position α with respect to the carbonyl group, it cannot self-condense but reacts readily with acetone in the reaction medium. The initial aldol adduct cannot be isolated because it dehydrates spontaneously under the reaction conditions, but the α, β-unsaturated ketone thus obtained also contains active hydrogen, which can be condensed with another molecule of benzaldehyde. Depending on the amount of ketone used, the formation of *mono-* and *bis*-adducts, respectively, can be optimized, and both adducts can be differentiated by their physical and spectroscopic properties. In the first part of this experiment, acetone is used in great excess to minimize the second step of condensation. In the second part, benzaldehyde is present in excess, and EtOH is added to the reaction to maintain the initial condensation product in solution long enough to react with a second molecule of benzaldehyde.

8.10.3. Procedure

A) *E*-4-Phenylbut-3-en-2-one (benzylideneacetone) preparation:
Pour 4.25 g of benzaldehyde in a 100 ml Erlenmeyer flask with magnetic stirring, and add 8 ml of acetone dropwise followed by 1 ml of aqueous NaOH (10%). Place the flask in a water bath at 25–30 °C (see p. 83), and stir the reaction for 90 min. Afterward, slowly add diluted HCl to an acidic pH, transfer the mixture to a separatory funnel, and extract with CH_2Cl_2 (2 × 15 ml). Separate the layers, collect the organic extracts, wash with 15 ml of water, and dry over anhydrous Na_2SO_4. Filter the solution by gravity, wash with 5 ml of the desiccant CH_2Cl_2, and remove solvent by rotary evaporation to yield a solid of low-melting-point solid (estimated yield 77% and m.p. = 42 °C; see ref. [3], p. 1033).

B) 1,5-Diphenyl-(*E,E*)-1,4-pentadien-3-one (benzylideneacetone) preparation:
Dissolve 5 g of NaOH in 25 ml of water, add 25 ml EtOH (95%), and cool the mixture under running water to r.t. In another 100 ml Erlenmeyer, pour 10.5 ml of benzaldehyde and add 2.9 g of acetone with a pipette and then add the alkaline EtOH solution previously prepared and stir the mixture for 15 min at 20–25 °C (may require external cooling). Filter the yellow solid under vacuum, and wash with cold water to remove the alkali. Leave the product to dry at r.t. on a filter paper. Recrystallize from ethyl acetate (2.5 ml per g) (estimated yield 90%, m.p. = 122 °C; see ref. [3], p. 1033).

Table 8.10: Physico-chemical properties of the reagents used.

Compound	M_w	M.p. (°C)	B.p. (°C)	Density (g·ml^{-1})	Danger[a] (GHS)
Benzaldehyde	106.12	−26	178–179	1.044	!
CH_2Cl_2	84.93	−97	40.0	1.33	
Ethyl acetate	88.11	−84	77.1	0.902	!
Acetone	58.08	−94	56	0.791	!
NaOH	40.00	318	1,390	2.130	
Benzylideneacetone	146.19	34–40	260–262	1.008	!
trans,trans-Dibenzylideneacetone	234.29	104–107	-	-	Non-hazardous

[a] For brevity, only GHS icons are indicated. The information offered in the Material Safety Data Sheet (MSDS) should be consulted.

8.11. Saponification reaction: preparation of soap from vegetable oil

Estimated time	Difficulty	Basic lab operations
1 h	⚗	✔ Vacuum filtration (see p. 103).

8.11.1. Goal

To make soap by reacting a vegetable oil with sodium hydroxide (saponification reaction).

Triacylglycerol

R = large hydrocarbon change

8.11.2. Background

Soap (metallic or carboxylate ester) forms as a product of a saponification reaction. It is called the reaction between a carboxylic acid and a strong base such as sodium hydroxide NaOH or potassium hydroxide KOH. The main feature is the presence of soap in the molecule having two zones of different polarity: a hydrophilic (or lipophobic) and a lipophilic (or hydrophobic). The hydrophilic region located around the carboxyl group is strongly polarized and also forms hydrogen bonds with water molecules. The lipophilic region, which is very polar, is kept away from the water molecules and corresponds to the hydrocarbon chain. Because of its dual amphiphilic character (hydrophilic-lipophilic), soap molecules have the property of solubilizing polar and non-polar molecules. Soap molecules show a strong tendency to migrate to the interfaces, so that the polar part is within the water and the apolar part is located facing an apolar part such as air or fat medium. This tendency to orient its structure with respect to water soap molecules lowers the surface tension at an air-water or oil-water interface, and therefore such molecules are called tensoactive agents.

8.11.3. Procedure

In a 100 ml beaker, dissolve 9 g of NaOH in 12 to 15 ml of EtOH/water solution (50%). In another 100 ml beaker, weigh 5 g of sunflower oil and add to the solution containing NaOH. Gently heat the mixture and stir continuously with a glass rod for 15 min. Remove the beaker from the hot plate if foam forms, and stir until the foam subsides. Cool the mixture and pour, while stirring, into a cold solution of 15 g NaCl in 60 ml of water.[9] Afterward, cool to r.t., and then cool in an ice bath or place in the freezer. Cool the solution to precipitate the soap, which is isolated by vacuum filtration. Wash the resulting solid three times with cold water, dry, weigh, and calculate the yield (see ref. [5,6]).

Table 8.11: Physico-chemical properties of the reagents used.

Compound	M_w	M.p. (°C)	B.p. (°C)	Density (g·ml^{-1})	Danger[a] (GHS)
Vegetable oil					🔥
NaOH	40.00	318	1,390	2.130	☢
EtOH	46.07	−114.1	78.5	0.790	🔥
NaCl	58.44	801	1,413	2.165	Non-hazardous

[a] For brevity, only GHS icons are indicated. The information offered in the Material Safety Data Sheet (MSDS) should be consulted.

---

[9]This solution must be prepared in a 250 ml beaker and heating if necessary until the salt is dissolved.

8.12. Esterification: synthesis of aspirin

Estimated time	Difficulty	Basic lab operations
2.5 h		✔ Gravity filtration (see p. 100). ✔ Vacuum filtration (see p. 103). ✔ Recrystallization from water (see p. 105). ✔ Reflux (see p. 89).

8.12.1. Goal

The purpose of this practice is to produce aspirin (acetylsalicylic acid) from salicylic acid and acetic anhydride.

Salicilic acid   Acetic anhydride   Aspirin   Acetic acid

8.12.2. Background

Acetylsalicylic acid, marketed under the trade name aspirin by the Bayer company, is one of the most widely consumed drugs in the world. It was synthesized by the German chemist Felix Hofmann at the end of the 19th century. The preparation consists of an esterification reaction catalyzed by acid (H_2SO_4 or H_3PO_4), where salicylic acid treated with acetic anhydride gives acetylsalicylic acid (aspirin). In this reaction, a hydroxyl group is converted to an ester, with acetic acid as a byproduct. Essentially, aspirin acts as an antipyretic and analgesic. As an antipyretic, it exerts its effect on two levels: it increases heat dissipation by vasodilation (insignificant action) and acts on the hypothalamus, which is the regulatory center of body temperature. It is taken orally as it is well absorbed by the gastrointestinal system. Acetylsalicylic acid may be partially hydrolyzed; this, plus the easily noticeable smell of acetic acid, can be recognized by a trial with $FeCl_3$ and observing whether the violet color of $FeCl_3$ is produced.

8.12.3. Procedure

Place 3 g (0.022 mol) of salicylic acid in a 100 ml round-bottom flask. Add 6 ml of acetic anhydride and then 6–8 drops of phosphoric acid at 85%. Stir gently to mix the layers, and immerse the flask and reflux equipment in a hot water bath (70–80 °C) for 15 min. Take the flask out of the bath and cool in an ice-water bath; while still hot, carefully add dropwise about 1 ml of water, stirring after each addition. After adding the first ml of water, another 20 ml can be added rapidly. Cool the flask in an ice bath, so that the product should start to crystallize (estimated yield 70%; and m.p. = 134–136 °C; see ref. [2]).

Table 8.12: Physico-chemical properties of the reagents used.

Compound	M_w	M.p. (°C)	B.p. (°C)	Density (g·ml^{-1})	Danger[a] (GHS)
Glacial acetic acid	60.05	16.2	118	1.049	
H$_2$SO$_4$ 96–98%	98.08	3	-	1.80–1.84	
Acetic anhydride	102.09	−73.1	139.8	1.080	
Salicylic acid	138.12	158–161	211	1.440	
Acetylsalicylic acid	180.16	134–136	-	-	
Ethyl acetate	88.11	−84	77.1	0.902	
Phosphoric acid	98.00	40	158	1.685	

[a] For brevity, only GHS icons are indicated. The information offered in the Material Safety Data Sheet (MSDS) should be consulted.

Noticeboard 8.12.1

⚠ DANGER!

Acetic anhydride reacts violently with water so that the mixture can splash.

 Spectroscopic data (aspirin):

- IR (KBr): 3006 (bm), 1754 (s), 1693 (s), 1606 (s), 1308 (s), 1190 (s), 91.8 (m) cm^{-1}.

- ^{1}H-NMR (CDCl$_3$): δ_H = 11.0 (bs, 1H), 8.12 (m, 2H), 7.62 (m, 1H), 7.35 (m, 1H), 7.14 (m, 1H) ppm.

- ^{13}C-NMR (CDCl$_3$): δ_C = 170.0, 169.7, 151.3, 134.9, 132.5, 126.2, 124.0, 122.3, and 21.0 ppm.

- MS: m/z = 180.2 (M$^+$), 138, 120, 92, 43.

8.13. Difunctional compound chemoselectivity: reduction of 4-nitroacetophenone

Estimated time	Difficulty	Basic lab operations
2 h		✔ Gravity filtration (see p. 100). ✔ Vacuum filtration (see p. 103). ✔ Liquid-liquid extraction (see p. 111). ✔ Recrystallization from water (see p. 105). ✔ Rotary evaporator (see p. 117). ✔ Reflux (see p. 89).

8.13.1. Goal

To illustrate the chemoselective reduction of 4-nitroacetophenone. This compound has two reducible groups (carbonyl and nitro), and, depending on the reducing agent used, this will achieve one without having to protect the other.

4-Aminoacetophenone   4-Nitroacetophenone   1-(4-Nitrophenyl)ethanol

8.13.2. Background

Chemoselectivity is the selective reactivity of a functional group in the presence of others. Protecting groups can be used instead of chemoselectivity but with the disadvantage of having to perform two additional synthesis steps (protection and deprotection). However, if the reagent and reaction conditions are well chosen, the chemoselectivity can be more effective than protecting groups. In this experiment, both functional groups will hold two chemoselectivity reductions of 4-nitroacetophenone, a compound with two reducible groups (nitro and carbonyl). In the first part, the aromatic nitro group is reduced to aromatic amine using tin and HCl, which does not reduce carbonyl groups. In the second part, the ketone is reduced using the smooth transfer agent hydride, $NaBH_4$.

8.13.3. Procedure

A) Reduction using tin and HCl of 4-nitroacetophenone:

Cut 3.3 g tin into small pieces (or use granulated tin) and place in a 100 ml round-bottom flask equipped with a reflux condenser and a stir bar. Add 1.65 g of 4-nitroacetophenone and then 24 ml of water and 9 ml of HCl (conc.). Stir the mixture and heat the flask to reflux for 1.5 h. Cool the reaction mixture to r.t. Filter the reaction crude mixture under vacuum if there is solid tin. Slowly add to the mother liquor ammonia until a precipitate appears, basic pH. Vacuum filter and wash with water. Recrystallize from water (m.p. = 106 °C; see ref. [7]).

B) Reduction to 1-(4-nitrophenyl)ethanol using $NaBH_4$:
Dissolve 1.65 g of 4-nitroacetophenone with 20 ml of warm EtOH in a 100 ml Erlenmeyer flask. Stir and cool the flask in an ice-water bath, if the reaction is very hot. Add 0.45 g of $NaBH_4$ in small portions for 5 min and stir the mixture at r.t for 15 min. Add a solution of diluted HCl dropwise until the bubbling of hydrogen ceases. Add 40 ml of water to the reaction crude, transfer to a 100 ml separatory funnel, and extract with 2 × 20 ml of CH_2Cl_2. Combine the organic extracts and dry over Na_2SO_4. Filter the drying agent by gravity, and remove the solvent under reduced pressure (rotary evaporator). The final reaction product is a liquid (see ref. [7]).

Table 8.13: Physico-chemical properties of the reagents used.

Compound	M_w	M.p. (°C)	B.p. (°C)	Density (g·ml^{-1})	Danger[a] (GHS)
HCl	36.46	−30	>100	1.200	
NH_3	17.03	−78	−33	0.590	
Tin	118.71	231.9	2,270	7.310	Non-hazardous
4′-Nitroacetophe-none	165.15	75–78	202	-	Non-hazardous
4′-Aminoacetophe-none	135.16	103–107	293	-	
1-(4-Nitrophenyl)-ethanol	167.16	-	-	-	See MSDS
$NaBH_4$	37.8	400	-	1.07	
EtOH	46.07	−114.1	78.5	0.790	
CH_2Cl_2	84.93	−97	40.0	1.33	

[a] For brevity, only GHS icons are indicated. The information offered in the Material Safety Data Sheet (MSDS) should be consulted.

Spectroscopic data (4-nitroacetophenone):

- IR (KBr): 1694 (s), 1527 (s), 1346 (s), 857 (s), 750 cm^{-1}.
- ^{1}H-NMR (CDCl$_3$): δ_H = 8.31 (m, 2H), 8.14 (m, 2H), 2.70 (s, 3H) ppm.
- ^{13}C-NMR (CDCl$_3$): δ_C = 196.4, 150.5, 141.6, 129.4, 123.9, 26.9 ppm.
- MS: m/z = 165.1 (M$^+$), 150 (BP), 104.43.

8.14. Perkin reaction: cinnamic acid synthesis

Estimated time	Difficulty	Basic lab operations
4 h	⚗️	✔ Gravity filtration (see p. 100). ✔ Vacuum filtration (see p. 103). ✔ Recrystallization from water (see p. 105). ✔ Steam distillation (see p. 127). ✔ Reflux (see p. 89).

8.14.1. Goal

To perform a condensation between an acid anhydride (acetic anhydride) and an aromatic aldehyde (benzaldehyde) (Perkin condensation reaction) to synthesize cinnamic acid.

8.14.2. Background

Cinnamic acid together with cinnamaldehyde are natural products from cinnamon oil. In this experiment, cinnamic acid is synthesized by the Perkin condensation reaction, which involves the reaction between an acid anhydride (acetic anhydride) and an aromatic aldehyde (benzaldehyde) catalyzed by a base (sodium acetate). The anhydride generates a carbanion due to the influence of the base (sodium acetate), which attacks the carbonyl group of the aldehyde. Then the anhydride group undergoes a process of dehydration and hydrolysis.

8.14.3. Procedure

In a 250 ml round-bottom flask provided with a reflux condenser and a drying tube, put 5 g of benzaldehyde 7.5 g of acetic anhydride and 2.5 g of sodium acetate (anhydride), and heat to reflux for 3 h. Allow to cool to r.t. Cool the reaction crude, and add 100 ml of water. Then perform a steam distillation (internal vapor source) until all the unreacted benzaldehyde (one volume of approximately 75 ml of distillate) separates; then discard. Vacuum filter the remaining residue to remove resinous solids that have been formed. The filtrate is acidified by slowly adding concentrated HCl. Cool in an ice bath and isolate the resulting solid by vacuum filtration. Recrystallize from water; then dry, weigh, and determine the yield (estimated yield is 60%; see ref. [3], p. 1038; and ref. [4]).

Table 8.14: Physico-chemical properties of the reagents used.

Compound	M_w	M.p. (°C)	B.p. (°C)	Density (g·ml^{-1})	Danger[a] (GHS)
Benzaldehyde	106.12	−26	178–179	1.044	!
Acetic anhydride	102.09	−73.1	139.8	1.080	🔥 ⚗ !
Glacial acetic acid	60.05	16.2	118	1.049	⚗ 🔥
Cinnamic acid	148.16	132–135	300	1.248	!
Sodium acetate	82.03	328	-	1.528	Non-hazardous

[a] For brevity, only GHS icons are indicated. The information offered in the Material Safety Data Sheet (MSDS) should be consulted.

8.15. Synthesis of imide: preparation of N-(p-chlorophenyl)-maleimide

Estimated time	Difficulty	Basic lab operations
2 h	⚗⚗	✔ Gravity filtration (see p. 100). ✔ Vacuum filtration (see p. 103). ✔ Recrystallization from organic solvent (see p. 105). ✔ Reflux (see p. 89).

8.15.1. Goal

To prepare a cyclic amide from p-chloroaniline and maleic anhydride.

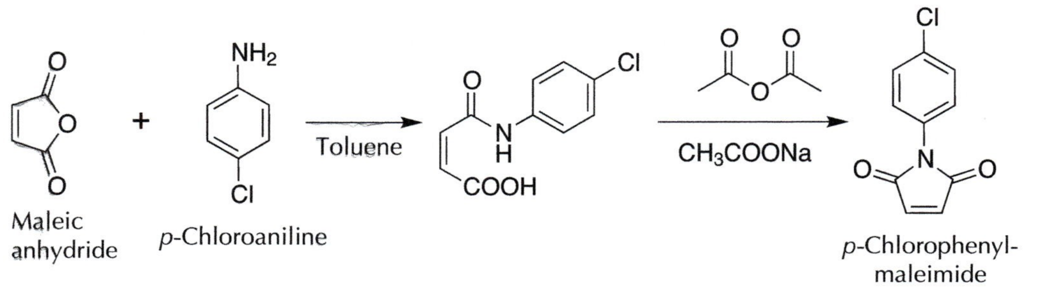

8.15.2. Background

Obtaining N-(p-chlorophenyl)-maleimide from the p-chloroaniline and maleic anhydride is a two-step process. First, a nucleophilic attack of the nitrogen (aromatic amine), catalyzed by the acid medium on the carboxylic carbon of the cyclic anhydride, occurs. In the second step, the N-(p-chlorophenyl)-maleamic acid is heated, and nitrogen attacks the other carboxylic carbon to form the N-(p-chlorophenyl)-maleimide.

8.15.3. Procedure

A) Formation of N-(p-chlorophenyl)-maleamic acid:

In a 100 ml round-bottom flask, pour 3.8 ml of p-chloroaniline and 6 ml of acetic acid. Heat the mixture slightly with magnetic stirring until complete dissolution of the solid. Then slowly add 2.94 g of maleic anhydride while stirring until a precipitate appears. Isolate the resulting solid by vacuum filtration and wash with acetic acid (3 ml). Purify the product in this step by recrystallization from EtOH. Weigh the solid and calculate the yield for this step (see ref. [8]).

B) Formation of *N*-(*p*-chlorophenyl)-maleimide:

In a 100 ml round-bottom flask, place 3.3 g of *N*-(*p*-chlorophenyl)-maleamic acid, 6 ml of acetic anhydride, and 0.15 g of sodium acetate.[10] Attach a reflux condenser to the flask and heat the mixture (bath, see p. 83), at a temperature of 85–95 °C, for 1 h with magnetic stirring. Remove the condenser and allow the flask to cool to r.t. Then insert into an ice bath and shake the flask occasionally until the appearance of a precipitate. Filter the maleimide formed by vacuum, and recrystallize from EtOH, using activated carbon to remove colored impurities. Dry the solid and then weigh and calculate yields (at this stage and the overall reaction; see ref. [8]).

Table 8.15: Physico-chemical properties of the reagents used.

Compound	M_w	M.p. (°C)	B.p. (°C)	Density (g·ml^{-1})	Danger^a (GHS)
N-(4-Chloro-phenyl)-maleamic acid	225.019	176.28	466.16	1.449	See MSDS
p-Chloroaniline	127.57	72.5	232	1.140	
Glacial acetic acid	60.05	16.2	118	1.049	
Maleic anhydride	98.06	51–56	200	1.480	
EtOH	46.07	−114.1	78.5	0.790	
N-(4-Chlorophenyl)-maleimide	207.61	110–112	350.63	1.46	See MSDS
Acetic anhydride	102.09	−73.1	139.8	1.080	
Sodium acetate	82.03	328	-	1.528	Non-hazardous
Active carbon	12.01	3,550	-	0.25–0.60	Non-hazardous

^{a} For brevity, only GHS icons are indicated. The information offered in the Material Safety Data Sheet (MSDS) should be consulted.

---

[10]Adjust quantities to the mass obtained in the previous stage.

8.16. Condensation polymerization reaction: synthesis of nylon 6,6

Estimated time	Difficulty	Basic lab operations
1 h		-

8.16.1. Goal

To familiarize the student with the reactions that produce organic polymers, a polyamide such as nylon. In this case a very simple technique called "interfacial polymerization" is used.

Adipoyl chloride   Hexamethylenediamine

Nylon 6,6

8.16.2. Background

Nylon is a synthetic polymer of the family of polyamides having great strength, providing the manufacture of fabrics for carpets, clothing, bands, ropes, hoses, conveyor belts, etc. It is synthesized by a polycondensation reaction of a diacid (or a derivative) with a diamine. Depending on the carbon atom number of the diacid and diamine, nylon is given the initials PA, the abbreviation for polyamide, and the number of carbons that the polymer gets from each of the two components. One of the most popular is PA 6,6 coming from adipic acid (hexanedioic acid) or its derivatives and hexamethylenediamine. In this experiment, PA polymer is prepared from hexamethylene diamine and adipoyl chloride in heterogeneous phase, forming the polymer at the interface.

8.16.3. Procedure

In a 50-ml beaker, put 10 ml of aqueous hexamethylenediamine 5%, and then add 10 drops of a solution of 20% NaOH. Subsequently, add 10 ml of a solution (5%) of adipoyl chloride in cyclohexane, pouring very carefully over the slightly inclined wall of the beaker. Two other organic and aqueous immiscible layers (cyclohexane) are formed, and immediately a polymer film appears at the interface. This technique is called "interfacial polymerization."

With the help of a glass rod (or copper wire hook) gently peel the polymer from the walls of the beaker. The polymer is gathered in the center and is slowly raised using the glass rod, so that the polyamide formed continuously goes high and a large thread of nylon is produced. The thread may break if it stretches too quickly. Wash with water several times and let dry on filter paper. This operation is repeated as often as necessary. Finally, strongly agitate with the glass rod the two phases to produce greater amount of polymer. Let it dry for 24 h and weigh (see ref. [9]).

Table 8.16: Physico-chemical properties of the reagents used.

Compound	M_w	M.p. (°C)	B.p. (°C)	Density (g·ml^{-1})	Danger[a] (GHS)
Hexamethylenedi-amine	116.20	42–45	204–205	0.890	
Adipoyl chloride	183.03	105–107	112	1.259	
Cyclohexane	84.16	4–7	80.7	0.779	
NaOH	40.00	318	1,390	2.130	

[a] For brevity, only GHS icons are indicated. The information offered in the Material Safety Data Sheet (MSDS) should be consulted.

8.17. Radical polymerization: producing polystyrene

Estimated time	Difficulty	Basic lab operations
1 h		.

8.17.1. Goal

To produce an organic polymer (polystyrene) so that students become familiar with the polymerization reactions initiated by radicals (organic peroxides).

Benzoyl peroxide   Benzoyl radical

8.17.2. Background

Polystyrene is a polymer that can be prepared by a monomer addition process. The addition reaction is catalyzed by radical cations or anions, although the most common are the free-radical processes. In this experiment, a radical process with benzoyl peroxide as a radical initiator is used. This organic peroxide is a relatively unstable compound that decomposes between 80 and 90 °C, with homolytic breakage of the oxygen–oxygen bond.

If there is in the middle a monomer having unsaturated bonds, the radical will join it, leading to a new radical and a chain reaction. The chain continues to grow and ends by two radicals combining, either both radical polymers or a polymer and a radical initiator, or by the abstraction of a hydrogen atom of another molecule.

8.17.3. Procedure

Place 10 ml of styrene in a test tube and add 0.7 g of benzoyl peroxide (Luperox R A75).[11] Heat the mixture over a bath until the mixture takes on a yellow color. With the disappearance of the color, bubbles begin to form; immediately remove the tube from the bath, since the reaction is exothermic. When the bubbling subsides, return the tube with styrene on the bath and continue heating until the liquid becomes very sticky.

With a wire, tease a thread of polymer out of the test tube. If the thread, after a few seconds (on cooling), is easily broken, polystyrene is ready to pour. If the filament does not break, continue heating the mixture and repeat the above operation until the filament easily breaks. Pour the sticky fluid into a watch glass. After cooling, the polystyrene can be removed from the surface with a spatula (see ref. [10]).

Table 8.17: Physico-chemical properties of the reagents used.

Compound	M_w	M.p. (°C)	B.p. (°C)	Density (g·ml^{-1})	Danger^a (GHS)
Styrene	104.15	−31	145–146	0.906	!
Polystyrene	35,000	-	240	1.050–1.130	
Benzoyl peroxide	242.23	105	-	-	🔥 !

^{a} For brevity, only GHS icons are indicated. The information offered in the Material Safety Data Sheet (MSDS) should be consulted.

Noticeboard 8.17.1

⚡ **DANGER!**

Benzoyl peroxide is an oxidizing agent and therefore prone to explode by heating, percussion, or friction.
HANDLE WITH EXTREME CAUTION !

---

[11]With water, wash all residues with remaining benzoyl peroxide, including disposable weigh substances.

8.18. Multistep synthesis of a drug: paracetamol

Estimated time	Difficulty	Basic lab operations
2 h	🧪🧪	✔ Gravity filtration (see p. 100). ✔ Vacuum filtration (see p. 103). ✔ Recrystallization from water (see p. 105). ✔ Rotary evaporator (see p. 117). ✔ Reflux (see p. 89).

8.18.1. Goal

To familiarize the student with the multi-step synthesis (three steps) of acetaminophen (paracetamol) from nitrobenzene.

Phenol $\xrightarrow[\text{NaNO}_3]{\text{H}_2\text{SO}_4 \text{ (dil.)}}$ o-Nitrophenol (36%) + p-Nitrophenol (25%) $\xrightarrow[\text{Pd(C)/NaOH}]{\text{NaBH}_4}$ p-Aminophenol (74%) $\xrightarrow{\text{Ac}_2\text{O}}$ Paracetamol

8.18.2. Background

Paracetamol is a crystalline solid (m.p. = 169–171 °C) that is used as an analgesic and antipyretic drug, but has no significant anti-inflammatory action. Commercially, it appears under various trade names. For example, in 1955 it was put on sale in the United States under the name Tylenol and in 1956 in the United Kingdom under the name of Panadol. It was first synthesized by J. von Mering in 1893 although it was not marketed until 1953, considered since then as safer than aspirin, especially for children and patients with an active stomach ulcer. Acetaminophen can be synthesized in three steps starting from the phenol. The first step consists of the nitration of phenol, and the isolation of the isomers obtained. In the second, the nitro group is reduced to an amino group, and in the third one, the amino group is selectively acetylated with acetic anhydride versus the phenolic -OH group.

8.18.3. Procedure

A) Nitration of phenol —synthesis of *p*-nitrophenol:

To a 250 ml flask, add 15 g of sodium nitrate ($NaNO_3$) and 40 ml water, and cool in an ice bath. Stir the mixture until complete dissolution of the solid, and then gradually add 25 g (13.6 ml) of H_2SO_4 (conc.). Then slowly add a spatula with 9.4 g of phenol, so that the mixture temperature does not exceed 20 °C, which takes approximately 20 min. After the addition, stir the reaction for 2 h at r.t. Under these conditions, a mixture is formed of *o*- and *p*-nitrophenol.

Using a Pasteur pipette, remove the supernatant, add 30 ml of water, and set up a steam distillation with an internal steam source. Heat the mixture, adding water from the dropping funnel at the same rate as the distillation vapors occurs, until it ceases to distill the product (2-nitrophenol). When the mixture remaining in the flask is cooled, the crystallization of the other isomer (4-nitrophenol) occurs and can be recrystallized using a solution of HCl (0.5 mol·100 ml^{-1}; see ref. [11]).

B) Reduction of nitro group to amine —synthesis of *p*-aminophenol:

In a 100 ml Erlenmeyer, add 4 g NaOH and 10 ml of deionized water. Once the base is dissolved, cool the solution to r.t. and add 0.56 g (14.7 mmol) of $NaBH_4$ and 50 mg of 5% Pd(C). Cool the mixture in an ice-salt bath (-12 °C), and when cold, slowly add 1 g (7.2 mmol) of *p*-nitrophenol with magnetic stirring for 30 min, keeping the reaction temperature around 15 °C.[12] After the addition, maintain stirring at r.t. for an additional 15 min.

Then acidify the crude reaction with HCl 6 M. Filtrate the catalyst under vacuum, and adjust the filtrate pH to 8.0 by adding $NaHCO_3$ in small portions. Filtrate the precipitate under vacuum, and then dry and weigh to use in the next step (estimated yield 75%; see ref. [11]).

C) Amide formation —producing acetaminophen:

In an Erlenmeyer containing 3.3 g of *p*-aminophenol and 9 ml of water, add dropwise with caution 3.6 ml of acetic anhydride, and stir the mixture constantly. Then heat in a water bath at 60 °C until complete dissolution of the solid. Keep stirring for an additional 10 min, and then cool the solution

---

[12]Check the temperature inside the Erlenmeyer with a thermometer without the stir bar striking the bulb.

in an ice bath until the appearance of a slightly pink crystalline product. Under vacuum, filter the crystals that appear in the reaction liquid in a Büchner (see ref. [11]).

Table 8.18: Physico-chemical properties of the reagents used.

Compound	M_w	M.p. (°C)	B.p. (°C)	Density (g·ml^{-1})	Danger[a] (GHS)
NaNO$_3$	84.99	306	380	2.261	☣ !
Phenol	94.11	40–42	182	1.07	☠ ☣ ⚠
o-Nitrophenol	139.11	45	214	-	!
p-Nitrophenol	139.11	110–115	279	1.480	☣ !
4-Aminophenol	109.13	185–189	-	-	☣ ! ☣
Pd(C)	106.42	-	-	-	Non-hazardous
NaCl	58.44	801	1,413	2.165	Non-hazardous
NaBH$_4$	37.8	400	-	1.07	☠ ⚠
H$_2$SO$_4$ 96–98%	98.08	3	-	1.80–1.84	⚠
HCl	36.46	−30	>100	1.200	⚠ !
Acetic anhydride	102.09	−73.1	139.8	1.080	☠ ⚠ !
NaHCO$_3$	84.01	300	-	2.160	Non-hazardous
p-Acetamidophenol	151.16	168–172	-	-	!

[a] For brevity, only GHS icons are indicated. The information offered in the Material Safety Data Sheet (MSDS) should be consulted.

Spectroscopic data (paracetamol):

- IR (KBr): 3226 (s), 1667 (s), 1611 (s), 1567 (s), 1516 (s), 1509, 1444 cm^{-1}.

- ^{1}H-NMR (DMSO-d$_6$): δ_H = 9.66 (bs, 1H), 9.14 (s, 1H), 7.35 (d, 2H, J = 8.88 Hz), 6.69 (d, 2H. J = 8.8 Hz), 1.99 (s, 3H) ppm.

- ^{13}C-NMR (DMSO-d$_6$): δ_C = 167.4, 153.2, 131.0, 126.9, 115.0, and 23.6 ppm.

- MS: m/z = 151.2 (M$^+$), 109 (BP), 80, 43.

8.19. Multistep synthesis of an anesthetic: preparation of benzocaine

Estimated time	Difficulty	Basic lab operations
4 h	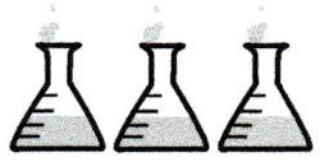 🜊🜊🜊	✔ Gravity filtration (see p. 100). ✔ Vacuum filtration (see p. 103). ✔ Liquid-liquid extraction (see p. 111). ✔ Recrystallization from organic solvent (see p. 105). ✔ Rotary evaporator (see p. 117). ✔ Reflux (see p. 89).

8.19.1. Goal

To provide an example of multi-step synthesis (using protecting groups) by preparing an organic compound that is present in many consumer products.

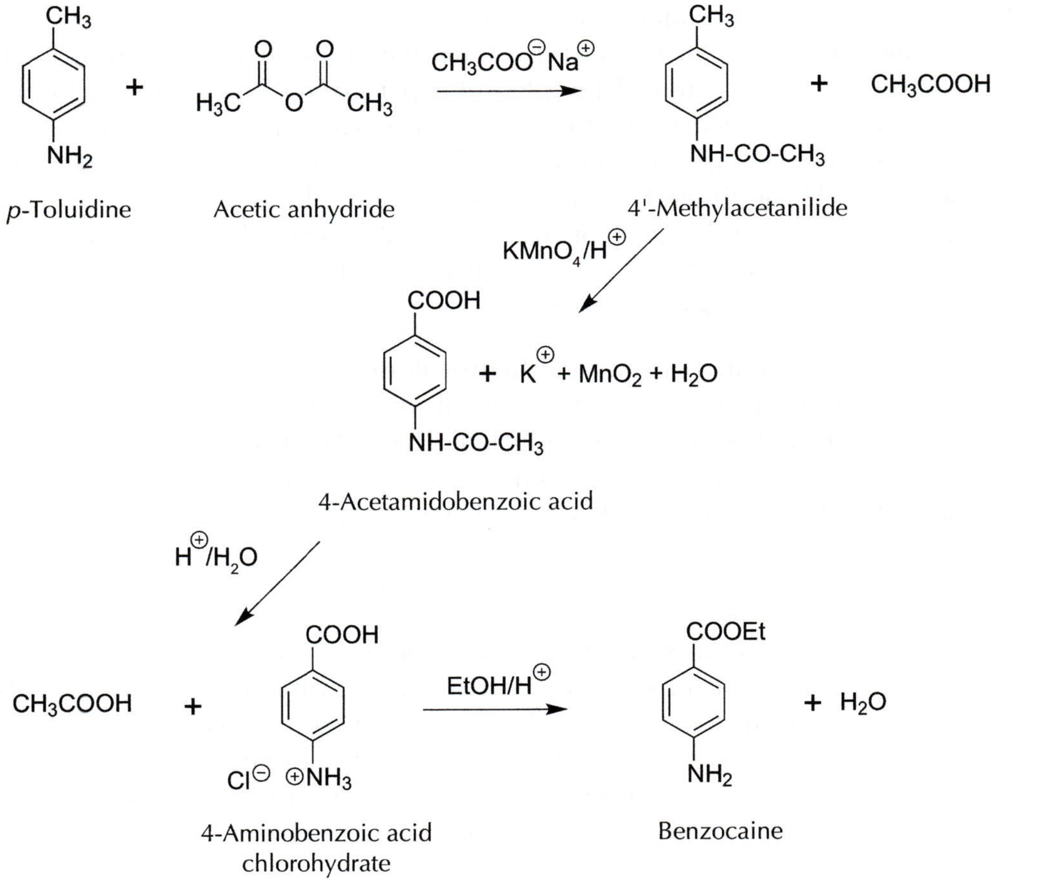

p-Toluidine Acetic anhydride 4'-Methylacetanilide

4-Acetamidobenzoic acid

4-Aminobenzoic acid chlorohydrate

Benzocaine

8.19.2. Background

Benzocaine (ethyl 4-aminobenzoate) is an anesthetic that is also used to relieve the pain of injuries and burns. Benzocaine is the derivative of p-aminobenzoic acid and can be synthesized via the Fischer esterification reaction. It can be prepared from p-toluidine by a four-step synthesis. The problem of a synthesis of this type is that the overall yield of the final product is usually quite low.

8.19.3. Procedure

A) Synthesis of N-acetyl-p-toluidine:

Place 4 g of p-toluidine in a 100 ml Erlenmeyer flask. Then, in a fume hood, add 10 ml of acetic anhydride and slowly stir. The reaction is highly exothermic. Allow the reaction mixture to stand for 10 min and then pour the reaction crude onto 50 ml of water/ice. If the reaction mixture crystallizes in the flask, drag the solid with a small amount of cold water. Stir the suspension of the product in water with a rod, and then collect the resulting solid by vacuum filtration. Keep the product to carry out the next stage of the synthesis (see ref. [4]).

B) Synthesis of p-acetamidobenzoic acid:

In a 400 ml beaker place N-acetyl-p-toluidine (4′-methylacetanilide), 200 ml of water, and an amount of 1.8 g $KMnO_4$ per gram of N-acetyl-p-toluidine. In a water bath with magnetic stirring, heat the mixture until it acquires a distinct brown (approximately 30 min). Vacuum filter the resulting hot solution.[13] If the filtered solution has a violet color, add EtOH dropwise, and heat gently until the color disappears. When the filtrate is colorless or slightly yellow, allow to cool and acidify with H_2SO_4 (20%). Vacuum filter the resulting white solid (see ref. [4]).

C) Synthesis of p-aminobenzoic acid hydrochloride:

Assemble the reflux in a flask of 100 ml and provide with a gas outlet to a trap with a diluted NaOH solution. Heat the flask for 30 min with a mixture of 2.5 g of p-acetamidobenzoic acid and 25 ml of HCl (conc.). Cool and collect the precipitate of p-aminobenzoic acid hydrochloride by vacuum filtration. (see ref. [4]).

D) Synthesis of benzocaine:

Into a 100 ml flask, mix 1.25 g of p-aminobenzoic acid hydrochloride, 10 ml of EtOH, and 0.5 ml H_2SO_4 (conc.). Heat the mixture at reflux for 2 h. After cooling the mixture, neutralize with an aqueous sodium carbonate solution (10%). Extract with CH_2Cl_2 (3 × 5 ml), dry the organic layer with anhydrous sodium sulfate, and evaporate in a rotary evaporator. Recrystallize the solid resulting from the EtOH/water (see ref. [4]).

---

[13]On the filter paper of a Büchner funnel, deposit a silica layer of 2 cm thick.

Table 8.19: Physico-chemical properties of the reagents used.

Compound	M_w	M.p. (°C)	B.p. (°C)	Density (g·ml^{-1})	Danger[a] (GHS)
H$_2$SO$_4$ 96–98%	98.08	3	-	1.80–1.84	⚠
EtOH	46.07	−114.1	78.5	0.790	🔥
HCl	36.46	−30	>100	1.200	⚠ ❗
NaOH	40.00	318	1,390	2.130	⚠
CH$_2$Cl$_2$	84.93	−97	40.0	1.33	☣
Acetic anhydride	102.09	−73.1	139.8	1.080	🔥 ⚠ ❗
KMnO$_4$	158.04	240	-	2.71	⭕ ❗ 🌿
CaCO$_3$	100.09	800	-	2.93	Non-hazardous
Na$_2$SO$_4$	142.04	884	-	2.630	Non-hazardous
p-Toluidine	107.15	41–46	200	-	☠ ☣ 🌿
p-Aminobenzoic acid	137.14	187–189	-	1.374	❗
Benzocaine	165.19	88–90	-	-	Non-hazardous
p-Aminobenzoic acid hydrochloride	173.60	-	-	-	❗
4′-Methylacetanilide	149.19	149–151	307	-	❗
p-Acetamidobenzoic acid	179.17	259–262	-	-	Non-hazardous

[a] For brevity, only GHS icons are indicated. The information offered in the Material Safety Data Sheet (MSDS) should be consulted.

8.20. Using protecting groups: multistep synthesis of *p*-nitroaniline

Estimated time	Difficulty	Basic lab operations
2 h	⚗⚗⚗	✔ Gravity filtration (see p. 100). ✔ Vacuum filtration (see p. 103). ✔ Recrystallization from water (see p. 105). ✔ Reflux (see p. 89).

8.20.1. Goal

To teach the students the concept of the protecting group. Particularly, protection/deprotection reactions of an amino group are performed, in order to use reactions such as nitration, which demand strongly acidic conditions, to which an amino group is highly sensitive.

Aniline   Acetanilide   *p*-Nitroacetanilide   *p*-Nitroaniline

8.20.2. Background

Aniline cannot be directly nitrated, since the amino group is very sensitive to the acidic conditions of the nitration. Therefore, a protecting group for the amino group should be used; then the nitration reaction will take place, and finally the protecting group should be removed, in order to produce the desired product. The product produced in a previous step should be used in the following one, so that the amounts indicated in the procedure must be adjusted with those determined in each step.

8.20.3. Procedure

A) Preparation of acetanilide:

In a 250 ml round-bottom flask, place 9 ml (0.1 mol) of aniline, 15 ml of glacial acetic acid, and 15 ml of acetic anhydride. Fit a reflux condenser to the flask, and heat the solution to boiling for 10 min. Then allow the flask to cool. Pour the reaction crude into a beaker with 50 ml of water and 40–50 g of ice. Mix well and collect the acetanilide crystals by filtration in a Büchner, stirring. Recrystallize the product from water, approximately 200 ml, and decolorize with activated charcoal. Dry, weigh, and determine the yield (see ref. [3], p. 919).

B) Nitration of acetanilide:

In a 100 ml beaker, place 15 ml of concentrated H_2SO_4, and add 6.75 g (0.05 mol) of acetanilide in small portions with magnetic stirring. As soon as all the acetanilide, or virtually all, is dissolved, place the glass into an ice bath, and using an addition funnel, add a solution of 6 ml of HNO_3 in 6 ml of H_2SO_4 (conc.). Add the mixture dropwise with gentle stirring and regulating the addition so that the temperature of the reaction mixture does not exceed 35 °C. After the addition, remove the beaker from the ice bath, and allow to stand at r.t. for 5 min. Pour the acetanilide nitrated solution in a 600 ml beaker (with approximately 100 ml of water and 30 g ice). Stir the mixture and collect the precipitate of p-nitroacetanilide by vacuum filtration in a Büchner. Wash, in the same Büchner, with two portions of 50 ml of cold water, pressing well, passing a stream of air to dry. Recrystallize from EtOH. Dry, weigh, and determine the yield (see ref. [3], p. 919).

Noticeboard 8.20.1

 DANGER!

Perform the experiments in a fume hood for both nitration (step B) and the addition of ammonia (step C).

C) p-Nitroaniline:

Place the wet p-nitroacetanilide in a 400 ml beaker, and with it form a thin paste, add 100 ml of water, and stir. Transfer this mixture to a 250 ml round-bottom flask, add 35 ml of HCl (conc.), and adapt a reflux condenser to the flask. Then heat to boiling for 35 min. Allow the reaction crude to cool to r.t. After cooling the reaction crude, pour into a 500 ml beaker, with approximately 50–75 g of crushed ice. The p-nitroaniline precipitates by basifying the solution with the addition of ammonia. Filter the precipitate under vacuum in a Büchner, and wash with small quantities of water. Recrystallize from water, decolorizing with activated carbon (approximately 0.5 g of substance 40–50 ml of water). Dry, weigh, and determine the (partial and total) yields (see ref. [3], p. 919).

Table 8.20: Physico-chemical properties of the reagents used.

Compound	M_w	M.p. (°C)	B.p. (°C)	Density (g·ml^{-1})	Danger[a] (GHS)
Aniline	93.13	−6	184	1.022	
Glacial acetic acid	60.05	16.2	118	1.049	
Acetic anhydride	102.09	−73.1	139.8	1.080	
Acetanilide	135.16	113–115	304	-	
H_2SO_4 96–98%	98.08	3	-	1.80–1.84	
HNO_3 68–70%	63.01	-	120.5	1.413	
HCl	36.46	−30	>100	1.200	
NH_3	17.03	−78	−33	0.590	
p-Nitroaniline	138.12	146–149	260	1.440	
p-Nitroacetanilide	180.16	213–215	-	-	
EtOH	46.07	−114.1	78.5	0.790	
Active carbon	12.01	3,550	-	0.25–0.60	Non-hazardous

[a] For brevity, only GHS icons are indicated. The information offered in the Material Safety Data Sheet (MSDS) should be consulted.

Spectroscopic data (p-NO$_2$-aniline):

- IR (KBr): 3484 (s), 3365 (s), 1633 (s), 1603, 1482, 1471, 1313 (s), 1116 (s) cm^{-1}.

- ^{1}H-NMR (DMSO-d$_6$): δ_H = 7.97 (m, 2H), 6.71 (bs, 2H), 6.64 (m, 2H) ppm.

- ^{13}C-NMR (DMSO-d$_6$): δ_C = 155.6, 135.8, 126.3, and 112.4 ppm.

- MS: m/z = 138.1 (M$^+$), 108, 92, 65 (BP), 44.

8.21. References

1. A. I. Vogel and B. S. Furniss, *Vogel's Textbook of Practical Organic Chemistry*, Longman, 1989.

2. D. B. Brown and L. B. Friedman, *The Aspirin Project. Laboratory Experiments for Introductory Chemistry*, Journal of Chemical Education **50** (1973), no. 3, 214, DOI 10.1021/ed050p214.

3. R. E. Buckles, *The use of the Perkin reaction in organic laboratory classes*, Journal of Chemical Education **27** (1950), no. 4, 210–211, DOI 10.1021/ed027p210.

4. P. Demare and I. Regla, *Synthesis of two local anesthetics from toluene: an organic multistep synthesis in a project-oriented laboratory course*, Journal of Chemical Education **89** (2012), no. 1, 147–149, DOI 10.1021/ed100838a.

5. J. A. Poce-Fatou, *A superficial overview of detergency*, Journal of Chemical Education **83** (2006), no. 8, 1147, DOI 10.1021/ed083p1147.

6. C. S. Judd, *News from online: cleaning up-soap, detergent, and more*, Journal of Chemical Education **79** (2002), no. 10, 1179, DOI 10.1021/ed079p1179.

7. A. G. Jones, *The selective reduction of meta- (and para-) nitroacetophenone*, Journal of Chemical Education **52** (1975), no. 10, 668, DOI 10.1021/ed052p668.

8. M. P. Cava, A. A. Deana, K. Muth, and M. J. Mitchell, *N-Phenylmaleimide*, Organic Synthesis; Collective Volume **5** (1973), 944.

9. M. R. Dintzner, C. R. Kinzie, K. Pulkrabek, and A. F. Arena, *The cyclohexanol cycle and synthesis of nylon 6,6: green chemistry in the undergraduate organic laboratory*, Journal of Chemical Education **89** (2012), no. 2, 262–264, DOI 10.1021/ed2000878.

10. D. W. Armstrong, J. N. Marx, D. Kyle, and A. Alak, *Synthesis and a simple molecular weight determination of polystyrene*, Journal of Chemical Education **62** (1985), no. 8, 705, DOI 10.1021/ed062p705.

11. F. Ellis, C. Osborne, and M. J. Pack, *Paracetamol: A Curriculum Resource*, Royal Society of Chemistry, 2002.

Chapter 9

Advanced Organic Synthesis Experiments

In this chapter a total of 24 experiments of increasing complexity (compared to the previous chapter) are discussed. This greater complexity is due to several factors —for example, the reactions have more complex mechanisms; the procedures required to develop the corresponding syntheses are more sophisticated, and therefore requires the student to have basic training in the Organic Chemistry laboratory; the nature of starting materials and/or reagents require very careful handling; or the initial and/or final products are not covered in the contents of an elementary Organic Chemistry course.

In this chapter, the multistep processes are more active than in the previous chapter, and various methods of forming C–C bonds are also described, including by Wittig, Diels-Alder reactions, or by the use of organometallic reagents. Also, the chemical preparation and modification of some heterocycles and natural products are described.

In some cases, these reactions require special conditions, such as an inert atmosphere, the gradual addition of reagents, and/or the use of anhydrous solvents, which adds difficulty to these reactions.

9.1. Reductive amination: producing (±)-α-methylbenzylamine

Estimated time	Difficulty	Basic lab operations
2 h	🜊🜊🜊	✔ Liquid-liquid extraction (see p. 111). ✔ Simple distillation (see p. 120). ✔ Reflux (see p. 89).

9.1.1. Goal

To perform a reductive amination of a carbonyl compound, emulating the biosynthesis of an amino acid but without using enzyme catalysis.

Acetophenone   (±)-1-Amino-1-phenylethanol   1-Phenylethanimine

(±)-1-Phenylethanamine

9.1.2. Background

When reagents, products, and non-chiral catalysts are used in a chemical reaction, the overall result is an optically inactive mixture. However, if a center or chiral carbon has been generated, the product will invariably appear as a racemic mixture (50% of the two possible enantiomers). In nature, optically active compounds are generated by means of enzymatic catalysis, producing only one of the two enantiomers (enantiospecific reaction). In this experiment, a reductive amination is performed by reacting a carbonyl compound (acetophenone) with ammonia, which acts as the nucleophile being added to the carbonyl group and generates an aminoalcohol that dehydrates, giving an imine. This imine is subsequently reduced with formate ion to the corresponding amine (Leuckart reaction).

9.1.3. Procedure

In a 100 ml round-bottom flask, add 25 g of acetophenone, 43 g of ammonium formate (NH_4HCO_2), and a magnetic stir bar. Assemble simple distillation equipment. Heat the mixture with a silicone bath, ensuring that the bath temperature does not exceed 200 °C.[1] Continue heating until no distillate is collected. Cool the flask and then add 25 ml of water. Transfer the reaction mixture to a separatory funnel and extract with toluene (2 × 25 ml). Mix the toluene extracts with the above distillate and discard the aqueous layer. Place in a flask equipped for simple distillation. Add 20 ml of concentrated HCl. Heat the mixture until all the toluene is distilled, taking care to collect on a glass with enough water so that the distillation gases bubble. Then, continue to heat for 40 min more, and allow the mixture to reach r.t. If some crystals form, add the minimum amount of water to dissolve. Extract the mixture with toluene (3 × 20 ml) and transfer the aqueous layer to a round-bottom flask. Add 43 ml of a solution of NaOH 50% and conduct simple distillation. Continue to distill until the pH of the distillate has no basic pH value (approximately 150 ml). Extract the distillate containing the amine with toluene (3 × 15 ml). Dry the toluene extract with NaOH pellets. Once dry, submit the solution to a simple distillation and the product distilled between 180–190 °C is collected (see ref. [10]).

Table 9.1: Physico-chemical properties of the reagents used.

Compound	M_w	M.p. (°C)	B.p. (°C)	Density (g·ml^{-1})	Danger[a] (GHS)
α-Methylbenzyl-amine	121.18	−60	197–200	0.962	
Acetophenone	120.15	19-20	202	1.03	
NH_4HCO_2	63.06	119–121	-	1.260	
NH_3	17.03	−78	−33	0.590	
NaOH	40.00	318	1,390	2.130	
Toluene	92.14	−93	110.6	0.867	
HCl	36.46	−30	>100	1.200	

[a] For brevity, only GHS icons are indicated. The information offered in the Material Safety Data Sheet (MSDS) should be consulted.

---

[1]On the neck of the Erlenmeyer collector, place cotton impregnated with H_2SO_4 diluted to prevent the escape of ammonia vapors.

9.2. Chiral resolution: (±)-α-methylbenzylamine

Estimated time	Difficulty	Basic lab operations
24 h	⚗⚗	✔ Vacuum filtration (see p. 103). ✔ Vacuum distillation (see p. 125). ✔ Reflux (see p. 89).

9.2.1. Goal

To acquaint students with the concept of chiral resolution: separation of the two enantiomers of a chiral compound. In this case, the different solubility of two diastereomeric salts in water will be used.

9.2.2. Background

When products, reagents, or non-chiral catalysts are used in a chemical reaction in which there are chiral centers, the overall result is the same as producing an optically inactive mixture of stereoisomers. If a chiral carbon or center is generated, the product will always appear as a racemic mixture (50% of the two possible enantiomers).

For the separation of the two enantiomers, a process called chiral resolution is carried out, reacting the racemic mixture with an optically active compound (chiral resolving agent). This results in the formation of two diastereomers, which possess different physical and chemical properties and therefore are capable of being separated by different procedures.

After the α-phenylethylamine racemic is synthesized (or purchased), the next procedure is to produce one of the optically pure enantiomers. In practice, only one of the two enantiomers separated is optically pure, the levorotatory, because it can be isolated more easily than can the other enantiomer. The chiral resolving agent to be used is (+)-tartaric acid, which forms two diastereomeric salts with α-methylbenzylamine (α-phenylethylamine) racemic.

The optically pure (+)-tartaric acid is very abundant in nature and is yielded mainly as a byproduct of winemaking. The (–)-amine-(+)-tartrate salt is less soluble than the salt (+)-amine-(+)-tartrate salt and precipitates from solutions in crystalline form. The crystals are filtered and purified. The salt is then treated with a diluted base, which is isolated with the free (–)-amine. On the other hand, the mother liquor, which preferably contains (+)-amine-(–)-tartrate salt, may be purified to produce the other diastereomeric salt. The hydrolysis of this salt would give the (+)-amine.

9.2.3. Procedure

Add 6.25 g of *L*-(+)-tartaric acid with 90 ml of MeOH to a 250 ml Erlenmeyer flask. Heat the mixture in a water bath until almost boiling. Add 5 g of racemic α-methylbenzylamine slowly because the mixture generates foam and can spill. Close the flask with a septum, and cool slowly for a day for good crystallization of the (+)-amine-(-)-tartrate salt.

In case of the formation of needle-shaped crystals, they can be redissolved by gentle heating and re-cooled. By this crystallization method, crystals of the desired shape can be produced.[2]

Filter the crystals under a vacuum and wash with a few milliliters of cold MeOH. A second amount of crystals can be obtained by concentrating the mother liquor, which will give an additional amount of approximately 1.5 g of the salt. Once the sufficient amount of crystals has been produced, partially dissolve 10 g salt[3] in 40 ml of water. Add 6 ml of NaOH 50% solution, and extract the mixture with diethyl ether (2×20 ml).

Dry the organic layer (upper) on Na_2SO_4 (anhydrous). Then remove the desiccant by gravity filtration. The solvent is removed in a rotary evaporator to give a residue that is purified by vacuum distillation. The estimated yield is 55%, and b.p. is 184–186 °C.

Measure the optical rotation of the product. For this, accurately weigh 20 mg of amine and dissolve in 2 ml MeOH (measured by pipette). Calculate the total volume of the solution considering the sum of the volumes of the components (the amine density is 0.9395 g/ml).

Transfer the solution to the polarimeter tube and determine the optical rotation ($[\alpha]_D^{20} = -40.3$ °C). Calculate the specific optical rotation and optical purity of the solution (see ref. [23]).

---

[2]Note that the needle-shaped crystals do not have a sufficiently high optical purity to then achieve good resolution of the enantiomers.

[3]If the procedure fails to provide 10 g of salt, reagent amounts should be adjusted to the mass of salt obtained.

Table 9.2: Physico-chemical properties of the reagents used.

Compound	M_w	M.p. (°C)	B.p. (°C)	Density (g·ml^{-1})	Danger[a] (GHS)
α-Methylbenzyl-amine	121.18	−60	197–200	0.962	
L-(+)-Tartaric acid	150.09	170–172	-	-	
MeOH	32.04	−98	64.7	0.791	
NaOH	40.00	318	1,390	2.130	
Diethyl ether	74.12	−116	34.6	0.71	
MgSO$_4$	120.37	1124	-	1.070	Non-hazardous

[a] For brevity, only GHS icons are indicated. The information offered in the Material Safety Data Sheet (MSDS) should be consulted.

9.3. Stereospecific synthesis of glycols: preparation of *trans*-cyclohexane-1,2-diol

Estimated time	Difficulty	Basic lab operations
4 h	⚗⚗⚗	✔ Gravity filtration (see p. 100). ✔ Vacuum filtration (see p. 103). ✔ Liquid-liquid extraction (see p. 111). ✔ Rotary evaporator (see p. 117). ✔ Simple distillation (see p. 120). ✔ Reflux (see p. 89).

9.3.1. Goal

To perform a stereospecific synthesis, as well as to show the relevance of the epoxides (oxacyclopropanes) as versatile intermediates in Organic Synthesis.

9.3.2. Background

In this experiment, the *trans*-cyclohexane-1,2-diol is obtained from cyclohexene, prior to the transformation of the latter to *trans*-2-bromocyclohexanol through a stereoselective process (*anti* addition to a double bond), followed by the formation of an epoxide and further its opening.

Formally, the reaction produces the anti addition of the component of hypobromous acid (HOBr) to the double bond by reaction of cyclohexene with *N*-bromosuccinimide (NBS) in an aqueous medium. The intermediate (bromonium ion) initially formed reacts with water, which is the most nucleophilic species present (due to almost no bromide ion concentration possible). In the experiment described, the intermediate bromohydrin is not isolated but immediately becomes an epoxide. Treating the *trans*-2-bromocyclohexanol with a base leads to the formation of the corresponding epoxide by an intramolecular nucleophilic substitution process. The available antiperiplanar disposition of the OH and Br groups allows the formation of 1,2-epoxycyclohexane.

The greater reactivity of the epoxides compared with acyclic or cyclic ethers, the larger the ring tension allows for relatively easy breakage of the C−O bond. The opening of the epoxide is performed in acid medium, whereby a steric control over the process is exercised, since the epoxide is protonated and induces the attack of water on the rear side of the epoxide. The result is the formation of a *trans* diol.

9.3.3. Procedure

A) Preparation of 2-bromocyclohexan-1-ol:

To an Erlenmeyer with a stir bar, add (in this order): 7.6 ml of cyclohexene, 20 ml of deionized water, and 25 ml of THF. Place the flask into an ice bath, set it on a stir plate, and vigorously mix the reaction [4]. Next, add a solution of 14.7 g of NBS in 40 ml of THF gradually for approximately 20 min, so that the reaction temperature does not exceed 30 °C. After the addition of NBS, maintain the stirring for 30 min more. After 30 min of reaction, pour the reaction crude into a separatory funnel. Add 30 ml of diethyl ether and 30 ml of brine. Shake the mixture and decant. Separate the organic (upper) layer, and extract the aqueous layer again with diethyl ether (2×20 ml). Combine the diethyl ether extracts, and wash again with brine (2×20 ml) to make a solution of *trans*-2-bromocyclohexanol to be used in the next step.

B) Preparation of 7-oxabicyclo[4.1.0]heptane:

To a 250-ml round-bottom flask with ground-glass stoppers and Claisen adapter (with stir bar) or otherwise a two-necked one, add 25 ml of a solution of NaOH 5M. Set up a reflux condenser, and slowly add for 40 min, from the addition funnel, a solution of 2-bromocyclohexan-1-ol in ether, prepared in the previous step; while stirring, maintain the temperature of reaction at approximately 40 °C. After the addition, stir the mixture for 30 min. When the addition is complete, transfer the mixture to a separatory funnel. Separate the two layers, and dry the organic (upper) layer over sodium hydroxide pellets, with occasional stirring for 15 min. Decant the solution and remove the solvent under reduced pressure (rotary evaporator). Distill the residue at atmospheric pressure (124–134 °C), and then weigh and calculate the yield of this step.

C) Preparation of *trans*-cyclohexane-1,2-diol:

In a 100-ml round-bottom flask, place 2 ml (1.95 g, 15 mmol) 7-oxabicyclo-[4.1.0]heptane.[5] Add 10 ml of water and 1 ml of H_2SO_4, and vigorously stir the mixture for 1 h (cover the flask). During this time period, observe that the flask is heated and that the reaction mixture becomes transparent. After stirring for 1 h, adjust the pH of the solution to 7 by dropwise adding the NaOH solution. Extract the aqueous solution with ethyl acetate (3×15 ml). Dry the organic extracts over anhydrous sodium sulfate, remove the desiccant by gravity filtration, and concentrate by rotary evaporation to 1/3 of the volume. Then, introduce the ethyl acetate solution into an ice bath

---

[4]The temperature is controlled throughout the entire process by inserting a thermometer in the reaction carefully so as not to strike the stir bar.

[5]Adjust quantities to the result obtained in the previous step.

until the appearance of a solid. Filter the solid under vacuum, dry with an air flow, weigh, and calculate the yield.

Table 9.3: Physico-chemical properties of the reagents used.

Compound	M_w	M.p. (°C)	B.p. (°C)	Density (g·ml^{-1})	Danger[a] (GHS)
Cyclohexene	82.14	−104	83	0.779	
H$_2$SO$_4$ 96–98%	98.08	3	-	1.80–1.84	
NBS	177.98	175–180	-	-	
THF	72.11	−108.0	65–67	0.89	
NaOH	40.00	318	1,390	2.130	
2-Bromocyclohexan-1-ol	179.055	11.82	225.52	1.520	See MSDS
trans-Cyclohexane-1,2-diol	116.16	101–104	-	-	Non-hazardous
1,2-Epoxycyclohex-ane	98.14	-	129–130	0.97	

[a] For brevity, only GHS icons are indicated. The information offered in the Material Safety Data Sheet (MSDS) should be consulted.

9.4. Reactivity of carbenes: preparation of 7,7-dichlorobicyclo[4.1.0]heptane

Estimated time	Difficulty	Basic lab operations
1 h		✔ Liquid-liquid extraction (see p. 111). ✔ Rotary evaporator (see p. 117). ✔ Vacuum distillation (see p. 125).

9.4.1. Goal

To perform an electrophilic addition of a carbene to an alkene in order to allow the simultaneous formation of two C−C bonds, producing a cyclopropane derivative.

Cyclohexene   (1*R*,6*S*)-7,7-Dichlorobicyclo[4.1.0]heptane

9.4.2. Background

Electrophilic additions to the double bonds are one of the most common transformations in an Organic Chemistry laboratory. They are also among the most powerful reactions in Organic Chemistry. Using this type of reaction, a relatively simple compound can represent a large amount of molecular complexity in one step. Many carbenes are electrophilic and and are able to be added to alkenes. Most carbenes (R_2C :) are prepared by a α-elimination, where the electrophilic and nucleophilic groups are from the same carbon atom. Generally, it is a base-catalyzed elimination.

9.4.3. Procedure

To a 50 ml Erlenmeyer, add 2 g (0.05 moles) of NaOH and 2 ml of water. Close the flask with a stopper and shake to dissolve the base. A strong exotherm reaction is observed, so the solution must be cooled using an external ice bath until r.t. is reached. Then add 0.4 g (0.001 mol) tricaprylmethyl ammonium chloride (Aliquat 336®), 2 ml (1.62 g, 0.02 moles) of cyclohexene and 2 ml (2.98 g, 0.025 moles) of chloroform, and stopper again. Shake the mixture vigorously for 20 min, ensuring that an emulsion is formed.

At the end of the reaction time, transfer the oil to a separatory funnel with the aid of an additional 15–20 ml water. Drag the contents of the flask with 5 ml CH_2Cl_2. Stir the mixture and decant. Repeat with another 5 ml of CH_2Cl_2.

Collect and dry the organic phase with anhydrous sodium sulfate. Filter (solid disposable), and transfer the filtrate to a tared flask. Remove the solvent together with the remains of cyclohexene and any remaining chloroform on a rotary evaporator. A liquid is the final product (see ref. [11].)

Table 9.4: Physico-chemical properties of the reagents used.

Compound	M_w	M.p. (°C)	B.p. (°C)	Density (g·ml^{-1})	Danger[a] (GHS)
Aliquat®336	-	−6	-	-	
CH_2Cl_2	84.93	−97	40.0	1.33	
Cyclohexene	82.14	−104	83	0.779	

[a] For brevity, only GHS icons are indicated. The information offered in the Material Safety Data Sheet (MSDS) should be consulted.

9.5. Regioselective halogenations: bromation of α-methylstyrene

Estimated time	Difficulty	Basic lab operations
2 h	🧪	✔ Vacuum filtration (see p. 103). ✔ Simple distillation (see p. 120).

9.5.1. Goal

To demonstrate the regioselectivity of electrophilic addition to a double bond. This characteristic (regioselectivity) is not possible in the addition reaction to cyclohexene, because it is symmetrical.

α-Methylstyrene   NBS   1-Bromo-2-phenylpropan-2-ol   or   2-Bromo-2-phenylpropan-1-ol   Succinimide

9.5.2. Background

Electrophilic additions to double bonds are ones of the most common as well as the most powerful chemical transformations available. These reactions allow, starting from a relatively simple compound, the addition of a large amount of molecular complexity through a single synthetic step. The reaction chosen here is the bromohydrin formation (addition of bromine and water to the double bond) of styrene. The bromine source employed is the NBS because it is easier to handle than solid elemental bromine (which is liquid and toxic). NBS produces small concentrations of bromine, under reaction conditions that will react with the double bond, forming a bromonium ion. The product of the reaction is succinimide, which can be precipitated from the reaction crude and be removed by filtration. The water of the reaction comes from the commercial acetone containing approximately 5% of water. Reaction of this intermediate with the bromonium ion may result in the formation of regioisomers according to the attack position.

9.5.3. Procedure

In a flask, place 5.9 g (50 mmol) of α-methyl styrene and 5 ml of acetone. Add 1.025 equivalents of NBS and stir the mixture at r.t. until all the NBS has dissolved. Then stir the solution for an additional 30 min. Remove the solvent by simple distillation (do not use a rotary evaporator due to the low product b.p.) below 60 °C. Add 25 ml of hexane to the flask, and dry this solution with anhydrous sodium sulfate. Cool the solution in an ice bath, and stir for 15 min. Filter and wash the solid with a small additional amount of hexane. Place the organic extract in a flask and distil hexane (do not use a rotary evaporator because of the low b.p. of products). The residue is the product of the reaction (see ref. [12]).

Table 9.5: Physico-chemical properties of the reagents used.

Compound	M_w	M.p. (°C)	B.p. (°C)	Density (g·ml^{-1})	Danger[a] (GHS)
NBS	177.98	175–180	-	-	☠ !
Hexane	86.18	−95	69	0.659	🔥 ☠ ! 🌿
Na$_2$SO$_4$	142.04	884	-	2.630	Non-hazardous
Acetone	58.08	−94	56	0.791	🔥 !
α-Methylstyrene	118.18	−24	165–169	0.909	🔥 ! 🌿

[a] For brevity, only GHS icons are indicated. The information offered in the Material Safety Data Sheet (MSDS) should be consulted.

9.6. Oxidative coupling of alkynes: the Glaser-Eglinton-Hay coupling

Estimated time	Difficulty	Basic lab operations
2 h		✔ Gravity filtration (see p. 100). ✔ Vacuum filtration (see p. 103). ✔ Recrystallization from water (see p. 105). ✔ Recrystallization from organic solvent (see p. 105). ✔ Rotary evaporator (see p. 117). ✔ Simple distillation (see p. 120). ✔ Reflux (see p. 89).

9.6.1. Goal

To study the oxidative coupling of alkynes.

$$R-C{\equiv}C-H \quad \xrightarrow[\substack{\text{Base:}\\ \text{NH}_4\text{OH/EtOH}\\ \text{Cu(AcO)}_2/\text{Pyridine}}]{\text{CuCl/O}_2} \quad R-C{\equiv}C-C{\equiv}C-R$$

Terminal alkyne   Conjugated di-alkyne

9.6.2. Background

The Glaser-Eglinton-Hay coupling is the oldest of the acetylenic couplings. It involves the reaction of the ethylenic compound with cuprous salts in the presence of oxygen and a base (ammonia initially). Subsequent modifications to this reaction changed the reaction conditions (reagents and bases used). It is assumed that the reaction mechanism proceeds via radical and copper(I) acetylide. Thus, the Glaser-Eglinton-Hay coupling is a modification in which two terminal alkynes are coupled directly, using a copper(II) salt, such as copper acetate and pyridine as a base.

9.6.3. Procedure

In a two-necked flask provided with agitation, put 30 ml of propan-2-ol, 100 mg of CuCl, and 20 drops of TMEDA. Put in one of the openings a reflux condenser, and in the other an adapter provided with a pipette bubbling oxygen into the solution. Add 2 g of 1-ethynylcyclohexanol and reflux the reaction for 30–40 min. Verify the absence of starting material by TLC, hexane/ethyl acetate (7:3).

Remove all propane-2-ol in a rotary evaporator. Then add 20 ml of water containing 1 ml of concentrated HCl. Cool and filter the product. If the product is blue or green, return to add HCl/water to remove all copper salts. If it has a dark color, dissolve in hot ethyl acetate and discolor with activated carbon. If it is almost white, recrystallize with ethyl acetate, and dry with an air stream through the same Büchner (see ref. [24], pp. 142–151).

Table 9.6: Physico-chemical properties of the reagents used.

Compound	M_w	M.p. (°C)	B.p. (°C)	Density (g·ml^{-1})	Danger[a] (GHS)
HCl	36.46	−30	>100	1.200	
Ethyl acetate	88.11	−84	77.1	0.902	
TMEDA	116.20	−55	120–122	0.775	
1-Ethynylcyclo-hexan-1-ol	124.18	30.0-33.5	180	-	
CuCl	99.00	430	1490	4.140	
Oxygen	32.00	−218	-183	-	
Propan-2-ol	60.10	−89.50	82	-	

[a] For brevity, only GHS icons are indicated. The information offered in the Material Safety Data Sheet (MSDS) should be consulted.

9.7. Diels-Alder reactions: butadiene and maleic anhydride

Estimated time	Difficulty	Basic lab operations
2 h		✔ Gravity filtration (see p. 100). ✔ Vacuum filtration (see p. 103). ✔ Recrystallization from organic solvent (see p. 105). ✔ Reflux (see p. 89).

9.7.1. Goal

To perform a Diels-Alder cycloaddition between a diene and an alkene to produce a highly functionalized cyclohexene derivative.

9.7.2. Background

The Diels-Alder is one of the most remarkable chemical reactions discovered in the 20th century. The authors of this reaction won the Nobel Prize in 1950. It is a powerful tool to build 6-membered rings and has been used for the synthesis of many complex molecules. A cycloaddition [4+2], it is usually performed at moderate or high temperatures and even under pressure. This reaction is stereospecific with respect to both the diene and dienophile. In this experiment, two examples of Diels-Alder reactions are described, in the first case the diene is prepared in situ from 3-sulfolene, and in the second case a functionalized diene is employed.

9.7.3. Procedure

A) Cycloaddition [4+2] between the maleic anhydride and buta-1,3-diene:
In a 100 ml round-bottom flask, place 10 g of 3-sulfolene (buta-1,3-diene sulfone) and 5 g of maleic anhydride powder. Next, add 10 ml of xylene and keep the mixture refluxed for 45 min with magnetic stirring. A gas trap is attached at the end of the condenser with a solution containing 5% NaOH, because the reaction releases SO_2. Cool the solution to r.t. and add 50 ml of toluene and activated carbon to discolor the solution. Filter by gravity. To the filtrate, add 25–30 ml of hexane, and heat to 80 °C. First cool the solution to r.t. and then place it in an ice bath. Then filter the crystals formed (see refs. [1,2]).

B) Cycloaddition [4+2] between maleic anhydride and *E,E*-2,4-hexadien-1-ol:
Reflux a mixture of 0.40 g of maleic anhydride and 0.40 g of *E,E*-2,4-hexadien-1-ol in 5 ml of toluene for 5 min. Let the solution cool slowly to r.t.; a solid will start to appear on the walls of the flask. To ensure a complete precipitation of the product, chill in an ice bath for 10 min. Vacuum filter and air dry. It can be recrystallized from toluene (m.p. = 156–159 °C, lit. 161 °C; see refs. [1,2]).

Table 9.7: Physico-chemical properties of the reagents used.

Compound	M_w	M.p. (°C)	B.p. (°C)	Density (g·ml^{-1})	Danger[a] (GHS)
E,E-hexa-2,4-dienol	98.14	27–33	80	0.871	
Maleic anhydride	98.06	51–56	200	1.480	
Toluene	92.14	−93	110.6	0.867	
3-Sulfolene	118.15	64–67	-	-	
p-Xylene	106.17	13.0	138.4	0.860	
SO_2	64.06	−73	−10	-	
NaOH	40.00	318	1,390	2.130	
cis-1,2,3,6-Tetrahydrophthalic anhydride	150.147	97–103	-	-	

[a] For brevity, only GHS icons are indicated. The information offered in the Material Safety Data Sheet (MSDS) should be consulted.

9.8. Wittig reaction: 4-vinylbenzoic acid synthesis

Estimated time	Difficulty	Basic lab operations
2 h		✔ Gravity filtration (see p. 100). ✔ Vacuum filtration (see p. 103). ✔ Recrystallization from organic solvent (see p. 105). ✔ Reflux (see p. 89).

9.8.1. Goal

To synthesize an alkene from a phosphonium salt and a carbonylic compound by a Wittig reaction.

9.8.2. Background

The Wittig reaction is so named for Georg Wittig (Nobel Prize in 1979). In this reaction, alkenes are produced from phosphonium salts and carbonylic compounds. The double bond is selectively formed on the carbon with a carbonylic group. Hydrogens in the α position of phosphonium salts are acidic because of the stabilization resonance effect of the corresponding carbanion, when they are treated with a base.

Phosphonium salts are prepared by the reaction of trisubstituted phosphines, usually triphenylphosphine with alkyl halides. Their deprotonation takes place under strong basic conditions, but, in the case of stabilized ylides with adjacent electron-withdrawing groups, it is possible to use weaker bases. It is assumed that the reaction begins with the nucleophilic attack on the carbonylic carbon of ylide to give a polar betaine, followed of triphenylphosphine elimination, to form an alkene. This cleavage involves a cyclic oxaphosphetane, which can be considered to be the result of a [2+2] cycloaddition to the carbonylic compound of the corresponding ylide. Due to the strength of the P=O bond, the final step of the reaction, the extrusion of this compound, is favored thermodynamically to yield the corresponding alkenes not available by other methods such as tensioned alkenes, non-conjugated alkenes, or terminal alkenes.

In this experiment, the ylide is generated in the presence of a considerable excess of the starting aldehyde. The ylide is stabilized by the electron-withdrawing effect of the carboxylic acid in the aromatic ring. The phosphonium salt is prepared from 4-bromomethylbenzoic acid, which is not a volatile substance and therefore is relatively safe to handle but can be somewhat irritating. 4-Bromomethylbenzoic acid is synthesized from methylbenzoic acid by radical reaction with *N*-bromosuccinimide initiated by benzoyl peroxide.

9.8.3. Procedure

A) Preparation of 4-bromomethylbenzoic acid:
   Put 2.7 g of methylbenzoic acid and 3.6 g of *N*-bromosuccinimide in a 100 ml round-bottom flask. Then carefully add the benzoyl peroxide (Luperox R A75), avoiding the adherence of the solid to the joint. Finally, add 25 ml of chlorobenzene, dragging the solids on the flask. Heat the mixture under reflux with stirring for 1 h. Cool the reaction to r.t., remove the condenser, and put the flask into an ice bath for 10 min to produce a white solid. Filter the precipitate under vacuum and wash it with petroleum ether (ligroin) (3 × 10 ml). Transfer the solid to a beaker with a stir bar and add 50 ml of water. Stir the mixture for 10 min to dissolve the succinimide formed. Filter the solid again under vacuum and wash it successively with water (2 × 10 ml) and petroleum ether (2 × 10 ml). Dry the solid under an air stream in the filter funnel. Once dry, the solid is used in the next step (see ref. [18]).

Noticeboard 9.8.1

 DANGER!

Benzoyl peroxide may explode if subjected to heat, impact, blows, shock, friction, or static discharge. Handle with care!

B) Preparation of 4-carboxybenzyltriphenoxophosphonium bromide:
   In a 100 ml round-bottom flask, dissolve 4-bromomethylbenzoic acid (4.3 g, 20 mmol) and triphenylphosphine (5.2 g, 20 mmol) in 60 ml of acetone. Add a stir bar, fit a reflux condenser, and reflux the mixture for 45 min with magnetic stirring. Once the reaction is finished, remove the condenser and cool the reaction to r.t. If necessary put the flask into an ice bath to separate a precipitate. Filter the solid under vacuum to give the corresponding phosphonium salt. Wash the solid with diethyl ether and dry it under an air stream. Weigh the solid and use it for the next step (see ref. [18]).

C) Preparation of 4-vinylbenzoic acid:
   Put 4-carboxybenzyltriphenoxophosphonium bromide (3.76 g, 8 mmol), aqueous formaldehyde (32 ml), and water (15 ml) in a 250 ml round-bottom flask

with a stir bar and adjust an appropriate clamp to the flask. Stir the mixture vigorously, and add an NaOH solution (2.5 g in 15 ml of water). When the base addition is finished, continue stirring for 45 min. Tirphenylphosphine is formed during the reaction. Separate it from the crude by vacuum filtration and wash it with water (disposable solid). Using HCl (conc.), treat the filtrate and the water used to wash the triphenylphosphine to produce a white solid corresponding to 4-vinylbenzoic acid. Recrystallize from aqueous EtOH (m.p. = 142–144 °C; see ref. [18]).

Table 9.8: Physico-chemical properties of the reagents used.

Compound	M_w	M.p. (°C)	B.p. (°C)	Density (g·ml^{-1})	Danger[a] (GHS)
Acetone	58.08	−94	56	0.791	🔥 !
Diethyl ether	74.12	−116	34.6	0.71	🔥 !
Petroleum ether	-	<-30	30–60	0.640	🔥 ☠ ! 🌿
4-Bromo-methylbenzoic acid	215.04	224–229	-	1.69	!
4-Methylbenzoic acid	136.15	177–180	274–275	-	Non-hazardous
Benzoyl peroxide	242.23	105	-	-	🔥 !
NBS	177.98	175–180	-	-	☠ !
Triphenylphosphine	262.29	79–81	377	-	☠ !
4-Vinylbenzoic acid	148.16	142–144	-	-	!
Chlorobenzene	112.56	−45	132	1.106	🔥 ! 🌿

[a] For brevity, only GHS icons are indicated. The information offered in the Material Safety Data Sheet (MSDS) should be consulted.

9.9. Grignard reagents: synthesis of triphenyl-methanol (triphenylcarbinol)

Estimated time	Difficulty	Basic lab operations
6 h		✔ Vacuum filtration (see p. 103). ✔ Liquid-liquid extraction (see p. 111). ✔ Liquid-liquid extraction with reaction (see p. 111). ✔ Washing in a separatory funnel (see p. 111). ✔ Rotary evaporator (see p. 117). ✔ Reflux (see p. 89). ✔ Inert atmosphere (see p. 130).

9.9.1. Goal

To familiarize the student with the preparation of Grignard reagents and their reaction with carbonyl compounds to form alcohols.

9.9.2. Background

One of the most popular methods to form C−C bonds in Organic Synthesis is by using Grignard reagents. The formation of these involves the reaction of an alkyl, vinyl, or aryl halide with magnesium and involves a change in the electronic nature of the carbon atom, which is changed from electrophile in the halide to strongly nucleophile in organomagnesium compounds. Grignard compounds are highly reactive, reacting with water, oxygen, CO_2, etc. Therefore, these compounds must be prepared under anhydrous conditions using an inert atmosphere. They have a strongly nucleophilic (and basic) character and produce additions to carbonyl groups, with the formation of a new C−C bond, yielding alcohols whose nature depends on the type of starting carbonyl compound used. In this experiment, the reactive chosen is an ester (methyl benzoate), so that the product formed is a tertiary alcohol, in this particular case the triphenylcarbinol.

9.9.3. Procedure

A) Preparation of phenylmagnesium bromide:

Place 2 g of magnesium turnings and 15 ml of anhydrous ether in a dry round-bottom flask. The necessary elements are coupled for reflux under anhydrous conditions and with the addition of reagents (see Figure 4.14 in p. 101). Prepare a solution of 9 g of bromobenzene in 10 ml of anhydrous ether, and place in the addition funnel. An inert atmosphere is then established by passing argon through the system for a few minutes and placing a balloon filled with argon over a septum. Add approximately 2 ml of the bromobenzene solution on the magnesium, and vigorously stir the reaction. After a few minutes, bubbles and turbidity will appear (heating also occurs).[6] Then continue adding the bromobenzene dropwise for 30 min. Once the addition is complete, wash the funnel with a few milliliters of ether. Maintain the reflux for approximately 15–20 min more, heating gently if necessary. A whitish reaction crude forms, which should not contain magnesium. Allow to cool to r.t. and use without isolation for the next step (see ref. [15]).

B) Synthesis of triphenylcarbinol:

The same assembly is used as in the previous section. Prepare a solution of 5 g of methylbenzoate in 15 ml of anhydrous ether, and transfer this solution to the addition funnel. Start adding this solution dropwise to the organomagnesium compounds. When the addition is finished, reflux another 30 min. Afterward, pour the reaction crude into a beaker containing 50 ml of H_2SO_4 (10%) and approximately 25 g of ice. Drag the remains of the flask with small portions of a cold solution of H_2SO_4. Stir the mixture, and when all the ice is melted, transfer to a funnel and decant (add a small amount of ether if there is a significant decrease in volume). Re-extract the aqueous layer with ether (15 ml). Combine the ether extracts and dry on anhydrous sodium sulfate; then filter and decant to a flask with 25 ml of hexane. The mixture is transferred to a rotary evaporator to yield a solid. Filter, wash with hexane, dry, and weigh the solid (see ref. [15]).

---

[6]If after a few minutes the reaction does not begin, add few crystals of iodine or a few drops of CCl_4.

Table 9.9: Physico-chemical properties of the reagents used.

Compound	M_w	M.p. (°C)	B.p. (°C)	Density (g·ml^{-1})	Danger[a] (GHS)
H$_2$SO$_4$ 96–98%	98.08	3	-	1.80–1.84	
Diethyl ether	74.12	−116	34.6	0.71	
Iodine I$_2$	253.81	113	184	4.930	
Magnesium (Mg)	24.31	648	1,090	1.740	
CCl$_4$	153.82	−23	76–77	1.594	
Methyl benzoate	136.15	−12	198–199	1.094	
Na$_2$SO$_4$	142.04	884	-	2.630	Non-hazardous
Hexane	86.18	−95	69	0.659	
Bromobenzene	157.01	−31	156	-	
Triphenylcarbinol	260.33	160–163	360	-	
Phenylmagnesium bromide	181.31	-	-	-	

[a] For brevity, only GHS icons are indicated. The information offered in the Material Safety Data Sheet (MSDS) should be consulted.

9.10. Organometallic compounds: synthesis of acetylferrocene

Estimated time	Difficulty	Basic lab operations
3 h	⚗⚗	✔ Vacuum filtration (see p. 103). ✔ Liquid-liquid extraction (see p. 111). ✔ Rotary evaporator (see p. 117). ✔ Chromatography (see p. 135).

9.10.1. Goal

To prepare acetylferrocene by electrophilic substitution on one of the cyclopentadiene rings of ferrocene (bis(η^5-cyclopentadienyl) iron(II)).

Ferrocene   Acetylferrocene

9.10.2. Background

The serendipitous discovery, in 1951, of ferrocene opened an entire, new area of research and contributed greatly to the technological advance of Chemistry. Among the uses and applications of ferrocene, the following bear noting:

- Aviation fuel catalyst, improving the rate of burning of up to 4-fold and lowering the temperature of the exhaust pipes.

- Additive to gasoline, for antiknock, instead of lead compounds.

- Additive to fuel oil to eliminate fumes and waste.

- Catalyst in ammonia synthesis under mild conditions.

- Additive in the manufacture of polymers as a protective agent and stabilizer against heat or UV radiation.

- Component of photosensitive materials, replacing the silver in films, photocopying, and printing with high resolution.

- Fertilizer production.

Ferrocene as any aromatic substance can undergo electrophilic aromatic substitution (S_EAr) reactions.

9.10.3. Procedure

In a test tube, mix 0.4 g of ferrocene, 2 ml of acetic anhydride, and 0.5 ml phosphoric acid. Heat the mixture in a water bath at 60 °C for 5 min.

Noticeboard 9.10.1

 DANGER!

Acetic anhydride reacts violently with water, and the mixture can splash.

Add a quantity of crushed ice equal to the volume of the reaction, stirring until the ice melts; remove the suspension adding hexane (2 × 5 ml); stir the mixture using a Pasteur pipette; and separate the organic layer. Collect the organic layer in another test tube and dry by adding anhydrous sodium carbonate.

At this point, make a TLC using as eluent hexane/diethyl ether (1:1). An orange stain will appear with an observed R_f value of approximately 0.3, generated by the reaction product, and probably another yellow one with R_f value of approximately 0.8, corresponding to unreacted ferrocene. Purify the reaction crude by CC using 5 g of silica gel. For this, prepare 75 ml of eluent hexane/diethyl ether (4:1) and a rack with 15 test tubes.

Prepare the sample to be purified by adding 0.5 g of silica gel to the reaction crude and remove the solvent under reduced pressure (rotary evaporator). This will produce an orange solid, which is to be placed into the CC column. Collect the eluent in fractions of approximately 5 ml in each test tube.

Check the nature and purity of the fractions collected, using TLC under the same conditions described above. Collect the tubes containing the acetylferrocene in a tared round-bottom flask, and remove the solvent under reduced pressure (rotary evaporator) to give a solid of m.p. = 85–86 °C; see ref. [19].

Table 9.10: Physico-chemical properties of the reagents used.

Compound	M_w	M.p. (°C)	B.p. (°C)	Density (g·ml^{-1})	Danger[a] (GHS)
Acetic anhydride	102.09	−73.1	139.8	1.080	🔥 ⚗ ❗
Phosphoric acid	98.00	40	158	1.685	⚗
Hexane	86.18	−95	69	0.659	🔥 ☣ ❗ 🌿
Na$_2$CO$_3$	105.99	851	-	2.532	❗
Diethyl ether	74.12	−116	34.6	0.71	🔥 ❗
Silica gel (SiO$_2$)	60.09	-	-	-	Non-hazardous
Ferrocene	186.03	172–174	249	-	🔥 ❗
Acetylferrocene	228.07	81–83	-	-	☠

[a] For brevity, only GHS icons are indicated. The information offered in the Material Safety Data Sheet (MSDS) should be consulted.

9.11. Chemoselectivity: cinnamaldehyde reduction with LiAlH$_4$

Estimated time	Difficulty	Basic lab operations
4 h		✔ Gravity filtration (see p. 100). ✔ Liquid-liquid extraction (see p. 111). ✔ Rotary evaporator (see p. 117). ✔ Vacuum distillation (see p. 125). ✔ Reflux (see p. 89).

9.11.1. Goal

To transform of cinnamaldehyde (2E)-3-phenylprop-2-enal) into 3-phenyl-2-propen-1-ol and 3-phenyl-propan-1-ol, using a single reagent, the LiAlH$_4$, depending on the relative quantities and the addition order, in an inert atmosphere.

Cinnamaldehyde

3-Phenylpropan-1-ol

(E)-3-Phenylprop-2-en-1-ol

9.11.2. Background

The functional group interconversion (FGI) reactions are among the most common synthetic processes. In this experiment, the conversion of cinnamic aldehyde is performed by reduction with LiALH$_4$ into two different products, depending upon the order of addition of reactants.

In a first experiment, the reaction is carried out by adding a solution of the aldehyde in anhydrous THF, over a suspension of LiALH$_4$ in THF, also anhydrous (direct addition). Under these conditions, the simultaneous reduction of the carbonyl group and the double bond occurs because the reducing agent is in excess in the medium in which the reaction occurs.

In a second experiment, the addition of a suspension of LiALH$_4$ in dry diethyl ether to a solution of the aldehyde in the same solvent (inverse addition) is performed. With this protocol, only the reduction of the aldehyde group occurs, leaving the double bond unchanged.

Because of the reactivity of $LiALH_4$, the reaction should be performed under anhydrous conditions in an inert atmosphere. For the correct performance of this experiment, the glassware should be dry, the solvents should have undergone a previous treatment to remove traces of moisture, and the whole experiment should be conducted in an inert atmosphere with argon or nitrogen. On the other hand, when the experiment is finished, but before discarding waste, the remaining $LiALH_4$ reactive must be handled properly to avoid accidents.

9.11.3. Procedure

A) Cinnamaldehyde reduction to alcohol:
In this process, $LiAlH_4$ in excess is added to a solution of cinnamaldehyde, and both functional groups (double bond and carbonyl group) are reduced. In a dry vial, weigh 2.9 g of $LiAlH_4$ and transfer it, using approximately 20 ml of THF, to a dry three-neck flask and set up an addition funnel and a reflux condenser with a bubbler. Couple a gas inlet to the third neck of the flask, and inlet a gently stream of inert gas (nitrogen or argon). Gentle reflux and add to the addition funnel a solution of 5 g of cinnamaldehyde in 50 ml of THF, and add dropwise to the flask while the reflux is maintained. When the reaction is complete (about 15 min),[7] cool the flask in an ice bath and carefully add 12 ml of a saturated sodium sulfate to destroy the excess hydride, and after stirring a few minutes add another solution of 10% H_2SO_4 (95 ml). Transfer to a separatory funnel, separate the two layers, and extract the aqueous layer with ether (4 × 30 ml). Collect the ether extracts, dry over anhydrous sodium sulfate, and remove the solvent on a rotary evaporator. Vacuum distill the resulting oil: hydrocinnamic alcohol (b.p. = 120–121 °C/13 mm Hg; see ref. [8]).

B) Cinnamaldehyde reduction to cinnamic alcohol:
In this procedure, an equimolar amount of $LiAlH_4$ is used only to reduce the carbonyl group to alcohol. This procedure is similar to the previous assembly. Add to the flask a solution of 10 g of cinnamaldehyde and 25 ml of anhydrous ether. Weigh 0.72 g of $LiAlH_4$, and add to addition funnel suspended in 50 ml of anhydrous ether. Place in an ice-water bath, and start adding the suspension $LiAlH_4$ during an interval of 30 min, ensuring that the temperature does not exceed 10 °C.[8] Check the reaction progress by TLC. When the reaction is complete (disappearance of cinnamaldehyde in the TLC plate), add 3 ml of water (to destroy possible reagent excess), and then add 25 ml of H_2SO_4 10%. Transfer to a separatory funnel, decant, and extract twice with 50 ml of ether. Collect the ether extracts, dry over anhydrous sodium sulfate, and remove the solvent in the rotary evaporator. Vac-

---

[7]Follow the reaction by TLC, using CH_2Cl_2 as the eluent.
[8]The reflux condenser can be replace with an adapter with a thermometer.

uum distill the oil produced, cinnamic alcohol (b.p. = 139 °C/14 mm Hg). Can be recrystallized by distilling (m.p. = 33 °C; see ref. [8]).

Table 9.11: Physico-chemical properties of the reagents used.

Compound	M_w	M.p. (°C)	B.p. (°C)	Density (g·ml^{-1})	Danger[a] (GHS)
Cinnamaldehyde	132.16	−7.5	248	1.050	❗
Na$_2$SO$_4$	142.04	884	-	2.630	Non-hazardous
LiAlH$_4$	37.95	125	-	0.920	🔥
THF	72.11	−108.0	65–67	0.89	🔥 ❗
Nitrogen	28.01	−210	−196	0.970	
Argon	39.95	−189.2	−185.7	-	Non-hazardous
H$_2$SO$_4$ 96–98%	98.08	3	-	1.80–1.84	
Diethyl ether	74.12	−116	34.6	0.71	🔥 ❗
Cinnamyl alcohol	134.18	30–33	250		❗
3-Phenyl-propan-1-ol	136.19	−18	119–121	1.001	❗

[a] For brevity, only GHS icons are indicated. The information offered in the Material Safety Data Sheet (MSDS) should be consulted.

9.12. Electrophilic aromatic substitution (S_EAr): 1,4-di-*tert*-butylbenzene

Estimated time	Difficulty	Basic lab operations
3 h		✔ Gravity filtration (see p. 100). ✔ Vacuum filtration (see p. 103). ✔ Liquid-liquid extraction (see p. 111). ✔ Liquid-liquid extraction with reaction (see p. 111). ✔ Recrystallization from organic solvent (see p. 105). ✔ Rotary evaporator (see p. 117). ✔ Simple distillation (see p. 120).

9.12.1. Goal

To perform an S_EAr, particularly a double Friedel-Craft alkylation on benzene.

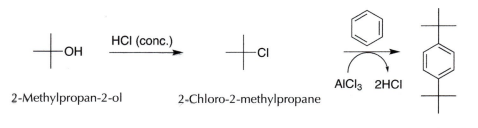

9.12.2. Background

The Friedel-Crafts alkylation of aromatic rings often uses alkyl halides and a Lewis acid compound as a catalyst. However, this reaction has some limitations since, first, the primary alkyl halides undergo rearrangement processes under the reaction conditions, and, second, they often generate polyalkylation products such as the benzene alkyl derivatives, which are more reactive than benzene itself.

In this experiment, the alkylating agent (*tert*-butyl chloride) is prepared from *tert*-butanol and HCl (conc.). In the second step, the product is used in a Friedel-Crafts reaction.

9.12.3. Procedure

A) Synthesis of *tert*-butyl chloride by reaction S_N1:

Fix a 250 ml separatory funnel to the corresponding ring stand inside a fume hood. Measure 25 ml of *tert*-butyl alcohol and pour into the funnel. Very slowly, add 100 ml of HCl 12 N and slowly mix the two layers. Then stir vigorously for 2 or 3 min continuously, sporadically opening the key funnel to release pressure. Then stir the funnel again, this time occasionally, for 10 min. Let the mixture stand until the two layers clearly separate.

Noticeboard 9.12.1

 DANGER!

Always work in a fume hood. Given the tear-inducing and irritant nature of *tert*-butyl chloride, measure the proper amount in the fume hood.

Next, discard the lower phase and wash the upper (organic) with a solution of K_2CO_3 5% (2×10 ml), making sure the funnel is open by the usual procedure, while stirring, to eliminate the overpressure. Then, transfer the organic phase to an Erlenmeyer, and proceed with the drying of the solution on anhydrous Na_2SO_4. Remove the desiccant by filtration. Transfer the solution to a dry 50 ml round-bottom flask that fits the components needed to perform a simple distillation; use a water bath as a heat source. Collect the two fractions: first, the material distilled to 48 °C and, second, the material distilled between 48 and 54 °C (previously tare the flask in which it is collected). Dispose of both the first fraction and the residue remaining in the flask. Despite the second fraction, calculate the yield, and use the compound to proceed to the next step.

B) Preparation of 1,4-di-*tert*-butylbenzene:

In a three-neck flask equipped with a thermometer, place 20 ml of *tert*-butyl chloride and 10 ml of benzene. Weigh on a rough paper 1 g of $AlCl_3$, and store it in a test tube with a septum. Connect the flask to a washing flask with sodium bicarbonate solution to trap the HCl that emerges with the formation of CO_2. Place the flask in an ice-water bath to cool it to a temperature of between 0 and 3 °C. Add approximately a quarter of the $AlCl_3$ into the necks of the flask. Remove the thermometer, cover, and shake the flask vigorously while in the ice bath. After a period of about 2 min, a strong reaction occurs, releasing HCl. Add the rest of the catalyst three times at 2 min intervals. When the reaction ends, a white solid will appear and the bubbling will cease. When this occurs, disconnect the wash bottle, preventing the return of the dissolution, and remove the flask from the ice-water bath. Let the reaction crude stand at r.t. for 5 min and add ice water to the reaction mixture so that when the ice melts the crude can be easily extracted with diethyl ether. Then transfer the contents to a separatory funnel and extract with diethyl ether. Combine the organic extracts (upper layer) and dry over sodium sulfate. Remove the solvent under reduced pressure (rotary evaporator). The resulting syrup may crystallize when cooled. Weigh the product, and calculate the yield. Recrystallize from MeOH (20 ml of MeOH for 15 g of product), and cool in an ice bath until the solid completely separates.

Table 9.12: Physico-chemical properties of the reagents used.

Compound	M_w	M.p. (°C)	B.p. (°C)	Density (g·ml^{-1})	Danger[a] (GHS)
tert-Butyl chloride	92.57	−25	51–52	-	
Benzene	78.11	5.5	80	0.874	
AlCl$_3$	133.34	100	120	2.440	
Diethyl ether	74.12	−116	34.6	0.71	
Na$_2$SO$_4$	142.04	884	-	2.630	Non-hazardous
MeOH	32.04	−98	64.7	0.791	
tert-Butyl alcohol	74.12	23–26	83	0.786	
1,4-Di-*tert*-butylbenzene	190.32	76–78	236	0.985	
HCl	36.46	−30	>100	1.200	
K$_2$CO$_3$	138.21	891	-	-	
NaHCO$_3$	84.01	300	-	2.160	Non-hazardous
CO$_2$	44.01	−78.5	-	-	

[a] For brevity, only GHS icons are indicated. The information offered in the Material Safety Data Sheet (MSDS) should be consulted.

9.13. Hofmann rearrangement: synthesis of 2-aminobenzoic acid (anthranilic acid)

Estimated time	Difficulty	Basic lab operations
2 h	⚗️⚗️	✔ Gravity filtration (see p. 100). ✔ Vacuum filtration (see p. 103). ✔ Recrystallization from water (see p. 105).

9.13.1. Goal

To produce anthranilic acid from phthalimide by the Hofmann rearrangement promoted by bromine in a basic medium.

Phthalimide → 1) NaOH/Br$_2$ → 2) Neutralization → Anthranilic acid

9.13.2. Background

The Hofmann rearrangement of an amide to form an amine with carbon dioxide loss is an example of reaction where alkyl or aryl groups migrate to electron-deficient nitrogen atoms. Other examples of very similar reactions include the Curtius, Lossen, and Schmidt rearrangements, in which a carboxylic acid or acid derivative is converted into an isocyanate group, yielding the amine. Also, a related process is the Beckmann rearrangement reaction, in which the oximes are converted to amides.

In the Hofmann rearrangement, an amide is subjected to an oxidation process with hypobromite to form an *N*-bromoamide intermediate, which in the presence of a base undergoes a deprotonation step followed by the migration of an alkyl group to the nitrogen atom, and simultaneous loss of bromine, whereby an isocyanate is generated.

Under the reaction conditions, hydrolysis of the isocyanate group occurs with the loss of CO_2 and formation of the corresponding amine, which has one carbon atom less than the starting amide. This reaction is especially useful for the preparation of aromatic amines.

In the experiment, the goal of this practice is to apply the Hofmann reaction to the synthesis of 2-aminobenzoic acid (anthranilic acid), which is used as a synthetic intermediate in the preparation of dyes or saccharin synthesis. Its esters are used in the preparation of perfumes, medications, or corrosion inhibitors in metals.

9.13.3. Procedure

Dissolve 8 g of NaOH in 30 ml of deionized water in a 100 ml Erlenmeyer containing a stir bar. Cool the solution in an ice bath and add at once 6.5 g of bromine (Br_2). Shake the mixture vigorously until the brown color disappears, which indicates that all the bromine has reacted. Then, while maintaining vigorous stirring, add 5.9 g of finely divided phthalimide and then add a solution of 5.5 g of NaOH in 20 ml water. Remove the ice bath and allow the temperature to rise spontaneously to approximately 70 °C, and stir for another 10 min. If there is turbidity, filter by gravity. Cool the reaction crude again in an ice bath and add dropwise HCl (conc.) using a dropper to neutrality (it will take about 15 ml), check the pH using indicator paper. If there is a slight excess of acid, adjust the pH by adding a base.

Transfer the reaction mixture to a 250 ml Erlenmeyer and add 5 ml of glacial acetic acid. Filter the precipitate by vacuum filtration, and wash with 10 ml of water. Recrystallize the 2-aminobenzoic acid (anthranilic acid) from water and a yellowish solid will form (m.p. = 146–148 °C; see ref. [20]).

Table 9.13: Physico-chemical properties of the reagents used.

Compound	M_w	M.p. (°C)	B.p. (°C)	Density (g·ml^{-1})	Danger[a] (GHS)
Phthalimide	147.13	232–235	310	1.210	Non-hazardous
NaOH	40.00	318	1,390	2.130	🧪
HCl	36.46	−30	>100	1.200	🧪 !
Br_2	159.81	7.2	58.8	-	☠️🧪🌿
Anthranilic acid	137.14	144–148	-	-	!
Glacial acetic acid	60.05	16.2	118	1.049	🧪🔥
Active carbon	12.01	3,550	-	0.25–0.60	Non-hazardous

[a] For brevity, only GHS icons are indicated. The information offered in the Material Safety Data Sheet (MSDS) should be consulted.

9.14. Radical coupling: synthesis of pinacol and pinacolone

Estimated time	Difficulty	Basic lab operations
2.5 h		✔ Vacuum filtration (see p. 103). ✔ Liquid-liquid extraction (see p. 111). ✔ Simple distillation (see p. 120).

9.14.1. Goal

To synthesize α-glycol (pinacol) from ketone by radical condensation and to dehydrate α-glycol to form pinacolone.

9.14.2. Background

Glycols can be synthesized by numerous methods, and this product can be used as starting olefin or unsaturated carbonyl compounds. For example, from olefins the synthesis can be carried out by adding halogen, by adding hypohalogenous acid (XOH), by epoxidation with peracid and subsequent hydrolysis of the epoxide formed, or by adding OsO_4 and/or MnO_4^-. Pinacol synthesis is induced by a reductive coupling of acetone; this ketone is treated with an active metal (Na or Mg), resulting in a transfer of an electron to the carbonyl producing a cetyl radical, which dimerizes, leading to C−C bond formation and yielding the diol. With aromatic ketones the synthesis can also be carried out photochemically. The pinacolinic transposition is formally a dehydration between two molecules ketone catalyzed by an acid medium.

9.14.3. Procedure

A) Preparation of pinacol (2,3-dimethyl-butane-2,3-diol):
   In a dry 100 ml round-bottom flask, place 3 g of magnesium and 0.5 g mercuric chloride. Add 10 ml of anhydrous acetone and stir until the reaction begins. Adapt the flask with a reflux condenser and gradually add another 40 ml of acetone. Stir for 15 to 20 min, and add another 25 ml of acetone. After 30 to 45 min, heat the mixture in a water bath to maintain a good reflux rate. After 1 h of reflux, end the reaction. Remove the flask and connect the condenser via an adapter to a vacuum line, while continuing to heat the flask, thereby removing the excess acetone (rotary evaporator can alternatively be used), and magnesium pinacolate is as a powder. Reconnect the reflux condenser and add 100 ml of water containing 5 g of trisodium phosphate through the condenser (mildly heating the reaction). Reflux the flask for 15 min. Vacuum filter the hot mixture and cool the filtrate in an ice bath with stirring. A precipitate will appear. When the crystallization is complete, vacuum filter and wash the resulting crystals with 25 ml of ice water or with 25 ml of petroleum ether (see ref. [7]).

Noticeboard 9.14.1

 DANGER!

The mercuric salts are toxic and therefore handle with caution, wearing gloves.

B) Preparation of pinacolone (2,3-dimethyl-butan-2-one):
   In a 100 ml Erlenmeyer flask, place 15 ml of water and add slowly and carefully 10 ml of H_2SO_4 (conc.). Then add 6 g pinacol previously prepared and dissolved, transfer to a flask, and distill. Allow the distillate to stand and then dry over anhydrous sodium sulfate and gravity filter (see ref. [7]).

Table 9.14: Physico-chemical properties of the reagents used.

Compound	M_w	M.p. (°C)	B.p. (°C)	Density (g·ml^{-1})	Danger[a] (GHS)
$HgCl_2$	271.50	277	302	5.440	☠☣⌇⌇
Magnesium (Mg)	24.31	648	1,090	1.740	🔥
Pinacol	118.17	41	174	0.967	🔥 !
Pinacol hexahydrate	226.27	-	-	-	See MSDS
Pinacolone	100.16	-	103–107	0.850	🔥 !
p-Xylene	106.17	13.0	138.4	0.860	🔥 !
Acetone	58.08	−94	56	0.791	🔥 !
H_2SO_4 96–98%	98.08	3	-	1.80–1.84	⌇
Na_3PO_4	163.94	75	-	1.620	⌇ !
Petroleum ether	-	<-30	30–60	0.640	🔥☣ ! ⌇

[a] For brevity, only GHS icons are indicated. The information offered in the Material Safety Data Sheet (MSDS) should be consulted.

9.15. Synthesis of a heterocyclic drug: *n*-butyl-barbituric acid

Estimated time	Difficulty	Basic lab operations
4 h	⚗⚗	✔ Vacuum filtration (see p. 103). ✔ Liquid-liquid extraction (see p. 111). ✔ Rotary evaporator (see p. 117). ✔ Vacuum distillation (see p. 125). ✔ Reflux (see p. 89).

9.15.1. Goal

To perform a two-step synthesis for the preparation of a derivative of barbituric acid. In the first step, the diethyl *n*-butylmalonate is prepared, which is subsequently reacted with urea in the presence of a strong base, such as NaOEt, to yield the desired compound.

1-Bromobutane   Diethyl 2-butylmalonate   *n*-Buthylbarbituric acid

9.15.2. Background

Barbiturates are drugs that are prescribed to treat severe nervous insomnia, some forms of epilepsy, and certain convulsive and psychological disorders. Structurally, they are derived from barbituric acid. They are legal substances of controlled prescription with drug action and addictive effects, and thus prescription is required for their sale. Their use leads to tolerance, and suppression causes the stop-eating syndrome. Ever since von Baeyer synthesized barbituric acid in 1863, which has no sedative effects, over 2,500 derivatives of the substance have been investigated. For a long period of time, barbiturates and opiates were the only substances available to relieve anxiety or agitation of some psychiatric patients.

9.15.3. Procedure

A) Preparation of diethyl *n*-butylmalonate:
    In a 100 ml flask place absolute EtOH (50 ml) and sodium (1.2 g). Cut sodium into small pieces or, if possible, use sodium wire. Adapt a reflux

condenser and fit a drying tube. Stir, at the beginning, at r.t. (the reaction generates heat due to a strongly exothermic character), and if necessary gently heat until the sodium disappears. Cool to r.t. and then add 8 g of diethyl *n*-malonate. Stir for about 15 min, and add 7.5 g of 1-bromobutane. Put the mixture to reflux for 1.5 h. After the reflux time, transfer the reaction crude to a separatory funnel, sliding the remaining crude with 30 ml of water. Stir the mixture vigorously, decanting and separating the aqueous layer. Extract the aqueous layer with diethyl ether (2 × 10 ml). Combine the organic extracts in an Erlenmeyer, and dry over anhydrous sodium sulfate. The desiccant is removed by gravity filtration. Next, remove the solvent under reduced pressure (rotary evaporator). Distill the residue under vacuum. Take into account the difference in b.p. of diethyl *n*-malonate (199.3 °C) and of ethyl *n*-butylmalonate (234–240 °C) at atmospheric pressure (see ref. [3], p. 682).

Noticeboard 9.15.1

 DANGER!

Sodium reacts violently with water; therefore, all materials should be dry. At the beginning of the reaction, material projections and breaks in the flask may occur. Ensure that the ensemble (flask and reflux condenser) is firmly fixed with clamp and clamp holder.

B) Preparation of *n*-butylbarbituric acid:
Pour 50 ml of absolute EtOH into a dry 250 ml round-bottom flask. Then add 0.46 g of sodium and adjust a reflux condenser. Maintain the mixture at r.t. until the complete dissolution of sodium. Add the solution of NaOEt formed at 4.32 g of ethyl *n*-butylmalonate dry, followed by a solution of 1.2 g of urea predried in 22 ml of absolute EtOH. Reflux the mixture for 2 h. Then stop the reflux, allow the crude to cool to r.t., and gradually acidify the reaction mixture with 20 ml of 10% HCl (check the acidity with indicator paper). After neutralization of EtONa, remove EtOH under reduced pressure; a precipitate will appear. Complete the crystallization by placing the flask in an ice bath. Collect crystals by vacuum filtering, and wash with hexane to remove the unreacted diethyl *n*-butylmalonate (detected by smell). Recrystallize the solid from water using a proportion of approximately 20 ml per gram of product. Weigh the product dry and calculate the yield (see ref. [3], p. 1177).

Table 9.15: Physico-chemical properties of the reagents used.

Compound	M_w	M.p. (°C)	B.p. (°C)	Density (g·ml^{-1})	Danger[a] (GHS)
Diethyl malonate	160.17	−50	199	1.055	!
1-Bromobutane	137.02	−112	100–104	1.276	🔥 ! 🌿
Hexane	86.18	−95	69	0.659	🔥 ☣ ! 🌿
Diethyl ether	74.12	−116	34.6	0.71	🔥 !
Na$_2$SO$_4$	142.04	884	-	2.630	Non-hazardous
EtOH	46.07	−114.1	78.5	0.790	🔥
Na	22.99	97.8	883	0.968	🔥 🜄
Diethyl n-butylmalonate	216.27	-	235–240	0.975	🜄 🔥
Urea	60.06	132–135	-	1.335	!
HCl	36.46	−30	>100	1.200	🜄 !
Barbituric acid	128.09	248–252	-	-	Non-hazardous

[a] For brevity, only GHS icons are indicated. The information offered in the Material Safety Data Sheet (MSDS) should be consulted.

9.16. Transformations of hydroxy ketone: synthesis and reactions of benzoin

Estimated time	Difficulty	Basic lab operations
48 h		✔ Gravity filtration (see p. 100). ✔ Vacuum filtration (see p. 103). ✔ Recrystallization from water (see p. 105). ✔ Recrystallization from organic solvent (see p. 105). ✔ Reflux (see p. 89).

9.16.1. Goal

To form the C−C bond via condensation reaction of benzaldehyde catalyzed by cyanide ion or thiamine, and to perform the stereoselective reduction of benzoin.

9.16.2. Background

One of the most important processes in Organic Synthesis is the formation of carbon-carbon bonds. Usually, this reaction requires the formation of a carbanion (carbon nucleophile) and its reaction with an electrophilic carbon (electron deficient).

The benzoin condensation involves the reaction of two molecules of benzaldehyde (or another non-enolizable aldehyde). As in this case, the carbonyl group has a partial positive charge needed to invert the polarity in one of these. To achieve this, normally, a catalytic amount of cyanide ion is often used, resulting in the formation of a carbanion, which can attack another carbonyl group, yielding the $C-C$ bond. Subsequent HCN loss occurs with the formation of benzoin.

Recently, it has been shown that thiamine (its hydrochloride, vitamin B_1) is a very efficient catalyst for producing the benzoin condensation. Benzoin allows an easy synthesis of glycol by reduction of the carbonyl group. This reduction could give a mixture of two diols, with *treo* and *erythro* configurations. However, the reduction with $NaBH_4$ produces mainly the isomer *erythro* (form *meso*), which can be determined by comparing the m.p. of the product compared to the pure isomers: *eritro* (m.p. 137 °C) and *treo* (m.p. 119 °C).

9.16.3. Procedure

A) Synthesis of benzoin:

Dissolve in a flask 0.7 g of thiamine hydrochloride (vitamin B_1) in 1.5 ml of water. Then add 6–7 ml of EtOH 95%, and cool the solution in an ice bath. To this solution under stirring, add slowly (for 7 to 10 min) 1.5 ml of a cold solution of NaOH 3 M. Add 4.24 g (4 ml) of benzaldehyde, shake well, and check the pH. If the pH is less than 8.0, add slightly more of NaOH, to a pH value in the 8.0–9.0 range.[9] Heat the solution in a water bath at 60–70 °C for 1 h. Remove the bath and allow the reaction to reach r.t. Cool in an ice bath whereupon a precipitate should appear. Filter the reaction crude, and wash with 10 ml of cold water. An oil may also appear, this being a mixture of benzaldehyde and the reaction byproducts. Recrystallize the product from EtOH/water (m.p. 137 °C; see ref. [6]).

B) Reduction of benzoin:

In a 100 ml Erlenmeyer, add 2 g of benzoin and 20 ml of absolute EtOH. Shake, and while shaking, add in small portions 400 mg of $NaBH_4$. When the addition is finished, stir for another 15 min at r.t. Cool in an ice bath and decompose the $NaBH_4$ excess, adding first 30 ml of water and then adding dropwise 1 ml of HCl 6 M. A foam could form if the acid is not added slowly enough. Add another 10 ml of water and stir for another 15 min. A white precipitate will appear. Filter the reaction product, and wash it with water.

---

[9]Be very careful on this point because if the pH is higher than 9.0, the Cannizzaro reaction occurs and, if the pH is too low, thiamine is not ionized.

Air dry and measure the m.p. If the m.p. range is too broad, recrystallize from acetone/hexane (see ref. [6]).

C) Benzoin oxidation:

In a round-bottom flask provided with a reflux condenser, place 2 g of benzoin, 3.75 g of cupric acetate monohydrate, 15 ml of acetic anhydride, and 5 ml of water. Heat by reflux for 15 min. Filter the hot mixture to remove the cuprous oxide. Cool the filtrate, and filter the precipitate that appears under vacuum, wash with a small amount of cold water, and air dry (m.p. = 95 °C; see ref. [6]).

Table 9.16: Physico-chemical properties of the reagents used.

Compound	M_w	M.p. (°C)	B.p. (°C)	Density (g·ml^{-1})	Danger[a] (GHS)
EtOH	46.07	−114.1	78.5	0.790	 🔥
NaOH	40.00	318	1,390	2.130	🜖
HCl	36.46	−30	>100	1.200	 🜖 ❗
Benzaldehyde	106.12	−26	178–179	1.044	❗
Benzoin	212.24	134–138	194	-	Non-hazardous
NaBH$_4$	37.8	400	-	1.07	 🔥☠🜖
Acetic anhydride	102.09	−73.1	139.8	1.080	 🔥🜖 ❗
Thiamine hydrochloride	337.27	250	-	-	Non-hazardous
Cu$_2$O	143.09	1,230	-	-	 ❗ 🌱
Copper(II) acetate monohydrate	199.65	115	-	1.882	 ❗ 🌱

[a] For brevity, only GHS icons are indicated. The information offered in the Material Safety Data Sheet (MSDS) should be consulted.

9.17. Enamines as reaction intermediates: producing 2-acetylcyclohexanone

Estimated time	Difficulty	Basic lab operations
3 h	🧪🧪	✔ Liquid-liquid extraction (see p. 111). ✔ Rotary evaporator (see p. 117). ✔ Vacuum distillation (see p. 125). ✔ Azeotropic distillation (see p. 127). ✔ Reflux (see p. 89).

9.17.1. Goal

To produce a β-dicarbonyl (2-acetylcyclohexanone) from cyclohexanone compound using an enamine as a synthetic intermediate.

Cyclohexanone   Pyrrolidine   2-Acetylcyclohexanone

9.17.2. Background

Organometallics and carbanions are not the only species with nucleophilic carbon atoms. The carbon in the β position of an enamine has a nucleophilic character and can be alkylated or acylated with an appropriate electrophilic reagent. This reactivity of enamines, which can easily be prepared from carbonyl compounds, is due to the relocation of the non-bonding electron pair of nitrogen through the carbon double bond in the β position. The alkylation or acylation of an enamine leads to the formation of an iminium ion, which by hydrolysis regenerates the carbonyl compound.

In the following experiment, cyclohexanone via pyrrolidinamine is acetylated. In a first step, the cyclohexanone is converted to the corresponding enamine by reaction with pyrrolidine in the presence of an acid as a catalyst, in toluene at the reflux using a Dean-Stark apparatus to remove water formed within the reaction. The enamine produced is not isolated but reacts immediately with acetic anhydride to yield the corresponding acetylation. The water treatment of the starting material hydrolyzes to 2-acetylcyclohexanone, which is purified by vacuum distillation. The compound generated exists as a mixture of keto and enol forms, and the percentage of each tautomer can be estimated by ^{1}H-NMR.

9.17.3. Procedure

Add 40 ml of toluene, 5 ml of cyclohexanone, 4 ml of pyrrolidine, and 0.1 g of
p-toluensulfonic acid to a 100 ml round-bottom flask. Connect the flask to a
Dean-Stark apparatus and this in turn to a reflux condenser with a drying
tube. Reflux the mixture for 1 h and then allow to cool to r.t.

The assembly is modified to perform an azeotropic distillation (to eliminate
pyrrolidine and water that may be left) to a maximum of 108–110 °C. At this
point, remove the heat source and allow the flask to cool to r.t. Remove the
distillation equipment, and add a solution of 4.5 ml of acetic anhydride in 10 ml
of toluene. Put this into the flask stopper, and allow the mixture to stand at
r.t. for at least 24 h.

Afterward, slowly pour 5 ml of water over the crude reaction, and heat the
mixture to reflux for 30 min. The reaction crude is allowed to cool to r.t. and is
transferred to a 50 ml separatory funnel containing 10 ml of water, separating
the two layers.

Wash the organic layer successively with HCl 3 M (3 × 10 ml) and water
(10 ml), transfer into an Erlenmeyer, and dry over anhydrous sodium sulfate.
Remove the solvent with a rotary evaporator under vacuum, transfer the residue
to a vacuum distillation apparatus of appropriate size and transfer the distilled
contents under reduced pressure using a water pump (or a pump that makes
a vacuum of approximately 15 mm Hg). Weigh the product and calculate the
yield.

Table 9.17: Physico-chemical properties of the reagents used.

Compound	M_w	M.p. (°C)	B.p. (°C)	Density (g·ml^{-1})	Danger[a] (GHS)
Cyclohexanone	98.14	−47	155	0.947	🔥 ☣ !
Pyrrolidine	71.12	−60	87–88	-	🔥 ☣ !
p-Toluenesulfonic acid (<5% H_2SO_4)	172.2	106–107	-	1.240	! ☣
Toluene	92.14	−93	110.6	0.867	🔥 ☣ !
Acetic anhydride	102.09	−73.1	139.8	1.080	🔥 ☣ !
HCl	36.46	−30	>100	1.200	☣ !
Na_2SO_4	142.04	884	-	2.630	Non-hazardous
2-Acetylcyclohex-anone	140.18	-	111–112	1.078	Non-hazardous

[a] For brevity, only GHS icons are indicated. The information offered in the Material Safety
Data Sheet (MSDS) should be consulted.

9.18. N-Heterocycle synthesis: producing benzo-triazol

Estimated time	Difficulty	Basic lab operations
2 h	⚗⚗	✔ Gravity filtration (see p. 100). ✔ Vacuum filtration (see p. 103). ✔ Recrystallization from water (see p. 105).

9.18.1. Goal

To produce a simple nitrogen heterocycle, in this case a triazole (five-membered ring with three nitrogens) fused to an aromatic six-membered ring, benzotriazole.

o-Phenylenediamine   Benzotriazol

9.18.2. Background

Benzotriazole is a heterocyclic compound that is used as the starting material in the preparation of some drugs in the pharmaceutical industry. Benzotriazoles such as vorozole and alizapride have inhibitory properties and act as agonists for many proteins.

9.18.3. Procedure

Dissolve 5.4 g of *o*-phenylenediamine in a mixture of 6 g (5.5 ml) of glacial acetic acid and 15 ml of water in a 250 ml Erlenmeyer. Heat gently, if necessary, to completely dissolve the mixture. Cool the solution in a bath of cold water and add all at once a solution of 3.8 g of sodium nitrite in 8 ml of water with manual stirring (avoid magnetic stirring).

When the reaction mixture reaches approximately 85 °C, cool again, and a change in color of the reaction mixture (dark red to light brown) could appear.[10] Continue to stir for 15 min and place the mixture in an ice bath for another 30 min.

---

[10]The sequence of cooling-heating-cooling is critical for good yield. If the initial or final cooling process is inefficient, the yield decreases.

Vacuum filter and wash three times with cold deionized water; and a solid will appear in the Büchner. The solid is benzotriazole. Recrystallize from water (see ref. [21]).

Table 9.18: Physico-chemical properties of the reagents used.

Compound	M_w	M.p. (°C)	B.p. (°C)	Density $(g \cdot ml^{-1})$	Danger[a] (GHS)
o-Phenylenediamine	108.14	100–102	256–258	1.030	
1*H*-Benzotriazole	119.12	94–99	-	-	
Glacial acetic acid	60.05	16.2	118	1.049	
NaNO$_2$	69.00	271	320	2.164	

[a] For brevity, only GHS icons are indicated. The information offered in the Material Safety Data Sheet (MSDS) should be consulted.

9.19. Synthesis of macrocycles: preparation of calix[4]pyrrole

Estimated time	Difficulty	Basic lab operations
2 h	⚗️⚗️	✔ Gravity filtration (see p. 100). ✔ Vacuum filtration (see p. 103). ✔ Recrystallization from organic solvent (see p. 105). ✔ Chromatography (see p. 135).

9.19.1. Goal

To produce a macrocycle, in this case a derivative of calix[4]pyrrole as an example of a sequestering agent anion.

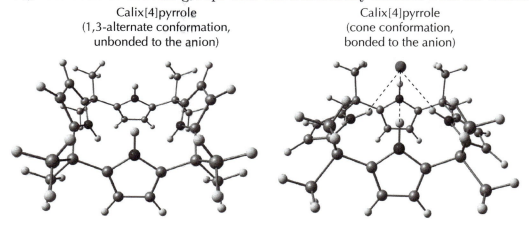

9.19.2. Background

In recent years, agents that selectively bind anions have attracted increasing interest because of their role in biological and ecological processes. Two strategies have been designed to produce such compounds. The first involves the inclusion in the molecule of cationic groups that electrostatically interact with the anion.

Calix[4]pyrrole
(1,3-alternate conformation,
unbonded to the anion)

Calix[4]pyrrole
(cone conformation,
bonded to the anion)

The second strategy is to introduce compounds capable of forming hydrogen bonds with certain anions. Porphyrins and other heterocyclic related macrocycles play key roles in many biological processes. The porphyrinogens, also known as calix[4]pyrroles, are excellent chelating agents[22]. The calix[4]pyrroles resulting from the condensation of pyrrole with aldehydes are compounds capable of easy oxidation. This is not true of the resulting condensation with ketones, which are very stable. This condensation is catalyzed by acid. An example is the *meso*-octamethylporphyrinogenic also known as "acetonepyrrole" described by Baeyer in 1886 [5].

9.19.3. Procedure

In a 10 ml Erlenmeyer provided with a stir bar, mix 5 ml of acetone and 1 ml pyrrole. Cover the flask with a septum and cool the mixture in an ice bath with magnetic stirring. Add dropwise to the reaction crude 0.5 ml of HCl using a syringe fitted with a needle piercing the septum. The solution, initially colorless, turns brown and shows slight effervescence. After some time, a white solid begins to precipitate. Stir for another 5 min and isolate the white solid by vacuum filtration using a Hirsch funnel, and then wash with cold acetone (15–20 ml) (can be dried at 40 °C for 30 min). Recrystallize the product of the reaction using approximately 5 ml of acetone per 100 mg product. If after 30 min the solid has not begun to crystallize, add water dropwise until turbidity appears (estimated yield 70%). Check the purity of the product by TLC using the eluent, hexane/ethyl acetate (7:3) (see ref. [13]).

Table 9.19: Physico-chemical properties of the reagents used.

Compound	M_w	M.p. (°C)	B.p. (°C)	Density (g·ml^{-1})	Danger[a] (GHS)
Acetone	58.08	−94	56	0.791	🔥 ❗
1*H*-Pyrrole	67.09	−23	131	0.967	🔥 ☠ 🜂
meso-Octamethyl-calix[4]pyrrole	428.61	-	-	-	❗
HCl	36.46	−30	>100	1.200	🜂 ❗

[a] For brevity, only GHS icons are indicated. The information offered in the Material Safety Data Sheet (MSDS) should be consulted.

9.20. Preparation of a sports dietary supplement: creatine synthesis

Estimated time	Difficulty	Basic lab operations
7 days	⚗⚗	✔ Gravity filtration (see p. 100). ✔ Vacuum filtration (see p. 103). ✔ Recrystallization from water (see p. 105).

9.20.1. Goal

To synthesize of creatine from sarcosine (*N*-methylglycine).

Cyanamide   Sarcosine   Creatine

9.20.2. Background

Creatine is a nitrogenous organic acid that occurs naturally in vertebrates and helps to supply energy to all cells in the body, primarily muscle by playing a key role in muscle energy metabolism. It is produced in the liver, pancreas, and kidneys and can also be derived from food and dietary supplements. Creatine provides the energy needed for muscle contraction and substantially improves performance in high-intensity exercise because it improves anaerobic capacity and protein synthesis. Therapeutically, it has been used to treat some types of muscular dystrophy, ocular atrophy, and some types of sclerosis. Creatine (a dietary supplement) is demanded by many athletes and is neither regulated by the Food and Drug Administration (FDA) nor prohibited by the International Olympic Committee (IOC). Creatine monohydrate can be synthesized in the laboratory (and commercially), starting from cyanamide and sarcosine (*N*-methylglycine).

9.20.3. Procedure

To an r.t. solution of *N*-methylglycine (sarcosine, 464 mg, 5.2 mmol) in deionized water (1 ml), add NaCl (304 mg, 5.2 mmol). In another flask, to a solution of cyanamide (412 mg, 9.8 mmol) in deionized water (0.26 ml), add two drops of concentrated ammonium hydroxide (approximately 1.4 mmol).

> ### Noticeboard 9.20.1
>
> **DANGER!**
>
> Cyanamide is extremely irritating and caustic when inhaled, ingested, or absorbed through the skin. Do the entire process in a fume hood.

Add this solution to the sarcosine and maintain magnetic stirring at r.t. for 1 h. Leave the reaction mixture at r.t. for 1 week. After a week, vacuum filter the creatine formed. Recrystallize from water (see ref. [4]).

Table 9.20: Physico-chemical properties of the reagents used.

Compound	M_w	M.p. (°C)	B.p. (°C)	Density (g·ml^{-1})	Danger^a (GHS)
Cyanamide	42.04	45–46	83	1.282	Non-hazardous
N-Methylglycine	89.09	208–212	-	-	Non-hazardous
Creatine	131.13	290	-	1.330	!
NaCl	58.44	801	1,413	2.165	Non-hazardous
NH$_4$Cl	53.49	340	-	-	!
NH$_4$OH (30%)	35.05	−60	38–100	-	🜛 ! 🜔

^{a} For brevity, only GHS icons are indicated. The information offered in the Material Safety Data Sheet (MSDS) should be consulted.

9.21. Claisen condensation: synthesis of ethyl acetoacetate

Estimated time	Difficulty	Basic lab operations
4 h	△△△	✔ Liquid-liquid extraction (see p. 111). ✔ Simple distillation (see p. 120). ✔ Vacuum distillation (see p. 125). ✔ Reflux (see p. 89).

9.21.1. Goal

To perform a Claisen condensation between two molecules of an ester (ethyl acetate) to form a β-ketoester (ethylacetoacetate).

9.21.2. Background

Ester enolates undergo addition-elimination reactions with other ester functional groups, resulting in β-ketoesters. These transformations, known as Claisen condensation, can occur between molecules of the same or different esters. Formation of enolates implies the use of strong bases compatible with the esters used.

9.21.3. Procedure

Place 50 g (55.5 ml) of pure ethyl acetate in a 250 ml round-bottom flask, and add 5 g of sodium wire. To the flask, connect a reflux condenser and on the upper end of this attach a drying tube. Heat the reaction mixture by reflux, so that the ethyl acetate is kept gently boiling.[11] Continue heating until the sodium is completely dissolved (approximately 3 h). If reflux is excessively prolonged, yield decreases.

---

[11]If the ethyl acetate contains only a small amount of alcohol or water, a violent reaction will not occur immediately, producing a slight release of hydrogen gas.

> **Noticeboard 9.21.1**
>
> **DANGER!**
>
> Sodium reacts violently with water.

After this time the hot plate is turned off, and with the reaction crude still warm, add slowly, with occasional stirring, a mixture composed of 14 ml of acetic acid in 16 ml of water until reaching an acid pH. Subsequently, add to the mixture an equal volume of cold saturated brine, and extract with the separatory funnel. The organic layer (top) is constituted by a mixture of acetic acid (bottom), ethyl acetate, and ethyl acetoacetate (majority). To remove ethyl acetoacetate, wash the funnel with saturated sodium bicarbonate solution, and purify the organic layer by simple distillation. Stop the distillation when the temperature reaches 95 °C. Distill the residue under vacuum. The first fraction (relatively small) is formed by ethyl acetate and water. In the second fraction, which distilled over a range of few degrees, the ethyl acetoacetate is obtained[12]; see ref. [3]).

Table 9.21: Physico-chemical properties of the reagents used.

Compound	M_w	M.p. (°C)	B.p. (°C)	Density (g·ml^{-1})	Danger[a] (GHS)
Ethyl acetate	88.11	−84	77.1	0.902	🔥 !
Ethyl acetoacetate	130.14	−43	181	1.029	!
Na	22.99	97.8	883	0.968	🔥 ⌖
CaCl$_2$	110.98	782	>1,600	2.15	!
Glacial acetic acid	60.05	16.2	118	1.049	⌖ 🔥
NaCl	58.44	801	1,413	2.165	Non-hazardous
NaHCO$_3$	84.01	300	-	2.160	Non-hazardous

[a] For brevity, only GHS icons are indicated. The information offered in the Material Safety Data Sheet (MSDS) should be consulted.

---

[12]Temperatures lower than specified may result, depending on the pressure reached by the vacuum system used —e.g., distillation at 14 mm Hg and 74 °C, at 29 mm Hg and 88 °C, at 45 mm Hg and 94 °C, and at 80 mm Hg and 100 °C.

9.22. Terpene transformation: citral conversion in α- and β-ionone

Estimated time	Difficulty	Basic lab operations
3 h	⚗⚗	✔ Liquid-liquid extraction (see p. 111). ✔ Washing in a separatory funnel (see p. 111). ✔ Rotary evaporator (see p. 117). ✔ Simple distillation (see p. 120). ✔ Vacuum distillation (see p. 125). ✔ Steam distillation (see p. 127).

9.22.1. Goal

To produce α- and β-ionone (violet perfume) from citral.

9.22.2. Background

Citral (3,7-dimethyl-2,6-octadienal) is the main constituent of the essential oil of lemon grass (citronella). This natural product readily undergoes aldol condensation with acetone to give ψ-ionone, which can be converted by cyclization under acidic conditions into a mixture of α- and β-ionones used as an artificial violet flavor. The nature of the catalyst, its concentration, and its temperature play decisive roles in the ratio between the two ionones.

9.22.3. Procedure

A) ψ-Ionone:

In a 100 ml round-bottom flask, prepare a solution of freshly distilled citral (5 g, b.p. = 91–93 °C at 2.6 mm Hg) in dry acetone (20 ml). Cool the mixture to −5 °C (ice/salt), and add dropwise a solution of sodium ethoxide prepared from 230 mg of sodium and 5 ml of absolute EtOH with magnetic stirring. Continue stirring for another 10 min at −5 °C, and afterward, neutralize the excess alkali with stirring by adding a solution of tartaric acid in water, prepared from 900 mg tartaric acid in 25 ml deionized water. Remove excess acetone and EtOH in a rotary evaporator. Isolate ψ-ionone from the remaining aqueous solution and extract with diethyl ether (2 × 5 ml).

Combine the organic extracts and dry over Na_2SO_4 anhydrous, remove the desiccant by gravity filtration, and the evaporation generates a liquid that is composed mainly of ψ-ionone. Weigh and calculate the yield. Purify by vacuum distillation (b.p. = 145–150 °C at 12 mm Hg; see ref. [14]).

B) α- and β-Ionone:

To a 100 ml round-bottom flask containing 15 g of a solution of phosphoric acid 85%, add dropwise 2 g of ψ-ionone with magnetic stirring. After the addition, maintain the stirring at r.t. for 1 h. After this time, pour the reaction crude into a 100 ml beaker containing 50 ml of cold water, and then extract the reaction mixture with diethyl ether (2 × 20 ml). Wash the combined ether extracts several times with water (until no acid pH reaction ocurrs).[13] Dry the organic layer over Na_2SO_4 anhydrous, separate the desiccant by gravity filtration, and remove the solvent on a rotary evaporator. Weigh the resulting liquid and calculate the yield (see ref. [14]).

Table 9.22: Physico-chemical properties of the reagents used.

Compound	M_w	M.p. (°C)	B.p. (°C)	Density (g·ml^{-1})	Danger^a (GHS)
Citral	152.23	-	229	0.888	!
α-Ionone	192.30	-	259–263	0.930	⬧
β-Ionone	192.30	−35	126–128	0.945	⬧
ϕ-Ionone	192.3	<25	114–116	0.900	Non-hazardous
H_3PO_4 (85%)	98.00	40	158	1.685	⬧
Acetone	58.08	−94	56	0.791	⬧ !
Tartaric acid	150.09	170–172	-	-	!
Diethyl ether	74.12	−116	34.6	0.71	⬧ !
Na_2SO_4	142.04	884	-	2.630	Non-hazardous
Na	22.99	97.8	883	0.968	⬧
EtOH	46.07	−114.1	78.5	0.790	⬧

^{a} For brevity, only GHS icons are indicated. The information offered in the Material Safety Data Sheet (MSDS) should be consulted.

---

[13]If emulsions are formed, add 10 ml of brine to the separatory funnel.

9.23. Carbohydrates: diisopropylidene derivative of glucose

Estimated time	Difficulty	Basic lab operations
4 h	⚗️⚗️	✔ Vacuum filtration (see p. 103). ✔ Rotary evaporator (see p. 117). ✔ Reflux (see p. 89).

9.23.1. Goal

To familiarize students with some of the most common reactions of protection of functional groups in carbohydrate chemistry, such as the protection of the hydroxyl groups.

D-Glucopyranose   1,2,5,6-Diisopropylidene-
                                                    α-D-glucofuranose

9.23.2. Background

Carbohydrates are compounds which present several hydroxyl groups that yield the characteristic reactions of alcohols. The key step in the synthesis of carbohydrates is the ways the different hydroxyls carry out different reactions on each. For example, acetone reacts with diols to form acetals. However, in the reaction with D-glucose catalyzed by iodine, it is possible to control the conditions for forming the primary and only anomeric acetals.

9.23.3. Procedure

Dissolve 1.5 g of iodine and 5 g of glucose in 250 ml of anhydrous acetone and place the mixture in a round-bottom flask, heat to reflux for 2 h. Then let the mixture cool to r.t.

Add saturated sodium thiosulfate to the cold crude reaction until complete discoloration of the aqueous solution occurs. Partially concentrate the resulting solution in the rotary evaporator to reach a third or quarter of the starting volume. Cool the solution and add 100 ml of water and 50 ml of chloroform. Transfer the resulting mixture to a separatory funnel and separate the organic layer (store). Extract the aqueous phase twice with 50 ml of chloroform. Joining the three organic extracts, dry over anhydrous sodium sulfate, filter, transfer the filtrate to a round-bottom flask, and evaporate to dryness on a rotary evaporator. The resulting white solid is the 1,2,5,6-di-*O*-isopropylidene-D-glucofuranose (see ref. [9]).

Table 9.23: Physico-chemical properties of the reagents used.

Compound	M_w	M.p. (°C)	B.p. (°C)	Density (g·ml^{-1})	Danger[a] (GHS)
1,2,5,6-Di-*O*-iso-propylidene-α-D-glucofuranose	260.28	109–113	-	-	Non-hazardous
D-(+)-Glucose	180.16	150–152	-	-	Non-hazardous
Acetone	58.08	−94	56	0.791	🔥 ❗
Sodium thiosulfate	158.11	52	-	1.010	Non-hazardous
Iodine I$_2$	253.81	113	184	4.930	❗ ☠
Chloroform (CHCl$_3$)	119.38	−63	60.5–61.5	1.492	☠ ❗
Na$_2$SO$_4$	142.04	884	-	2.630	Non-hazardous

[a] For brevity, only GHS icons are indicated. The information offered in the Material Safety Data Sheet (MSDS) should be consulted.

9.24. Preparation of a nucleoside: multistep synthesis of uridine derivative

Estimated time	Difficulty	Basic lab operations
2 days	⚗⚗⚗	✔ Rotary evaporator (see p. 117). ✔ Inert atmosphere (see p. 130). ✔ Chromatography (see p. 135).

9.24.1. Goal

To produce a derivative of uridine and use chromatographic techniques for their identification and isolation.

5'-*O*-Dimethoxytrityl-
2'-*O*-methyluridine

5'-*O*-Dimethoxytrityl-
3'-*O*-methyluridine

9.24.2. Background

Recent advances in the pharmaceutical industry have increased interest in the chemistry of nucleic acids —for example, the synthesis of AZT (3'-azido-2',3'-dideoxythymidine) used in treating AIDS. Furthermore, the multistep synthesis of these products requires the application of various chromatographic techniques to ensure the isolation and characterization of these products. Also these products are well characterized in the literature spectroscopically.

The proposed synthesis starts with uridine, which is first converted into its derivative dibutylstannylene to activate the secondary positions of the nucleoside. This is then treated with methyl iodide, which is used for the methylation of the 2' and 3' positions of the nucleoside. Finally, the methylated nucleosides are tritilated at the 5' position in order to facilitate the isolation and purification by chromatography. This last step can be avoided to shorten the experiment.

9.24.3. Procedure

A) 2',3'-*O*-(Dibutylestanilen)uridine:
In a 250 ml round-bottom flask, add 100 ml of anhydrous MeOH and then carefully add uridine (**1**, 488 mg, 2 mmol) and dibutyltin oxide (500 mg, 2 mmol). Heat the solution to reflux for 1 h, and then evaporate to dryness using a rotary evaporator. Store the resulting white powdery residue overnight at r.t. under an inert atmosphere, and use in the next step without further manipulation (see ref. [2]).

B) 2'-*O*-methyluridine and 3'-*O*-methyluridine (**2** and **3**):
Using a cannula, add 30 ml of anhydrous DMF to the flask from the previous step, and later with a syringe add 4 ml (64 mmol) of methyl iodide. The reaction is maintained at 40 °C under an inert atmosphere overnight.

> **Noticeboard 9.24.1**
>
> **DANGER!**
>
> Conduct the experiment in a fume hood because of the dangers of methyl iodide (CH$_3$I).

C) 5'-*O*-Dimethoxytrityl-2'-emph*O*-methyluridine and 5'-*O*-dimethoxytrityl-3'-*O*-methyluridine (**4** and **5**):
Evaporate the solution in a rotary evaporator until a gummy residue remains. The mixture can be characterized by TLC in ethyl acetate/MeOH (4:1) to give about R_f 0.60, 0.68, and 0.56, for products **1**, **2**, and **3**, respectively, or CH$_2$Cl$_2$/MeOH/triethylamine (96:3:1), and in this case the values would be 0.05, 0.17, and 0.14 (see ref. [2]). To the gummy residue from the previous step, add 50 ml of anhydrous pyridine and evaporate to dryness in

a rotary evaporator. Add back another 50 ml of anhydrous pyridine and then DMT-Cl (0.75 g, 2.2 mmol). Stir the mixture at r.t. under an inert atmosphere for 2 h. The progress of the reaction can be followed by TLC, taking care to elute first with diethyl ether to drag pyridine and subsequently with ethyl acetate/triethylamine (99:1). Under these conditions, **2** and **3** have an R_f of 0.06, while **4** and **5** have 0.48 and 0.36, respectively. When the reaction is complete, evaporate the solution to dryness, giving a syrup, which is resolved into its components by CC with CH_2Cl_2/triethylamine (99:1). Monitor the fractions as indicated above (see ref. [2]).

Table 9.24: Physico-chemical properties of the reagents used.

Compound	M_w	M.p. (°C)	B.p. (°C)	Density (g·ml^{-1})	Danger[a] (GHS)
Uridine	244.20	163–167	-	-	Non-hazardous
Diethyl ether	74.12	−116	34.6	0.71	🔥 ❗
MeOH	32.04	−98	64.7	0.791	🔥 ☠ ⬥
N,N-Dimethylformamide	73.09	−61	153	0.949	🔥 ⬥ ❗
Triethylamine	101.19	−115	88.8	0.600	🔥 ⬥ ❗
Ethyl acetate	88.11	−84	77.1	0.902	🔥 ❗
Dibutyltin(IV) oxide	248.94	300	-	1.580	☠ ⬥ ⬥
CH_3I	141.94	−64	41–43	2.280	☠ ⬥
Pyridine	79.10	−42	115	0.978	🔥 ❗
CH_2Cl_2	84.93	−97	40.0	1.33	⬥
2'-*O*-Methyluridine	258.23	-	-	-	Non-hazardous
3'-*O*-Methyluridine	258.23	-	-	-	Non-hazardous
4,4'-dimethoxytrityl chloride (DMT-Cl)	338.83	119-123	-	-	Non-hazardous

[a] For brevity, only GHS icons are indicated. The information offered in the Material Safety Data Sheet (MSDS) should be consulted.

9.25. References

1. A. I. Vogel and B. S. Furniss, *Vogel's Textbook of Practical Organic Chemistry*, Longman, 1989.

2. P. deLannoy and J. Howell, *An Undergraduate Organic Chemistry Laboratory Experiment: The Multistep Synthesis of a Modified Nucleoside*, Journal of Chemical Education **74** (1997), no. 8, 990, DOI 10.1021/ed074p990.

3. J. K. H. Inglis and K. C. Roberts, *Ethyl acetoacetate*, Organic Synthesis; Collective Volume **1** (1926), 235.

4. A. L. Smith and P. Tan, *Creatine synthesis: an undergraduate organic chemistry laboratory experiment*, Journal of Chemical Education **83** (2006), no. 11, 1654, DOI 10.1021/ed083p1654.

5. A. Baeyer, *Ueber ein condensationsproduct von pyrrol mit aceton*, Berichte der Deutschen Chemischen Gesellschaft **19** (1886), no. 2, 2184–2185, DOI 10.1002/cber.188601902121.

6. A. T. Rowland, *The determination of the stereochemistry of erythro-1,2-diphenyl-1,2-ethanediol: an undergraduate organic experiment*, Journal of Chemical Education **60** (1983), no. 12, 1084, DOI 10.1021/ed060p1084.

7. J. E. Weber and A. D. Boggs, *A simple procedure for preparing pinacol hydrate*, Journal of Chemical Education **29** (1952), no. 7, 363, DOI 10.1021/ed029p363.

8. N. G. Gaylord, *Reduction with complex metal hydrides*, Journal of Chemical Education **34** (1957), no. 8, 367, DOI 10.1021/ed034p367.

9. S. Pétursson, *Protecting groups in carbohydrate chemistry*, Journal of Chemical Education **74** (1997), no. 11, 1297, DOI 10.1021/ed074p1297.

10. A. W. Ingersoll, J. H. Brown, C. K. Kim, W. D. Beauchamp, and G. Jennings, *Extensions of the Leuckart synthesis of amines*, Journal of the American Chemical Society **58** (1936), no. 9, 1808–1811, DOI 10.1021/ja01300a089.

11. A. Ault and B. Wright, *2,2-Dichlorobicyclo[4.1.0]heptane from cyclohexene and dichlorocarbene by phase transfer catalysis*, Journal of Chemical Education **53** (1976), no. 8, 489, DOI 10.1021/ed053p489.

12. B. Andersh, K. N. Kilby, M. E. Urnis, and D. L. Murphy, *Regioselectivity in organic synthesis: preparation of the bromohydrin of α-methylstyrene*, Journal of Chemical Education **85** (2008), no. 1, 102, DOI 10.1021/ed085p102.

13. J. A. Shriver and S. G. Westphal, *Calix[4]pyrrole: synthesis and anion-binding properties. an organic chemistry laboratory experiment*, Journal of Chemical Education **83** (2006), no. 9, 1330, DOI 10.1021/ed083p1330.

14. G. A. Poulton, *Isomer analysis by spectral methods*, Journal of Chemical Education **52** (1975), no. 6, 397, DOI 10.1021/ed052p397.

15. R. D. Pointer and M. A. G. Berg, *Using a premade Grignard reagent to synthesize tertiary alcohols in a convenient investigative organic laboratory experiment*, Journal of Chemical Education **84** (2007), no. 3, 483, DOI 10.1021/ed084p483.

16. T. E. Sample and L. F. Hatch, *3-Sulfolene: a butadiene source for a Diels-Alder synthesis: an undergraduate laboratory experiment*, Journal of Chemical Education **45** (1968), no. 1, 55, DOI 10.1021/ed045p55.

17. L. C McKenzie, L. M. Huffman, J. E. Hutchison, C. E. Rogers, T. E. Goodwin, and G. O. Spessard, *Greener solutions for the organic chemistry teaching lab: exploring the advantages of alternative reaction media*, Journal of Chemical Education **86** (2009), no. 4, 488, DOI 10.1021/ed086p488.

18. S. H. Leung and S. A. Angel, *Solvent-free wittig reaction: a green organic chemistry laboratory experiment*, Journal of Chemical Education **81** (2004), no. 10, 1492, DOI 10.1021/ed081p1492.

19. H. T. McKone, *Acylation of ferrocene: effect of temperature on reactivity as measured by reverse phase high performance liquid chromatography*, Journal of Chemical Education **57** (1980), no. 5, 380, DOI 10.1021/ed057p380.

20. C. M. Dougherty, R. L. Baumgarten, A. Sweeney, and E. Concepcion, *Phthalimide, anthranilic acid, benzyne. An undergraduate organic laboratory sequence*, Journal of Chemical Education **54** (1977), no. 10, 643, DOI 10.1021/ed054p643.

21. R. E. Damschroder and W. D. Peterson, *1,2,3-Benzotriazol*, Organic Synthesis; Collective Volume **3** (1955), 106.

22. A. J. F. N. Sobral, *Synthesis of meso-octamethylporphyrinogen: an undergraduate laboratory mini-scale experiment in organic heterocyclic chemistry*, Journal of Chemical Education **82** (2005), no. 4, 618, DOI 10.1021/ed082p618.

23. A. Ault, *Resolution of D,L-α-phenylethylamine: an introductory organic chemistry experiment*, Journal of Chemical Education **42** (1965), no. 5, 269, DOI 10.1021/ed042p269.

24. K. M. Doxsee and J. E. Hutchison, *Green Organic Chemistry: Strategies, Tools, and Laboratory Experiments*, Thomson-Brooks/Cole, 2004.

Chapter 10

Microscale

10.1. Introduction

Organic Chemistry is an experimental science. This means that the laboratory work is essential for teaching the subject. Hence it is important for the students beginning to study Chemistry to acquire the necessary knowledge and appropriate habits for operating with ease in the laboratory or in a future professional occupation.

Working in the Organic Chemistry laboratory can be developed at three levels, according to the quantities of substance to be handled: macro (from quantities of 5–50 g of reagents and >50 ml of solvent); mini (in the 1–5 g range of reagents and 25 ml of solvent), and micro (<1 g of reagents and approximately 5 ml of solvent). The current trend in a practice laboratory is to use the miniscale, as reflected in most of the experiments described in the following chapters. However, the development of more sophisticated experiments and the use of more expensive products often make it advisable to minimize the costs, so there is a tendency to reduce the scale of the experiments. Also, elsewhere the microscale is the most common way of performing experiments at the research level. The biggest challenge of this scale is the adaptation of laboratory material as well as the basic operations. Both aspects are the focus of this chapter.

10.2. The scales in the Organic Chemistry laboratory

As stated above, work in the Organic Chemistry laboratory can be undertaken on three scales, according to the quantities of substance to be handled: macro-, micro-, and miniscale.

10.2.1. Macroscale

Laboratory experiments have long been carried out using amounts of reactants on the order of 5 to 50 g and volumes of solvent of 25 to 500 ml, which are known as macroscale or multigram-scale techniques. Currently, there is a tendency to dramatically decrease the amounts of substances handled, so that this methodology is progressively shifted to the teaching of experimental Organic Chemistry with much smaller amounts of substances.

However, research laboratories continue preparing large quantities of products when the need arises for starting material, for the development of a specific project, or the scaling of a process to the industrial synthesis of the substance. Note that the problem of industrial-scale synthesis of any product is very different from the techniques used in research laboratories (stirring, heating, purification, etc.), so that an adaptation of the processes are required when moving from the laboratory to industry.

10.2.2. Miniscale

Today, most of the Organic Chemistry experiments are performed on this scale. This method permits the isolation and characterization of products using conventional laboratory equipment and easily manageable quantities of material. The main features of these techniques are the following:

- Reagents: Amounts in the 1–5 g range are generally used. These quantities can be used with balances of, for example, a tenth of a gram accuracy. With these amounts, any errors made will not be significant.

- Solvent: Volumes of around 25 ml are often used for reactions as well as extraction processes. The amount of solvent is not critical to the performance of the experiment. For example, when 20 ml of diethyl ether is needed in a reaction, the same yield would probably result if 18 or 25 ml is used instead. For the measurement of the volumes, a graduated cylinder of 25 ml or a 10 ml pipette is adequate, although this does not mean that precise in measurements are unnecessary.

- Laboratory equipment: Conventional Organic Chemistry lab equipment is used as described in detail in Chapter 3. The student simply must adapt to the volume of solvent or quantities of reagents used in each particular experiment.

10.2.3. Microscale

Laboratory research is usually conducted on this scale. In recent years, there is a growing tendency to apply this methodology in the teaching of experimental Organic Chemistry, especially with beginning students.

- Reagents: Experiments performed in Organic Chemistry at microscale are carried out mainly with reactant amounts ranging from 0.005 to 0.5 g. Weighing measurements must be performed in microscale experiments with single-pan electronic balances, which are capable of weighing to ± 0.001 g. Note that these quantities with a deviation of 0.1 g in a reactive represents a significant percentage error in the proper proportions of the reagents used.

- Solvents: Amounts are usually below 100 microliters and 5 milliliters. Therefore, the student should use pipettes, micropipettes, dispensers, or syringes with adequate graduation and accuracy for each experiment.

- Lab equipment: An adjustment is required for the amounts used, especially when they are less than 100 mg. Such lab equipment may have various configurations, ranging from conventional lab equipment, with a size adapted to the specific needs of quantities and volumes used in this technique or specific design lab equipment (microscale kits).

10.3. Pros of working at the microscale

The advantages of using microscale techniques in Organic Chemistry laboratories are clear. Among the most notable ones are:

- Lower costs per student in teaching practices per experiment.

- Higher number and repertory of practical experiments with the same budget.

- Possibility of using more expensive reagents.

- Improved laboratory safety by reducing exposure to potentially toxic substances and risks of explosion or fire.

- Significantly lower amounts of reagents used and consequently less waste generated.

- Usually less reaction and experimentation time, enabling more time to analyze the results.

- Improved utilization of laboratories.

- Possibility of developing new technologies for using laboratory equipment.

- Need for less storage space for reagents and materials.

- Promotion of the principle of the three R's: Reduce, Recycle, Recover.

- Improved training of students, forcing them to be more careful at all stages of their work in practice.

10.4. Specific microscale lab equipment

There are various types of specific microscale equipment on the market designed to allow most of the basic laboratory operations. There are two main configurations arising from the publication of two books that describe experiments.

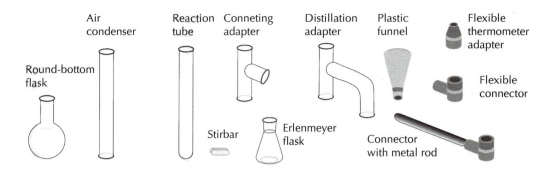

Figure 10.1: Characteristic elements of Kontes-Williamson microscale equipment.

10.4.1. Kontes-Williamson equipment

These kits are designed by K. L. Williamson and initially manufactured by Kontes Glass Co., as described in the book *Macroscale and Microscale Organic Experiments*[1]. It is a relatively inexpensive kit and fairly straightforward where similar elements are combined with a conventional apparatus but with a reduced size, such as Erlenmeyer flasks, filter flasks, centrifuge tubes or Hirsch funnels. These elements are specifically designed such as in the case of flasks with different types of necks, distillation heads and columns, reaction tubes, or chromatography columns. In this equipment the expensive ground-glass joints between elements have been removed and replaced by toric joints or flexible connectors (see Figure 10.1). However, recently this equipment has become available with the addition of unions based on screw caps.

10.4.2. ACE-Mayo equipment

This is the most popular and versatile equipment. The name comes from the company that first developed these kits (ACE Glass Inc.) following the publication of the book *Microscale Organic Laboratory* in 1986[2]. In this equipment, the different pieces are connected by ground or threaded joints. It is not recommended to use grease in the joints of the condensers to prevent contamination of samples and reactions, so rubber adapter O-rings are used to ensure sealing of the apparatus. Figure 10.2 depicts the main components of this equipment.

The most significant differences between components of this equipment and those normally used in practice (macroscale and miniscale) are the following:

- Conical vials: The use of conical vials is widespread both for reactions and liquid-liquid extractions (see Section 4.8, p. 111). This form is especially effective for handling liquid so that a disposable pipette can be used to empty one of these vials.

- Stir vanes or Teflon-coated vanes: Magnetic stir bars have also been adapted for use in conical vials. Teflon-coated vanes are triangular to rotate at any speed at full efficiency.

- Craig tube: Used for the recrystallization of solids.

- Hickman still head and Hickman column: They replace the traditional distillation apparatus (Claisen adapter, straight condenser and collector).

- Tube for gas vent: Used to measure the amount of gas produced in a reaction or for the collection of harmful gases.

10.5. Microscale laboratory techniques

10.5.1. Mass and volume measures

One of the first issues to resolve is to have electronic balances capable of weighing quantities within the required range. For weighing solids, the microspatula and weighing boats are used. These must be of adequate size for the order of magnitude of the substances used.

As for measuring volumes, pipettes (drain-out or blow-out), disposable-tip pipettes (Hamilton pipettes), and syringes (disposable or not) can be used, depending on the precision to be achieved.

For measuring liquids, when high accuracy is not needed, such as washing a solid with a solvent, or in liquid-liquid extractions, previously calibrated disposable Pasteur pipettes can be used. For calibrating a Pasteur pipette, 0.5 g of distilled water is weighed in a test tube of suitable size and, using a nipple connected to the Pasteur pipette, all the water is suctioned; using a glass marker, mark the water level reached by the 0.5 g of distilled water. The operation is repeated, this time for 1 g of distilled water. Thus, the pipette is calibrated to measure approximately 0.5 and 1 ml of another liquid (see Figure 10.3). Plastic pipettes graduated from factory can also be used.

10.5.2. Removal of small amounts of solvents

The usual way to remove a solvent is with a rotary evaporator. A solvent may be removed under reduced pressure with round-bottom flask of appropriate volume (e.g., 5 ml or 10 ml). However, there are other alternatives:

- Using a gas stream. The vessel is heated in a water bath and the solvent is forced to evaporate by a stream of gas. It is advisable to use an inert gas such as nitrogen, because the air oxygen can oxidize many compounds. This operation must be conducted in a fume hood to prevent accumulation of hazardous or harmful vapors in the workplace.

- By application of vacuum. Figure 10.4 lists options for removing a solvent in vacuo on a small scale.

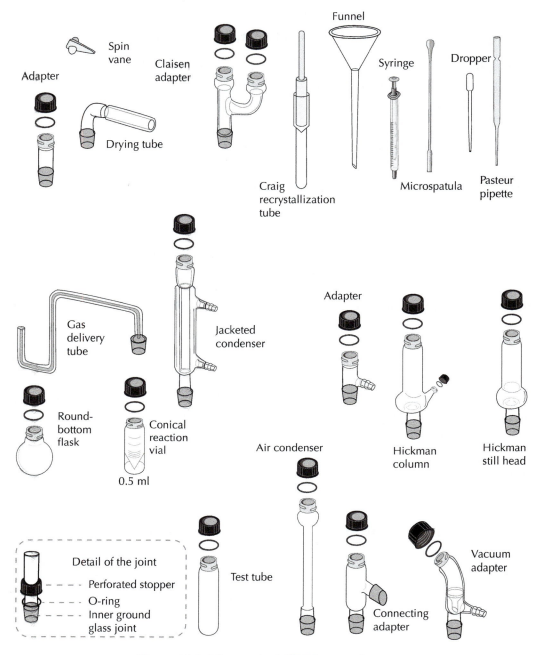

Figure 10.2: Microscale ACE-Mayo equipment.

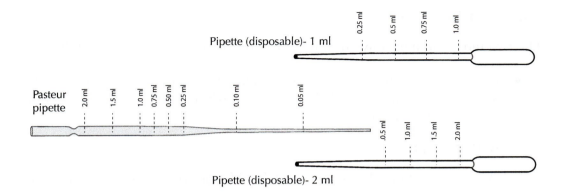

Figure 10.3: Approximate calibration of Pasteur and disposable pipettes.

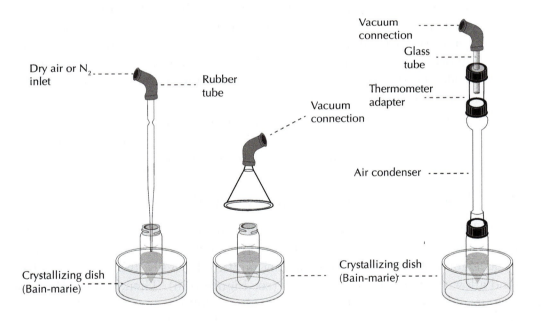

Figure 10.4: Different systems of solvent removal.

10.5.3. Filtration

The two most common ways to filter solids on the microscale are the following:

- By a Hirsch funnel, either glass or plastic, similar to the conventional vacuum filtration, assembled on a filter flask or a tube with a side outlet to connect the vacuum.

- By Pasteur pipette.

    A) For removing the supernatant liquid from the solid that is in a container (vial, Erlenmeyer flask, etc.), a Pasteur pipette can be used. The pipette (provided with a pippette bulb) is inserted to the bottom of the container (see Figure 10.5 (A)), so that it is completely attached and is suctioned so that the solid will remain in the bottom and the liquid in the pipette. If the solid is formed by very small particles, a cotton ball rolled into the pipette tip can help prevent the solid from passing through.

    B) For removing a solid, a pipette with a small amount of cotton is introduced into the solid with an applicator (e.g., a piece of wire is prepared; see Figure 10.5 (B)). Next, a filling material is introduced, such as celite or silica gel, and then the pipette is attached to a support with clamp. The liquid to be filtered is transferred using another pipette provided with a pipette bulb, and the filtering is collected in a container. The solid to remove is retained in the filling material.

    C) If the passage of liquid through the filter is very slow, a piece of rubber tubing can be used to couple a plastic syringe to the filter and force the drip by generating pressure with a syringe (see Figure 10.5 (C)).

10.5.4. Liquid-liquid extraction in conical vials

The liquid-liquid extraction at the microscale can be efficiently performed with a conical vial. Introduce the two layers into the conical vial, sealed with a screw cap, and vigorously stir the corresponding mixture for a few minutes. Open the cap and allow the two layers to separate (see Figure 10.6). Then, using a Pasteur pipette provided with a pipette bulb, separate as follows.

If the organic solvent is denser than water, such as CH_2Cl_2, insert the pipette directly to the bottom of the vial and gently suction to remove the entire organic layer. Then transfer to another container, which may be another vial, and thus the two layers are completely separated. If the organic layer is less dense than water, proceed similarly to the previous case. Slowly suction the organic layer while trying not to stir the mixture. The use of a small piece of cotton screwed into the pipette tip helps to prevent small amounts of water or interface from being transferred to the Pasteur pipette. Transfer the organic layer to another container and thereby the two layers.

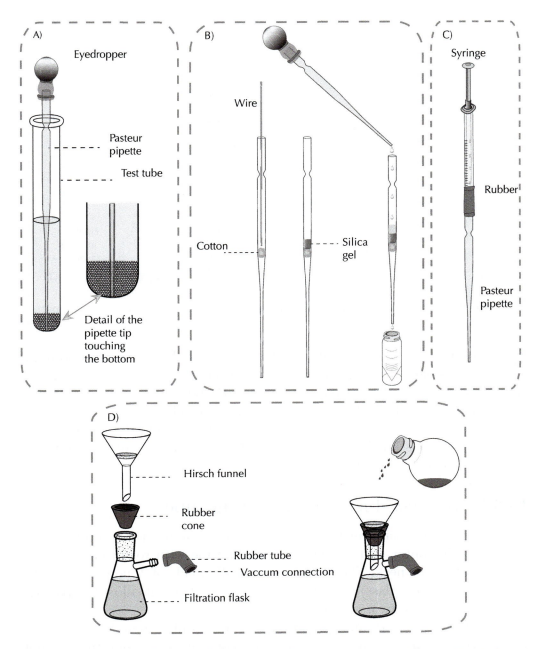

Figure 10.5: Different methods of filtering using Pasteur pipettes (A–C) and Hirsch funnel (D).

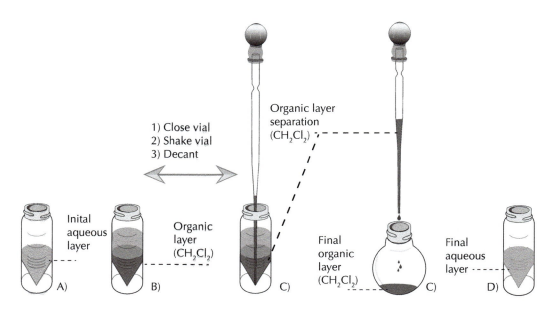

Figure 10.6: Microscale liquid-liquid extraction in conical vials using Pasteur pipettes.

Alternatively, conical centrifuge tubes can be used with this procedure, performing a similar operation. If the organic layer instead of CH_2Cl_2 consists of another solvent less dense than water, use a Pasteur pipette to remove the organic solvent from the vial (upper layer) almost entirely, and use another pipette for the aqueous layer. Use a small cotton ball in the tip of the pipette to avoid sucking up the possible interface that sometimes is formed, supporting it on the bottom of the conical vial.

10.5.5. Recrystallization in a Craig tube

The Craig tube consists of two separable parts, one inside the other (see Figure 10.7). The outer part is similar to a test tube with rough ground glass in the widening tube neck. The plunger is typically made of Teflon or glass (in the latter case, ground glass). The two parts fit together but allow the passage of liquids between the two, even when closed.

This device is particularly suitable to recrystallize quantities of solid of 100 mg or less, minimizing shifts between containers, so that the amount of solid lost during the handling is also minimal.

This tube is used to remove small amounts of crystals formed in microscale crystallizations, using volumes of less than 2 ml of solution. It is a special type of microscale filtering.

For recrystallization, the solid is placed in the glass tube, and the solvent or solvent mixture is added and heated in a water or sand bath to the hot solution of the crystals. This is allowed to cool at r.t. until the appearance of crystals and then is covered with a top. To easily remove the glass tube by hand, a thread with small knot (copper wire or other material inert to organic solvents) can be used and covered with a centrifuge tube (see Figure 10.7). The wire must

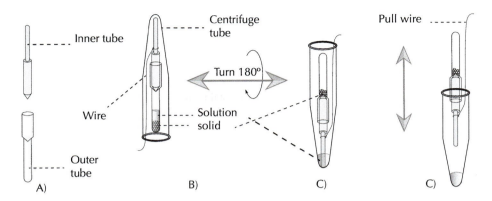

Figure 10.7: Craig tubes for microscale filtration.

be longer than the centrifuge tube. Then the centrifuge tube is reversed so that the filtrate from recrystallization falls to the bottom of the slide pipe through the plunger and the crystals will be deposited at the widest part. Remove the glass tube by hand, and using the cable, gently pull the plunger to retrieve the crystals.

10.5.6. Drying liquids and solids

To dry liquids at the microscale, a desiccant is added to a flask with the solution to be dried, and then it is separated by decantation or filtration or passing the solution through a Pasteur pipette filled with the desiccant and collected in a container (see Figure 10.8). Water will be retained in the desiccant. Solids can be dried by connecting a vacuum directly to the glass flask, stirring occasionally to facilitate operation. If the solid is wet with water or alcohols, a hot-water bath may help.

10.5.7. Sublimation

Sublimation consists of the evaporation of a solid from a hot surface and subsequent condensation on another surface at a lower temperature. Depending on the nature of the solid, sublimation can occur at atmospheric pressure or vacuum. It is a very effective microscale solid purification technique and involves minimal loss of substance and removal of solvent residues that may be caught in the solid with better results than in the case of conventional recrystallization. The efficiency of the sublimation process (see Section 4.7.4, p. 108) depends on the vapor pressure of the solid that is to be purified and the impurities that are to be removed. There are many ways to sublimate a solid at the microscale, ranging from the use of equipment specifically designed for this purpose (sublimators) to different adaptations of the material available to perform this operation. In the latter case, the use of a filter flask (side-arm flask), together with a centrifuge tube with a rubber cone, which is empty, can be effective. The centrifuge tube is filled with ice or carbonic ice at a distance of around 2 cm from the bottom of filter flask (side-arm flask), and this is placed on a heat source (see Figure 10.9).

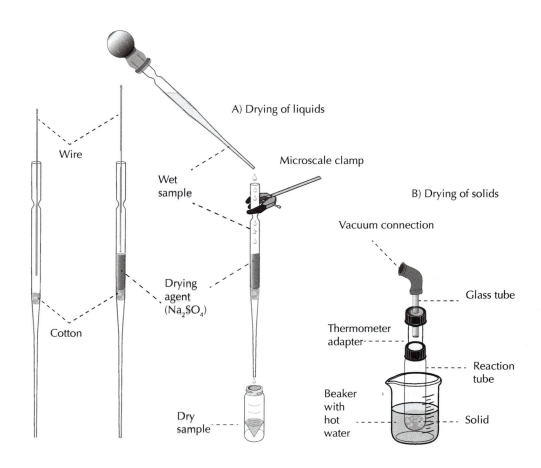

A) Drying of liquids

Wire

Microscale clamp

B) Drying of solids

Wet sample

Vacuum connection

Glass tube

Drying agent (Na$_2$SO$_4$)

Thermometer adapter

Cotton

Reaction tube

Beaker with hot water

Dry sample

Solid

Figure 10.8: Different types of drying. A) Liquids. B) Solids.

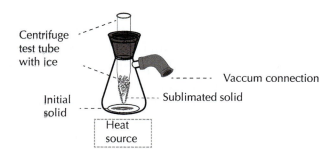

Centrifuge test tube with ice

Vaccum connection

Initial solid

Sublimated solid

Heat source

Figure 10.9: Microscale sublimation device.

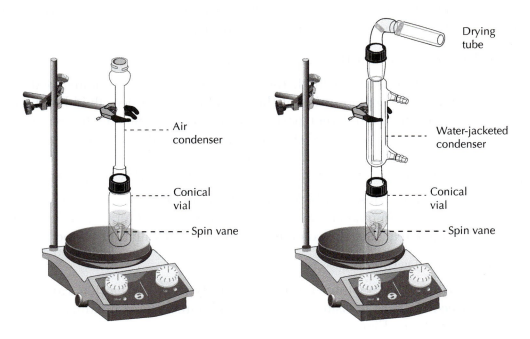

Figure 10.10: Different microscale reflux setup.

10.5.8. Reflux

There are different configurations to run a reaction under reflux at microscale. With the Kontes-Williamson equipment, reaction tubes can be used directly. Since the volume of solvent is very small, the mounting vapors along the tube are cooled on the wall and fall back to the bottom of the tube, thereby establishing the reflux.

Another possibility is to use a reaction tube, which consists of a flask or conical vial with an air condenser. If the amount of solvent used is not great and its b.p. is not too low, this can be a good option.

Finally, for highly volatile solvents, the safest way is to use a conical vial with a water-jacketed condenser. This condenser will prevent the vapors generated by heating the reaction from going into the atmosphere.

When anhydrous conditions are required in the reaction to reflux assemblies, it will be necessary either to couple a drying tube to prevent humidity or to adapt some of the components of the assembly by filling them with the corresponding desiccant (see Figure 10.10).

10.5.9. Simple and vacuum distillations and distillation with a Hickman still

Microscale distillation of a volatile liquid can be performed similarly to the cases described in Chapter 4 with assemblies consisting of a container, flask or vial, a distillation head, thermometer adapter to control the temperature of the vapors with a condenser, and a collector for the distillate. Figure 10.11 illustrates the components of these assemblies for Ace-Mayo equipment.

- Distillation with a Hickman still column.

  For small amounts of fluid, the Hickman column is recommended, as it simplifies the configuration of the assembly and reduces fluid loss by condensation in the different parts. This device is similar to the one in Figure 10.11. The bottom of the column has a small curvature in which the condensate collects. Some Hickman columns have a screw-top side opening to facilitate the transfer of the condensate with a Pasteur pipette.

  The column is inserted between the flask and condenser, as shown in the figure. A thermometer can be inserted so that the bulb is in contact with the vapor reaching the column to control the distillation temperature.

- Vacuum Distillation.

  By connecting vacuum assemblies this type of distillation can be performed. In case of using a Hickman column, a corresponding adapter is required (see Figure 10.10).

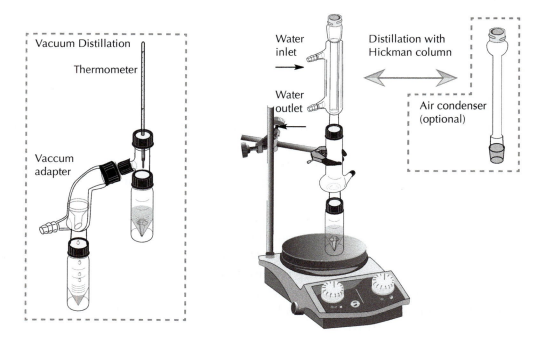

Figure 10.11: Adaptation to microscale of different distillation types.

10.5.10. Addition of reagents

Reagents can be added to a reaction in the simplest case by using a microspatula for solids, with an automatic pipette, or with a Pasteur pipette for liquids or solutions.

If the reaction must remain in a closed flask, the reagents should be added through septum-type plugs or pierced with a syringe fitted with a needle and adding the reagent. Sometimes, the puncture of a second needle is necessary to facilitate the process and prevent overpressure. In Figure 10.12, the simplest case is described, where a reagent is added to a reaction tube.

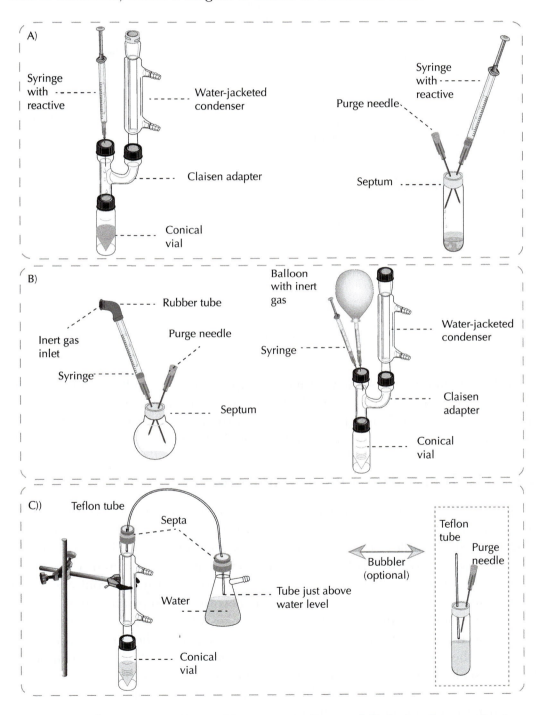

Figure 10.12: A) Microscale addition, B) inert atmosphere, and C) devices for removing harmful vapors.

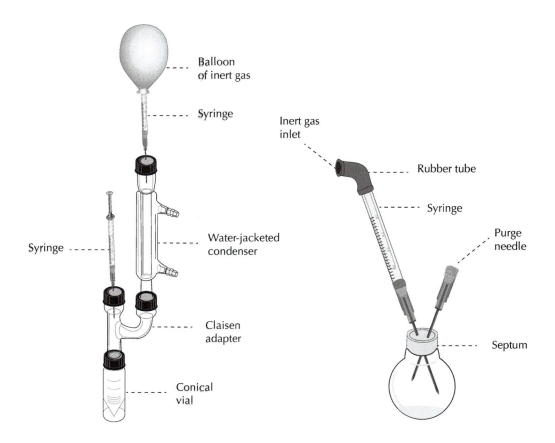

Figure 10.13: Inert atmosphere devices.

When assemblies are somewhat more complex, such as a reflux, septum plugs or similar ones are also used to add the reagents from a syringe. The septum must be placed directly on the flask in which the reaction is heated so that the contents of the syringe eject directly into the flask or conical vial.

10.5.11. Inert atmosphere reactions

In an inert atmosphere, microscale reactions can be performed as in miniscale reactions. The first possibility is to carry out the operation under a stream of nitrogen when available from a centralized gas system or a gas cylinder connected with a Teflon or silicone tube to a syringe without a plunger and a needle at the end. To check the flow, first a gas has to be bubbled into a container with an organic solvent. Afterward, the needle penetrates the septum that is used as cap of the container in which the reaction takes place, and the puncture of a second needle prevents overpressure. Nitrogen displaces the air and keeps the inside of the container free of oxygen. Another option, which is also used at the miniscale, is to puncture the rubber septum with a syringe. The syringe plunger has been replaced by a balloon filled with nitrogen, maintaining the inert atmosphere inside the assembly.

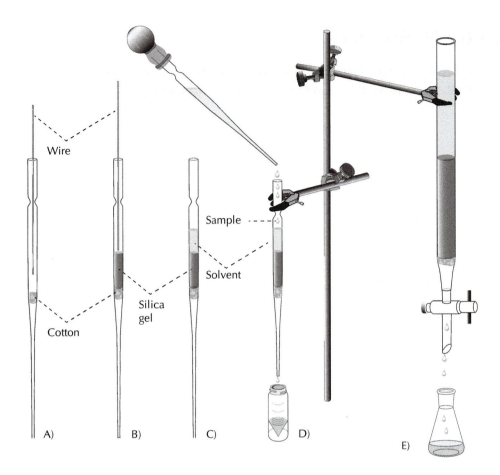

Figure 10.14: Column chromatography with Pasteur pipettes.

10.5.12. Removal of harmful vapors

As in the miniscale processes, it is sometimes necessary to remove, in a controlled way, harmful or dangerous vapors generated in a reaction. The following figures show assemblies that are often used for this purpose. Note the absolute parallelism of such assemblies with those described in the chapter on basic laboratory operations. The gas generated can be bubbled over a container with water or by applying vacuum to evacuate.

10.5.13. Column chromatography

A substance can be purified by column chromatography (CC) in two ways: Using a column adapted to the amount of substance to be purified or using a Pasteur pipette with a filling of alumina or silica gel prepared for this purpose (see Figure 10.14).

The Kontes-Williamson equipment has chromatography columns specially designed for the separation of small amounts of sample, but any column of the appropriate size for the amount of the substance to be separated can also be used.

10.6. References

1. K. L. Williamson, *Macroscale and Microscale Organic Experiments*, Lexington, 1989.
2. D. W. Mayo, R. M. Pike, and S. S. Butcher, *Microscale Organic Laboratory*, Wiley, 1986.

Chapter 11

Microscale Experiments

Microscale experiments are designed to perform basic operations with the usual small amounts both of starting materials and of final products (in all cases less than 0.5 g and/or 5 ml of solvent). These kinds of experiments are more present difficulties because the materials and methodology need to be adapted to such small amounts. Also, these amounts require greater measurement accuracy, whether for liquids or for solids, and careful attention because any handling error significantly compromises the results of the experiment.

The execution of such experiments by the students can be an ideal complement for their training. The operations will improve the skills of Chemistry students and will prepare them for transition to a next stage in research laboratories.

The experiments proposed in this section show how to perform the same types of reactions with similar methodologies as in conventional experiments but with specific microscale material (see Chapter 10). A total of 16 experiments including major types of reactions both at basic and advanced levels, are described. Usually less common reagents and products (more expensive) are used than in a miniscale experiments described in earlier chapters.

11.1. Addition of HX to alkenes: synthesis of 2-bromohexane

Estimated time	Difficulty	Basic lab operations
4 h	⚗⚗	✔ Microfiltration with Pasteur pipette (see p. 360). ✔ Microscale liquid-liquid extraction (see p. 360). ✔ Distillation with Hickman column (see p. 366).

11.1.1. Goal

To perform the addition of HBr to an alkene under phase-transfer catalyst conditions.

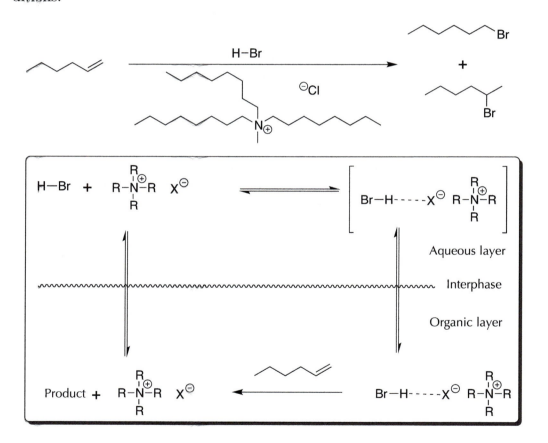

11.1.2. Background

Electrophilic addition ot HX to alkenes is a useful reaction for functional group interconversions. The reaction begins by a proton attack on the double bound and formation of a carbocation. These reactions are controlled by the Markovnicov rule, which indicates that the product is formed from the most stable carbocation. When this reaction is performed, certain practical problems arise. Most of alkenes are insoluble in such a polar medium as concentrated HBr, and water present in the reaction media causes the hydration reaction of the alkene to compete with the main reaction. This problem is avoided by using a phase-transfer catalyst, which improves the contact between the HBr and the alkene.

11.1.3. Procedure

Place 500 μl of hex-1-ene in a 5 ml conical vial with a spin vane, with 2 ml of 48% aqueous HBr and 150 mg of tricaprylmethylammonium chloride (Aliquat 336®). Set a water-jacketed condenser and reflux the reaction with magnetic stirring for 2 h. Cool the reaction mixture to r.t. and add 1 ml of hexane. Remove the condenser and set a stopper. Shake the vial by hand for 1 min, and then let the two layers separate. Remove the aqueous layer (lower) with a Pasteur pipette.

Wash the remaining organic layer with 2 ml of 10% solution of $NaHCO_3$, shaking the uncovered vial to avoid CO_2 accumulation. The operation is repeated until ensuring that no more acid remains in the organic layer (the pH paper test of the aqueous layer must be basic). With a microspatula, add anhydrous Na_2SO_4 until the solution becomes transparent, indicating that all water has been removed.

Finally, transfer the organic solution to a clean 5 ml conical vial with a Pasteur pipette, fit a Hickman column, and proceed to distill (b.p. = 166–167 °C). Carefully transfer the distillate to a tared clean and dry vial to calculate the yield (see ref. [16]).

Table 11.1: Physico-chemical properties of the reagents used.

Compound	M_w	M.p. (°C)	B.p. (°C)	Density (g·ml^{-1})	Danger[a] (GHS)
Hexane	86.18	−95	69	0.659	🔥💀❗🌿
$NaHCO_3$	84.01	300	-	2.160	Non-hazardous
Hex-1-ene	84.16	−140.0	60–66	-	🔥💀
Hydrobromic acid (48%) (HBr)	80.91	-	100	1.490	💀
Aliquat®336	-	−6	-	-	💀🌿
Na_2SO_4	142.04	884	-	2.630	Non-hazardous
CO_2	44.01	−78.5	-	-	

[a] For brevity, only GHS icons are indicated. The information offered in the Material Safety Data Sheet (MSDS) should be consulted.

11.2. Production of margarine: partial hydrogenation of a vegetable oil with cyclohexene and Pd(C)

Estimated time	Difficulty	Basic lab operations
2 h	⚗️⚗️	✔ Microfiltration with Pasteur pipette (see p. 360). ✔ Microscale reflux (see p. 365). ✔ Conventional rotary evaporator (see p. 117).

11.2.1. Goal

To perform a non-conventional hydrogenation of a vegetable oil (olive, palm, sunflower seed, etc.) to produce margarine.

11.2.2. Background

Margarine has traditionally been manufactured by hydrogenation of vegetable oils (olive, palm, sunflower seed, etc.) by an industrial process for converting these oils into a solid, stable, and spreadable substance.

The procedure followed in this experiment is the partial hydrogenation of a sample of olive oil using cyclohexene as the hydrogen source in the presence of Pd on carbon, which is used as catalyst. Under these conditions, cyclohexene is a hydrogen source that is converted in situ into benzene, which is added to the double bonds of the fatty acids comprising part of oil molecules. The reaction is the so-called heterogeneous catalytic transfer hydrogenation.

11.2.3. Procedure

In a conical 5 ml vial with a spin vane, add 600 mg of olive oil, 12 mmol of cyclohexene, and 40–50 mg of Pd (on C 10%). Reflux the mixture with stirring for 50 min.

Use as a filter a Pasteur pipette attached to a support by a clamp, with a small cotton ball in the narrowest part. Fill it with 2 or 3 cm of Celite. Moisten the Celite with 2 or 3 ml of solvent (diethyl ether or CH_2Cl_2). Filter the catalyst in the above-described device, forcing with compressed air filtration if necessary, passing 1 ml of additional solvent to a previously tared round-bottom flask, to collect the liquid that comes out the end of the pipette.

The catalyst should be retained in the Celite. Collect the filtrate in the round-bottom flask, and then remove the solvent under reduced pressure (rotary evaporator). Determine the mass of the product in order to calculate the yield (see ref. [2]).

Table 11.2: Physico-chemical properties of the reagents used.

Compound	M_w	M.p. (°C)	B.p. (°C)	Density (g·ml^{-1})	Danger[a] (GHS)
Vegetable oil					🔥
Cyclohexene	82.14	−104	83	0.779	🔥☠💥
Pd(C)	106.42	-	-	-	Non-hazardous
CH_2Cl_2	84.93	−97	40.0	1.33	💥
Diethyl ether	74.12	−116	34.6	0.71	🔥 !
Celite® S	-	-	-	-	!

[a] For brevity, only GHS icons are indicated. The information offered in the Material Safety Data Sheet (MSDS) should be consulted.

11.3. Isomerization of alkenes: derivatives of fumaric acid from maleic acid

Estimated time	Difficulty	Basic lab operations
1 h	⚗	✔ Vacuum microfiltration (see p. 360). ✔ Conventional recrystallization (see p. 105). ✔ Microscale reflux (see p. 365). ✔ Chromatography (see p. 135).

11.3.1. Goal

To perform an isomerization reaction of double bonds C=C in acidic medium.

11.3.2. Background

Double bonds are rigid structures where interconversion between isomers *E* and *Z* does not occur. However, in acid the protonation of the double bond may occur, with consequent change in hybridization. Then, the σ bond permits the rotation of the substituents around the bond, which through additional proton loss regenerates again the double bond. Under equilibrium conditions the thermodynamically stablest double bond forms. Maleic acid is isomerized to fumaric acid with acid catalysis.

11.3.3. Procedure

Into a conical vial provided with a spin vane, place 200 mg of maleic acid, 300 μl of deionized water, and 250 μl of concentrated HCl. Couple the vial to a condenser and reflux with stirring for 20 min. A precipitate will appear and become more visible when the reaction liquid is cooled after the reflux time is completed. After reaching r.t., completely separate the precipitate by vacuum filtration using a Hirsch funnel. Recrystallize the solid from water; dry, weigh, and determine the yield (m.p. = 130 °C) and calculate the reaction yield (estimated 85%). Check purity of the product by TLC using ethyl acetate/EtOH (absolute) (1:1) as the eluent.

Table 11.3: Physico-chemical properties of the reagents used.

Compound	M_w	M.p. (°C)	B.p. (°C)	Density (g·ml^{-1})	Danger[a] (GHS)
HCl	36.46	−30	>100	1.200	
Ethyl acetate	88.11	−84	77.1	0.902	
EtOH	46.07	−114.1	78.5	0.790	
Maleic acid	116.07	137–140	160	1.590	
Fumaric acid	116.07	287	-	1.635	

[a] For brevity, only GHS icons are indicated. The information offered in the Material Safety Data Sheet (MSDS) should be consulted.

11.4. Nucleophilic substitution reaction: ethyl iodide from ethanol

Estimated time	Difficulty	Basic lab operations
3 h	⚗⚗	✔ Microscale liquid-liquid extraction (see p. 360). ✔ Distillation with Hickman column (see p. 366). ✔ Microscale reflux (see p. 365).

11.4.1. Goal

To produce ethyl iodide from EtOH by a nucleophilic substitution reaction.

11.4.2. Background

Ethyl iodide is usually prepared in laboratories by the reaction of iodine, red phosphorus, and EtOH. The key step in this synthesis is the formation of the intermediate phosphorus triiodide, which reacts with EtOH to yield the desired product and phosphoric acid as a byproduct.

$$2P \ + \ 3I_2 \ \longrightarrow \ 2PI_3$$

$$3CH_3-CH_2-OH \ + \ 2PI_3 \ \longrightarrow \ 3CH_3-CH_2-I \ + \ H_3PO_3$$

11.4.3. Procedure

In a 10 ml round-bottom flask with an ice-water bath, add 500 mg of red phosphorus, and 5 ml of EtOH. Then, with magnetic stirring, add in 500 mg portions of iodine for approximately 15 min. When the addition is finished, remove from the ice-water bath and fit a water-jacketed condenser with a drying tube, and reflux the mixture for 2 h. After the reflux time, remove the condenser and fit a Hickman column, and separate the final product by distillation, giving a brownish liquid, ethyl iodide, with unreacted free iodine. Transfer the liquid to a conical vial that has been washed with water (several times) to eliminate the remaining EtOH. Then, with several drops of sodium bisulfite, treat the solution until the brown color disappears. Finally, treat the solution with the same volume of diluted NaOH. Add anhydrous $CaCl_2$, remove the liquid with a Pasteur pipette, and transfer it on a clean conical vial. Fit a clean Hickman column and distill (b.p. = 72 °C). Weigh and calculate the yield (see ref. [13]).

Table 11.4: Physico-chemical properties of the reagents used.

Compound	M_w	M.p. (°C)	B.p. (°C)	Density (g·ml^{-1})	Danger[a] (GHS)
Red phosphorous	30.97	280	-	-	🔥
EtOH	46.07	−114.1	78.5	0.790	🔥
NaOH	40.00	318	1,390	2.130	🧪
PI_3	411.69	61	-	4.180	🧪 ❗
Iodine I_2	253.81	113	184	4.930	❗ 🌳
$NaHSO_3$ (40%)	104.06	-	-	1.48	🧪 ❗
$CaCl_2$	110.98	782	>1,600	2.15	❗
Ethyl iodide	155.97	−108	69–73	1.950	☠ ❗

[a] For brevity, only GHS icons are indicated. The information offered in the Material Safety Data Sheet (MSDS) should be consulted.

11.5. Methylketone reactivity: acetophenone oxidation with sodium hypochlorite

Estimated time	Difficulty	Basic lab operations
2 h	🜊🜊	✔ Vacuum microfiltration (see p. 360). ✔ Microscale liquid-liquid extraction (see p. 360). ✔ Microscale reflux (see p. 365).

11.5.1. Goal

To modify of the haloform reaction using everyday oxidizing chemicals such as bleach.

11.5.2. Background

The haloform reaction is characteristic for compounds bearing a methyl group adjacent to a carbonylic group, these being the so-called methylketones. The reaction is carried out using chlorine in basic media to give a carboxylic acid. This reaction can be used as an analytical test to detect this family of compounds. On this occasion, the process is modified, chlorine being replaced by easy-to-handle commercial bleach for the methylketone transformation.

11.5.3. Procedure

Weigh 180 μl of acetophenone, place it in a 10 ml round-bottom flask with a spin vane, and add 6.3 ml of commercial bleach. Stir the mixture at r.t. while adding 500 μl of 10% NaOH solution. Fit a water-jacketed condenser and heat the reaction in a water bath at 70 °C for 20 min while stirring (see p. 83).

After the heating, cool the reaction at r.t., remove the condenser and add in small amounts 45 mg of sodium sulfite, while stirring to eliminate the remaining NaOCl. Transfer the reaction crude to a conical centrifuge tube and extract twice with 1.5 ml of diethyl ether. Add HCl (conc.) dropwise to the aqueous phase (lower) until reaching pH = 3.0 to give a precipitate. Filter it under vacuum in a Hirsch funnel, and dry it under an air stream. Weigh the benzoic acid and calculate the yield (see ref. [12]).

Table 11.5: Physico-chemical properties of the reagents used.

Compound	M_w	M.p. (°C)	B.p. (°C)	Density (g·ml^{-1})	Danger[a] (GHS)
NaOH	40.00	318	1,390	2.130	
Acetophenone	120.15	19-20	202	1.03	
Sodium hypochlorite (solution NaOCl)	74.44	−20	111	-	
NaOH	40.00	318	1,390	2.130	
Sodium sulfite	126.04	-	-	2.630	Non-hazardous
Diethyl ether	74.12	−116	34.6	0.71	
HCl	36.46	−30	>100	1.200	
Chloroform (CHCl$_3$)	119.38	−63	60.5–61.5	1.492	
Benzoic acid	122.12	125	249	1.08	

[a] For brevity, only GHS icons are indicated. The information offered in the Material Safety Data Sheet (MSDS) should be consulted.

11.6. Electrophilic aromatic substitution (S_EAr): preparation of 4-methyl-3- and 4-methyl-2-nitroacetanilide from 4-methylaniline

Estimated time	Difficulty	Basic lab operations
4 h	⚗️⚗️	✔ Vacuum microfiltration (see p. 360). ✔ Recrystallization with Craig tube (see p. 362). ✔ Microscale reflux (see p. 365).

11.6.1. Goal

To prepare and isolate two isomers in the nitration reaction of 4-methylaniline using two synthetic strategies.

11.6.2. Background

In electrophilic aromatic substitution reactions, the amino groups have a strong activating character directing *ortho* and *para*. However, when the amino group is protonated due to the positive charge on the nitrogen atom, it is strongly deactivating and directs *meta*. Therefore, to perform a nitration reaction at *ortho* with respect to the -NH$_2$ group, it is necessary first to form the acetylated derivative of the amine, and to release the amino group (deprotection reaction) after the nitration step. This is because the reagent used for nitrating is a mixture of H$_2$SO$_4$ and HNO$_3$, signifying that the group -NH$_2$ should be as -NH$_3^{\oplus}$. On the other hand, the possibility of oxidation of an amino group is minimized.

11.6.3. Procedure

A) Preparation of 4-methyl-3-nitroaniline:
Dissolve, in a 2 ml conical vial at r.t. with magnetic stirring, 107 mg of *p*-toluidine in 200 μl of concentrated H$_2$SO$_4$. Cool the solution in an ice bath and continue stirring while adding dropwise 400 μl of a solution (prepare the solution mixing 300 μl of concentrated HNO$_3$ and 100 μl of concentrated H$_2$SO$_4$). The addition must made while maintaining the temperature of the reaction mixture below 5 °C. After the addition, allow the reaction to reach r.t. and continue stirring for 15 min. Then pour the reaction crude into a 5 ml NaOH 40% solution and stir until a yellow precipitate appears. Isolate the solid by vacuum filtration using a Hirsch funnel and wash it with 10 ml of cold deionized water. Purify by recrystallization from 95% EtOH in a Craig tube (do not use more than 3 ml of solvent). The estimated yield is 75% (m.p. 78–79 °C; see ref. [3]).

B) Acylation of *p*-toluidine:
In a conical vial with a spin vane, prepare a solution of 107 mg (1 mmol) of *p*-toluidine in 500 μl acetic acid. Then add 200 μl of acetic anhydride. Fit a water condenser and reflux the mixture with magnetic stirring for 15 min. Upon completion of the reflux, cool the reaction crude to r.t. and pour it into 5 ml of ice water. Filter the solid under vacuum using a Hirsch funnel and wash it with 10 ml of deionized water. Dry and weigh the solid and calculate the yield (estimated yield 75%). Use this product for the following steps, adjusting in each case the amounts of reagents (see ref.[3]).

C) Nitration of 4-methylacetanilide:
In a conical vial, place 100 mg (0.67 mmol) of 4-methylacetanilide together with a mixture of 300 μl of concentrated HNO$_3$ and 100 μl of concentrated H$_2$SO$_4$. Fit a water condenser to the vial and heat the reaction mixture to 50 °C for 15 min with magnetic stirring. After this time, allow the reaction crude to cool to r.t., and pour it into a 5 ml Erlenmeyer containing an ice-water mixture. Vacuum filter the yellow solid that precipitates, and wash the crystals with 10 ml of cold water. Dry under an air steam, weigh, and calculate the yield (estimated 70%; m.p. = 91–93 °C; see ref. [3]).

D) Hydrolysis of 4-methyl-2-nitroacetanilide:
Prepare, in a conical vial with a spin vane, a suspension of 80 mg (0.41 mmol) of 4-methyl-2-nitroacetanilide in 500 μl of concentrated HCl. Couple a water condenser and reflux the mixture for 30 min. After the reflux period, cool

to r.t., and add dropwise a solution of NaOH 0.5 M until reaching a basic pH. A solid precipitates; filter under vacuum and recrystallize the product from EtOH in a Craig tube (estimated yield 70%; m.p. = 116–117 °C; see ref. [3]).

Table 11.6: Physico-chemical properties of the reagents used.

Compound	M_w	M.p. (°C)	B.p. (°C)	Density (g·ml^{-1})	Danger[a] (GHS)
p-Toluidine	107.15	41–46	200	-	☠☣🏵
H$_2$SO$_4$ 96–98%	98.08	3	-	1.80–1.84	🜨
HNO$_3$ 68–70%	63.01	-	120.5	1.413	🜨🔥
Glacial acetic acid	60.05	16.2	118	1.049	🜨🔥
Acetic anhydride	102.09	−73.1	139.8	1.080	🔥🜨 ❗
4′-Methylacetanilide	149.19	149–151	307	-	❗
4-Methyl-2-nitroaniline	152.15	115–116	-	-	☣☠🏵
4-Methyl-3-nitroaniline	152.15	74–77	-	-	☣☠🏵
NaOH	40.00	318	1,390	2.130	🜨
EtOH	46.07	−114.1	78.5	0.790	🔥
HCl	36.46	−30	>100	1.200	🜨 ❗
4′-Methyl-2′-nitro-acetanilide	194.19	-	-	-	Non-hazardous

[a] For brevity, only GHS icons are indicated. The information offered in the Material Safety Data Sheet (MSDS) should be consulted.

11.7. Sandmeyer reaction: 2-iodobenzoic acid synthesis

Estimated time	Difficulty	Basic lab operations
2 h		✔ Vacuum microfiltration (see p. 360). ✔ Recrystallization with Craig tube (see p. 362).

11.7.1. Goal

To produce iodobenzoic acid by the formation of an intermediate diazonium salt.

11.7.2. Background

The Sandmeyer reaction is a versatile synthetic tool by which an amino group on an aromatic ring is replaced with a wide range of substituents by converting an amino group attached to an aromatic ring into a diazonium salt that can be transformed into several functional groups. In this experiment, the 2-iodo-benzoic acid is synthesized from 2-aminobenzoic acid by this reaction.

11.7.3. Procedure

In a 5 ml conical vial provided with a spin vane, add 137 mg of 2-aminobenzoic acid (1 mmol), 1 ml of deionized water, and 250 μl of concentrated HCl. Stir the mixture until complete dissolution of the 2-aminobenzoic acid, gently heating if necessary. Place the vial with a septum in an ice bath, and slowly add with a syringe and a needle a solution of 75 mg of $NaNO_2$ dissolved in 300 μl of water with stirring. Wash the syringe with 100 μl of additional water to the reaction crude.

Stir the mixture for 5 min while keeping it cold, and then add, again with the help of a clean syringe and a needle, 170 mg of potassium iodide dissolved in 250 μl of water. Observe that a brown product is formed that partly precipitates.

After the addition is finished, the vial is quickly uncovered and fit to an air condenser to prevent spillage of the foam formed. Remove the ice bath, and then stir the reaction mixture for 5 min at r.t. Then introduce the conical vial into a water bath and heat at 40 °C for 10 min (see p. 83). At this point, gas (nitrogen) vigorously bubbles and a dark precipitate appears. Then allow the temperature of the bath to rise to 80 °C for another 10 min, considering at this point that the reaction is finished.

Cool the vial with an ice bath, and add several mg of sodium sulfite to destroy the iodine that may be present. Filter the resulting solid under vacuum and wash with 1 ml of deionized water. Recrystallize the solid from the water using a Craig tube (see ref. [3], pp. 931, 1073).

Table 11.7: Physico-chemical properties of the reagents used.

Compound	M_w	M.p. (°C)	B.p. (°C)	Density (g·ml^{-1})	Danger[a] (GHS)
HCl	36.46	−30	>100	1.200	⚠ !
NaNO$_2$	69.00	271	320	2.164	🔥☠🌱
KI	166.00	681	1,330	3.130	!
Anthranilic acid	137.14	144–148	-	-	!
2-Iodobenzoic acid	248.02	160–162	-	-	⚠ !
Nitrogen	28.01	−210	−196	0.970	⬳
Na$_2$SO$_3$	126.04	>500	-	2.630	Non-hazardous
NaCl	58.44	801	1,413	2.165	Non-hazardous
KCl	74.55	770	1,500	-	Non-hazardous

[a] For brevity, only GHS icons are indicated. The information offered in the Material Safety Data Sheet (MSDS) should be consulted.

11.8. Synthesis of a carbohydrate derivative: preparation of 2,3:5,6-di-*O*-isopropylidene-α-D-mannofuranose

Estimated time	Difficulty	Basic lab operations
2 h	🧪🧪	✔ Microfiltration with Pasteur pipette (see p. 360). ✔ Microscale reflux (see p. 365). ✔ Conventional rotary evaporator (see p. 117).

11.8.1. Goal

To prepare a mannose derivative both efficiently and under safe conditions, using commercially supplied solvents and reagents.

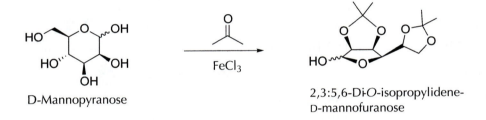

D-Mannopyranose

FeCl₃

2,3:5,6-Di-*O*-isopropylidene-D-mannofuranose

11.8.2. Background

Carbohydrates are a family of biomolecules that are essential for many biochemical processes and that are components of many natural and synthetic products with biological activity. A key question in Carbohydrate Chemistry is related to the protection-deprotection of hydroxyl groups with adequate protective groups, in order to transform them chemically. A common protective group for carbohydrate moieties is isopropylidene acetal derivatives (acetonides), which are produced by a reaction of acetone with two hydroxyl groups of the sugar unit. Isopropylidene groups are formed by acid catalysis and are stable under basic conditions. In the case of D-mannose, when it reacts with acetone in presence of an acidic catalyst, 2,3:5,6-di-*O*-isopropylidene-α-D-mannofuranose is formed. This protected intermediate is used, for example, in the synthesis of ovalicin, a molecule with antibiotic, antitumor, and immunosuppressive activity. Commercial D-mannose is constituted by a mixture of α- and β-anomers of D-mannopyranose. When it reacts with acetone 2,3:5,6-di-*O*-isopropylidene-α-D-mannofuranose is formed, according to the scheme.

11.8.3. Procedure

In a 5 ml conical vial containing a magnetic stir bar, place anhydrous ferric chloride (5 mg, 0.03 mmol), D-mannose (100 mg, 0.6 mmol), and anhydrous acetone (2 ml). Attach a water-cooled condenser and heat the reaction under reflux with stirring for 30 min.[1] After the reflux period, remove the condenser and cool until r.t. Add aqueous potassium carbonate solution (10%) until the reaction becomes clear and colorless. Then remove the excess acetone by gently heating in a fume hood. After evaporation the crude is cooled again to r.t. and extracted with portions of CH_2Cl_2 (2×1 ml). Remove the organic layer with a Pasteur pipette and dry by microfiltration with another Pasteur pipette using anhydrous sodium sulfate. Collect the organic solution in a tared vial, and eliminate the solvent under vacuum (rotary evaporator) to give the final product (m.p. 121–122; expected yield, 75%; see ref. [4]).

Table 11.8: Physico-chemical properties of the reagents used.

Compound	M_w	M.p. (°C)	B.p. (°C)	Density (g·ml^{-1})	Danger^a (GHS)
$FeCl_3$	162.20	304	-	2.800	
D-(+)-Mannose	180.16	132	-	1.539	Non-hazardous
Acetone	58.08	−94	56	0.791	
K_2CO_3	138.21	891	-	-	
Na_2SO_4	142.04	884	-	2.630	Non-hazardous
CH_2Cl_2	84.93	−97	40.0	1.33	

^{a} For brevity, only GHS icons are indicated. The information offered in the Material Safety Data Sheet (MSDS) should be consulted.

---

[1]If any solid material remains in the reaction, the crude is refluxed further until complete solution.

11.9. Free-radical halogenation: 9-bromoanthracene preparation

Estimated time	Difficulty	Basic lab operations
2 h		✔ Vacuum microfiltration (see p. 360). ✔ Recrystallization with Craig tube (see p. 362). ✔ Microscale reflux (see p. 365). ✔ Conventional rotary evaporator (see p. 117).

11.9.1. Goal

To acquaint students with the radical reactions of bromination using NBS.

11.9.2. Background

N-bromosuccinimide (NBS) is used as a highly specific bromination agent both in free-radical substitutions and in electrophilic additions of unsaturated systems, because releases small quantities of bromine.

Anthracene can be brominated at position 9 by a radical process. The reaction begins with Br_2 formation from NBS. Light or other initiating agents can produce $Br\cdot$, which attacks anthracene at position 9 to give a 9-anthracenyl radical and HBr. HBr reacts with new molecules of NBS to produce more new Br_2 molecules and succinimide. Bromine reacts with the formed 9-anthracenyl radical to give the substitution product and a new bromine radical, which promotes the reaction (propagation step).

11.9.3. Procedure

In a 25 ml Erlenmeyer prepare a solution of 200 mg of I_2 in 10 ml of CCl_4. Cap the flask containing the mixture with a septum and store in the fume hood for the next step. Mix 50 mg of anthracene (0.28 mmol) 50 mg and NBS (0.28 mmol) with 0.4 ml of CCl_4 in a conic vial with a spin vane. With a Pasteur pipette, add 1 drop of the I_2/CCl_4 solution. Fix a water-jacketed condenser with a dry tube and reflux for 1 h. A brownish color will appear and a solid will form, corresponding to succinimide. After reflux, cool the mixture to r.t. Filter the succinimide under vacuum with a Hirsch funnel, and then wash the solid with cold CCl_4 (2 × 1 ml). Transfer the organic solution to a tared 10 ml round-bottom flask, and eliminate the solvent under vacuum (rotary evaporator) to get a greenish yellow solid corresponding to 9-bromoanthracene. Once the solid is dry, weigh and calculate the yield. Purify the final product by recrystallization in a Craig tube in 95% EtOH (m.p. = 101–103 °C; see ref. [14]).

Table 11.9: Physico-chemical properties of the reagents used.

Compound	M_w	M.p. (°C)	B.p. (°C)	Density (g·ml^{-1})	Danger^a (GHS)
NBS	177.98	175–180	-	-	❖ !
EtOH	46.07	−114.1	78.5	0.790	🔥
CCl_4	153.82	−23	76–77	1.594	❖ ☠
Iodine I_2	253.81	113	184	4.930	! ☢
Succinimide	99.09	123–125	285–290	-	Non-hazardous
9-Bromoanthracene	257.13	97–103	-	-	Non-hazardous
Anthracene	178.23	210–215	340	-	! ☢
HBr	80.91	−87	−67	2.140	⚗ ! ▬
Br_2	159.81	7.2	58.8	-	☠ ⚗ ☢

^{a} For brevity, only GHS icons are indicated. The information offered in the Material Safety Data Sheet (MSDS) should be consulted.

11.10. Alkylbenzene oxidation: reactivity of alkyl groups in aromatic compounds with KMnO$_4$

Estimated time	Difficulty	Basic lab operations
2 h	⚗⚗	✔ Vacuum microfiltration (see p. 360). ✔ Microscale reflux (see p. 365).

11.10.1. Goal

To obtain benzoic acid by oxidation of alkyl functional groups from aromatic compounds.

Toluene → (KMnO$_4$, Oxidation) → Benzoic acid

11.10.2. Background

The alkyl chains attached to aromatic rings are likely to be oxidized to acid if they have at least one hydrogen at the benzyl position. Permanganate is an oxidizing agent widely used both in organic and inorganic reactions. Permanganate, which has a very distinctive purple color is reduced to MnO$_2$, which is brown and insoluble in water. If the reaction is carried out in a basic medium the acid is present as a salt, so it is soluble in an aqueous medium. If the aromatic ring presents more than one alkyl chain with at least one hydrogen at the benzyl position, it will be produced as many carboxylic groups as chains has the molecule.

11.10.3. Procedure

In a conical vial of 5 ml provided with a spin vane, add in this order 825 mg of potassium permanganate, 100 mg of sodium carbonate, and 3 ml of water. Couple a condenser and gently heat the mixture until all reagents are dissolved. Then disconnect the plate, allow to cool sightly and add 250 μl of toluene and 10 mg of liquid detergent. Reflux the mixture with stirring for 45 min. Afterward, the purple color of the permanganate will turn to brown from MnO$_2$. Upon reflux completion, allow the reaction to cool without reaching r.t., disconnect the condenser while still hot, and add dropwise a solution of NaHSO$_3$ until complete disappearance of the purple color. Filter the reaction crude under vacuum with a Hirsch funnel[2] on a 10 ml Erlenmeyer. Wash the solid with 1 ml additional water. Cool the filtrate with an ice-water bath and add HCl (conc.) dropwise until the pH reaches approximately 2.0 (test with indicator paper). A white precipitate corresponding to benzoic acid appears. Filter the crystals with a Hirsch funnel, dry weigh, and calculate the yield (see ref. [15]).

Table 11.10: Physico-chemical properties of the reagents used.

Compound	M_w	M.p. (°C)	B.p. (°C)	Density (g·ml^{-1})	Danger[a] (GHS)
$KMnO_4$	158.04	240	-	2.71	
Benzoic acid	122.12	125	249	1.08	
$NaHSO_3$ (40%)	104.06	-	-	1.48	
Toluene	92.14	−93	110.6	0.867	
$Na_2CO_3 \cdot 10\,H_2O$	286.14	32–34	-	1.44	
HCl	36.46	−30	>100	1.200	
MnO_2	86.49	535	-	5.026	
Celite® S	-	-	-	-	
Sodium dodecylbenzenesulfonate	348.48	-	-	-	

[a] For brevity, only GHS icons are indicated. The information offered in the Material Safety Data Sheet (MSDS) should be consulted.

---

[2]It is advisable to use some filter aid like diatomaceous earth or Celite.

11.11. Reduction of anthraquinone: anthrone synthesis

Estimated time	Difficulty	Basic lab operations
2 h		✔ Vacuum microfiltration (see p. 360). ✔ Microscale reflux (see p. 365).

11.11.1. Goal

To produce anthrone by reduction of one of the anthracene carbonyl groups.

Anthraquinone   Anthrone

11.11.2. Background

Anthrone is a tricyclic aromatic ketone that can be used, for example for calorimetric determinations of carbohydrates in biological fluids. It can be synthesized by partial reduction of anthraquinone with several reagents, as sodium hydrogen sulfite, tin chloride, or tin. In this experiment, the partial reduction is carried out with tin, HCl, and acetic acid as a solvent.

11.11.3. Procedure

Mix, in a conical vial with a spin vane, 104 mg (0.5 mmol) of anthraquinone, 100 mg of granulated tin, and 800 μl of acetic acid. Fit a water condenser, and reflux the mixture for 1.5 h, add dropwise 260 μl of concentrated HCl along this period to the boiling the mixture. Afterward, anthraquinone should be dissolved; if not, add more tin and HCl until complete dissolution. Remove the condenser and vacuum filter the reaction crude while hot using a Hirsch funnel to eliminate any solids in suspension. The anthrone will crystallize when the solution cools to 10 °C (use an ice-water bath to encourage the crystallization). Vacuum filter the white solid with a Hirsch funnel and dry the solid with an air stream in the funnel. Weigh the final product, calculate the yield (estimated yield 60%), and determine the m.p. (156 °C; see ref. [5]).

Table 11.11: Physico-chemical properties of the reagents used.

Compound	M_w	M.p. (°C)	B.p. (°C)	Density (g·ml^{-1})	Danger[a] (GHS)
HCl	36.46	−30	>100	1.200	
Glacial acetic acid	60.05	16.2	118	1.049	
Anthraquinone	208.21	284–286	379–381	1.440	
Anthrone	194.23	154–157	-	-	
Tin	118.71	231.9	2,270	7.310	Non-hazardous

[a] For brevity, only GHS icons are indicated. The information offered in the Material Safety Data Sheet (MSDS) should be consulted.

11.12. Compounds for the fragrance industry: ester synthesis

Estimated time	Difficulty	Basic lab operations
1 h ($\times$ reac.)		✔ Microscale liquid-liquid extraction (see p. 360). ✔ Distillation with Hickman column (see p. 366). ✔ Microscale reflux (see p. 365). ✔ Microscale solvent removal (see p. 357).

11.12.1. Goal

To prepare various esters at a microscale with applications in the fragrance industry by combining different carboxylic acids and alcohols that are selected because they are common in Organic Chemistry laboratory practice.

11.12.2. Background

Esters are compounds with the following general formula $R-\overset{\overset{\text{O}}{\|}}{C}-O-R'$. They result from by a reaction between a carboxylic acid and an alcohol in the presence of a mineral acid that is used as a catalyst. This process is the very well-known Fischer-Speier reaction (Fischer-Speier esterification). Most esters have pleasant odors, and in fact many of them produce the aromas of many fruits. This property implies that many esters are used in the perfumery industry or in food as fragrances or flavorings. In the case of the food industry, different legislation in different countries regulates their use, although the esters are produced in high purity and have the same properties as those isolated from natural sources.

The carboxylic acids and alcohols used in this experiment are frequently available in an undergraduate Organic Chemistry laboratory, and it is possible to prepare a wide range of esters with recognizable flavors or odors. Table 11.13 lists the products that result from esterification reaction between different carboxylic acids and alcohols generally available in an Organic Chemistry laboratory.

11.12.3. Procedure

In a 5 ml conical vial provided with a spin vane, place 2.5 mmol of carboxylic acid, 1.25 mmol alcohol, and 1 drop of concentrated H_2SO_4. Fit a water condenser, and heat the mixture for 30 min in a water bath at 70 °C (see p. 83).

After heating, allow the mixture to cool to r.t., remove the condenser, and add 1 ml of deionized water and 1 ml of diethyl ether. Set a stopper and shake by hand a few times for 30 s, open the stopper, and let the two layers separate. Using a Pasteur pipette, separate the aqueous layer (lower) and the organic one, and then wash the organic layer with a solution of $NaHCO_3$ 5% in portions of 0.5 ml to ensure that all acid has been removed. Dry the organic layer by passing it through a Pasteur pipette with anhydrous Na_2SO_4 on a tared 5 ml conical vial. Remove the solvent by gentle heating or by passing an air stream (perform this evaporation in a fume hood). Weigh the crude product and calculate the yield. The ester may be contaminated with traces of unreacted alcohol. If the b.p. of the ester is less than 150 °C, it can be purified by distillation with a Hickman column[3] (see ref. [8]).

---

[3]The smell of pure esters may be too strong, so it is best to smell them by placing a few drops in a test tube with 2 or 3 ml of deionized water.

Table 11.12: Physico-chemical properties of the reagents used.

Compound	M_w	M.p. (°C)	B.p. (°C)	Density (g·ml^{-1})	Danger[a] (GHS)
Butyl acetate	116.16	−78	124–126	0.880	🔥 !
Isopentyl acetate (Isoamyl acetate)	130.18	−78	142	0.876	🔥 !
Octyl acetate	176.26	-	211	0.867	Non-hazardous
Phenethyl acetate	164.20	-	238–239	1.032	Non-hazardous
Ethyl benzoate	150.17	−34	212	1.045	Non-hazardous
Butyl benzoate	178.23	-	249	1.010	!
Isopropyl benzoate	164.20	-	-	-	!
Isoamyl salicylate	208.25	-	277–278	1.05	Non-hazardous
Phenetyl salicylate	242.27	39–41	370	1.154	!
H_2SO_4 96–98%	98.08	3	-	1.80–1.84	🜤
Diethyl ether	74.12	−116	34.6	0.71	🔥 !
Na_2SO_4	142.04	884	-	2.630	Non-hazardous
$NaHCO_3$	84.01	300	-	2.160	Non-hazardous
Benzoic acid	122.12	125	249	1.08	🜤 !
Glacial acetic acid	60.05	16.2	118	1.049	🜤 🔥
Salicylic acid	138.12	158–161	211	1.440	🜤 !
Butan-1-ol	74.12	−90	116–118	0.810	🔥 🜤 !
3-Methyl-butan-1-ol	88.15	−117	131–132	-	🔥 !
Octan-1-ol	130.23	−15	196	0.824	!
Propan-2-ol	60.10	−89.50	82	-	🔥 !
Phenethyl alcohol	122.16	−27	219–221	-	☠ 🜤
EtOH	46.07	−114.1	78.5	0.790	🔥

[a] For brevity, only GHS icons are indicated. The information offered in the Material Safety Data Sheet (MSDS) should be consulted.

Table 11.13: Scent esters synthesized according to the procedure of this experiment.

Ester	Structure	Fragrance	B.p. (°C)	Carboxylic acid	Alcohol
Butyl acetate		Apple	126	Acetic acid	Butan-1-ol
Isoamyl acetate		Banana	142	Acetic acid	3-Methyl-butan-1-ol
Octyl acetate		Orange	210	Acetic acid	Octan-1-ol
Phenethyl acetate		Honey	231	Acetic acid	Methylbenzyl alcohol
Ethyl benzoate		Mint	211	Benzoic acid	EtOH
Butyl benzoate		Balsamic	249	Benzoic acid	Butan-1-ol
Isopropyl benzoate		Fruity	218.5	Benzoic acid	Propan-2-ol
Isoamyl salicylate		Pineapple	277	Salicylic acid	3-Methyl-butan-1-ol
Phenethyl salicylate		Balsamic	41[a]	Salicylic acid	Methylbenzyl alcohol

[a] M.p. of the solid.

11.13. Tandem transesterification-esterification: wintergreen oil from aspirin tablets

Estimated time	Difficulty	Basic lab operations
3 h		✔ Vacuum microfiltration (see p. 360). ✔ Microscale liquid-liquid extraction (see p. 360). ✔ Microscale reflux (see p. 365). ✔ Conventional rotary evaporator (see p. 117).

11.13.1. Goal

To prepare the high valued-added methyl salicylate from commercial aspirin tablets by two simultaneous transesterification-esterification reactions.

Aspirin   Methyl salicylate

11.13.2. Background

Methyl salicylate is an ester salicylic acid derivative found in many plants. Wintergreen *Gaultheria procumbens* is one of the species with the highest concentration of this compound, and therefore this molecule is known as Gaultheria essential oil. It can be used as an additive in food, because it has a flavor similar to mint. Gaultheria essential oil is used as an antiseptic for preparing creams and liniments against muscle pain and bruises. In this experiment, this compound is synthesized using commercial aspirin tablets as the starting material. Once the excipients and additives of the tablets are eliminated, the remaining acetyl salicylic acid is converted into methyl salicylate by two simultaneous processes of transesterification–Fischer-Speier esterification catalyzed by acid. Methyl salicylate and the volatile methyl acetate are formed at the same time, the latter being removed by evaporation when reaction crude is processed.

11.13.3. Procedure

Weigh three aspirin tablets and grind them with a mortar and pestle. Transfer the powder to a 5 ml conical vial, and add 3.5 ml of MeOH. Fit a stopper and shake vigorously for 2 min. Filter the mixture with a Pasteur pipette with a piece of cotton tip to eliminate all the additives and excipients. Transfer the clear solution to a clean conical vial with a spin vane, and add 300 μl of H_2SO_4 (conc.). Fit a water condenser with a drying tube, and reflux the mixture for 90 min. Afterward, cool the mixture to r.t.

Transfer the reaction mixture to a suitable separatory funnel, containing 5 ml of satured $NaHCO_3$ solution. Gently shake the mixture by hand without a stopper to avoid CO_2 overpressure (check that all the acid has been eliminated with pH test paper). Extract the solution with CH_2Cl_2 (2 $\times$ 5 ml). Join the organic layer in a clean 25 ml Erlenmeyer, and dry the organic solution with anhydrous Na_2SO_4. Filter the mixture to eliminate the desiccant on a tared round-bottom flask, and eliminate the solvent under vacuum (rotary evaporator) to give a liquid corresponding to methyl salicylate. Weigh the flask and determine the yield.[4] (see ref. [9]).

Table 11.14: Physico-chemical properties of the reagents used.

Compound	M_w	M.p. (°C)	B.p. (°C)	Density (g·ml^{-1})	Danger[a] (GHS)
MeOH	32.04	−98	64.7	0.791	
H_2SO_4 96–98%	98.08	3	-	1.80–1.84	
CH_2Cl_2	84.93	−97	40.0	1.33	
Acetylsalicylic acid	180.16	134–136	-	-	
$NaHCO_3$	84.01	300	-	2.160	Non-hazardous
Na_2SO_4	142.04	884	-	2.630	Non-hazardous
Methyl salicylate	152.15	(−)8–(−)7	222	1.174	
CO_2	44.01	−78.5	-	-	

[a] For brevity, only GHS icons are indicated. The information offered in the Material Safety Data Sheet (MSDS) should be consulted.

---

[4]It is strongly advisable to wash the desiccant with an additional 5 ml of CH_2Cl_2.

11.14. Polyfunctional molecule reactivity: vanillin transformations

Estimated time	Difficulty	Basic lab operations
3 h		✔ Microfiltration with Pasteur pipette (see p. 360). ✔ Vacuum microfiltration (see p. 360). ✔ Conventional recrystallization (see p. 105).

11.14.1. Goal

To apply microscale procedures to the synthesis of three vanillin derivatives by means of functional group transformations.

11.14.2. Background

Vanillin (3-methoxy-4-hydroxybenzaldehyde) is a white crystalline compound that can be isolated from vanilla plant (*Vanilla planifola*), the highest concentration of this compound being in the seeds. It is one of most commonly used flavorings and perfuming agents worldwide in the food and cosmetic industries, and because of its high volume of consumption, today most vanillin is produced by chemical synthesis. Vanillin is a polyfunctional compound with an aromatic ring bonded to an aldehyde group, -OH phenolic, and a methoxy group. In this experiment, three derivatives of vanillin will be prepared by relatively simple chemical reactions, leading to the formation of molecules with different aromas of the starting molecule.

11.14.3. Procedure

A) Vanillyl alcohol synthesis by reduction of the aldehyde group:
Dissolve 380 mg (2.5 mmol) of vanillin in a 5 ml conical vial provided with a spin vane in 2.5 ml of NaOH 1 M solution until a yellow solution is formed. Cool the mixture in an ice-water bath at 10–15 °C. Then add slowly 75 mg of $NaBH_4$. After addition, continue stirring for 30 min at r.t. Then return the vial to an ice-water bath and add HCl 2.5 M dropwise until slightly acid pH (check with pH test paper). Under these conditions, a precipitate that is the final reaction product should appear. If crystallization does not occur, scrape the flask wall with a microspatula gently until the first solid crystals appear, and after complete crystallization, let stand at r.t. The estimated yield is approximately 65%. Isolate the solid by vacuum filtration in a Hirsch funnel and purify it by recrystallization from ethyl acetate (m.p. = 108–109 °C) (see ref. [6]).

B) Synthesis of vanillic acid by oxidation of the aldehyde group:
In a conical vial with a spin vane, dissolve 170 mg $AgNO_3$ in 1 ml of deionized water. Add dropwise 0.5 ml of NaOH 2.5 M solution, stir the mixture until the precipitation of Ag_2O, and continue stirring for an additional 5 min. After this, stop stirring to let the solid deposit in the bottom of the vial. Then remove the supernatant liquid with a Pasteur pipette, and wash the remaining solid by adding 4×1 ml of deionized water. Stir and discard the Ag_2O, allowing it to settle to the bottom each time to eliminate the nitrates completely. After this step, add to the solid 2.5 ml of a solution of NaOH 2.5 M, and heat the vial with a water bath to 60 °C (see p. 83). When this temperature is reached, slowly add 152 mg (1 mmol) of vanillin using a microspatula for approximately 15 min. A metallic silver depot will form on the vial walls (silver mirror), which will subsequently fall to the bottom of the vial and form a yellow solution. Transfer the yellow solution to an Erlenmeyer and wash the remaining solid in the vial with 3×1 ml of deionized water. Wash with water and collect the yellow solution, and to this add HCl 6 M dropwise until reaching an acidic pH (check with pH test paper). Put the flask into an ice-water bath to trigger the precipitation of the final product. If this does not occur, gently scrape the wall of the vial with a microspatula until crystallization begins. Vacuum filter the solid in a Hirsch funnel, and, when dried, weigh and calculate the yield (estimated yield 60%). The product may be recrystallized from water (1.2 ml of water per 100 mg product; see ref. [6]).

C) Preparation of vanillin acetate (esterifying the phenolic -OH):
Dissolve 304 mg of vanillin (2 mmol) in a 50 ml Erlenmeyer containing 5 ml

of NaOH 10% with magnetic stirring. Add 6 g of crushed ice and 1.6 ml of acetic anhydride, cover the flask with a septum, and shake the mixture at r.t. for 15 min. Vacuum filter the solid with a Hirsch funnel. Dry, weigh, and calculate the yield (estimated yield 35%). Purify the product by recrystallization from EtOH/water (1:4) (see ref. [6]).

Table 11.15: Physico-chemical properties of the reagents used.

Compound	M_w	M.p. (°C)	B.p. (°C)	Density (g·ml^{-1})	Danger[a] (GHS)
Vanillyl alcohol	154.16	110–117	-	-	❗
Vanillin acetate	194.18	77–79	-	-	Non-hazardous
Vanillin	152.15	81–83	170	1.056	❗
Ag$_2$O	231.74	-	-	7.143	🔥 ❗
AgNO$_3$	169.87	212	440	4.350	🔥 ⚠
NaBH$_4$	37.8	400	-	1.07	🔥 ☠ ⚠
Acetic anhydride	102.09	−73.1	139.8	1.080	🔥 ⚠ ❗
NaOH	40.00	318	1,390	2.130	⚠
EtOH	46.07	−114.1	78.5	0.790	🔥
Vanillic acid	168.15	208–211	-	-	Non-hazardous
HCl	36.46	−30	>100	1.200	⚠ ❗
Ethyl acetate	88.11	−84	77.1	0.902	🔥 ❗

[a] For brevity, only GHS icons are indicated. The information offered in the Material Safety Data Sheet (MSDS) should be consulted.

11.15. Multistep synthesis of N-heterocycles: triphenylpyridine preparation

Estimated time	Difficulty	Basic lab operations
4 h		✔ Vacuum microfiltration (see p. 360). ✔ Conventional recrystallization (see p. 105). ✔ Microscale reflux (see p. 365).

11.15.1. Goal

To synthesize a heterocyclic compound (triphenylpyridine) by a sequence of reactions involving an aldolic condensation, a Michael addition, and a final condensation reaction to yield a pyridine ring.

2,4,6-Triphenylpyridine

11.15.2. Background

The aldol condensation reaction produces β-hydroxyaldehyde (aldol) or β-hydroxyketones from two carbonylic compounds with at least a hydrogen atom at the α-position with respect to such groups. Starting from two different carbonyl compounds the reaction is called a cross-aldol condensation. In such reactions, usually the primary condensation product is not isolated, since the loss of a water molecule occurs very easily to give α,β-unsaturated carbonylic compounds, active for a Michael addition on the double bond. The experiment described in this practice combines three different reactions that are performed sequentially: aldol reaction, Michael addition, and formation of a pyridine derivative.

11.15.3. Procedure

In a mortar with a pestle, grind and mix an NaOH lentil (they have a mass of between 75 and 95 mg) and 240 mg of acetophenone to form a smooth paste. Then add 110 mg of benzaldehyde. Continue grinding the three reagents for 15 min. Throughout the process, a thick paste forms first and then a solid. Stir and scrape the walls of mortar with a spatula frequently to enhance the mixing process. When finished mixing, let the crude reaction stand for 20 min. Meanwhile, place 150 mg of ammonium acetate and 10 ml of acetic acid into a 25 ml round-bottom flask with a spin vane, stir the mixture at r.t. for 5 min, and then add the solid prepared in the previous step. Fit a water condenser and reflux the mixture for 2 h. Afterward, allow the crude reaction to cool to r.t. To the reaction mixture, add 10 ml of water and cool the flask in an ice-water bath until crystals appear. Filter the resulting solid under vacuum in a Hirsch funnel, and wash it first with deionized water (2 × 3 ml) followed of a solution of 10% sodium bicarbonate (2 × 5 ml). Disconnect the vacuum while adding the sodium bicarbonate, and stir the solid thoroughly with a spatula. Reconnect the vacuum until the liquid disappears completely by suction. Transfer the solid to a Petri dish or watch glass, and when dry, weigh and calculate the yield. The final product can be purified by recrystallization from ethyl acetate (m.p. = 137–138 °C; see ref. [10,11]).

Table 11.16: Physico-chemical properties of the reagents used.

Compound	M_w	M.p. (°C)	B.p. (°C)	Density (g·ml^{-1})	Danger[a] (GHS)
NaOH	40.00	318	1,390	2.130	
Benzaldehyde	106.12	−26	178–179	1.044	!
Acetophenone	120.15	19-20	202	1.03	
Ammonium acetate	77.08	110–112	-	-	Non-hazardous
2,4,6-Triphenylpyridine	307.39	-	-	-	
Glacial acetic acid	60.05	16.2	118	1.049	
Ethyl acetate	88.11	−84	77.1	0.902	!
NaHCO$_3$	84.01	300	-	2.160	Non-hazardous

[a] For brevity, only GHS icons are indicated. The information offered in the Material Safety Data Sheet (MSDS) should be consulted.

11.16. Synthesis of five-membered heterocycles: 2,5-dimethyl-1-phenylpyrrole by Paal-Knorr reaction

Estimated time	Difficulty	Basic lab operations
3-4 h	⚗️⚗️	✔ Vacuum microfiltration (see p. 360). ✔ Conventional recrystallization (see p. 105). ✔ Microscale reflux (see p. 365).

11.16.1. Goal

To synthesize a heterocyclic compound, in this case 2,5-dimethyl-1-phenylpyrrole.

Aniline   Hexane-2,5-dione   2,5-Dimethyl-1-phenyl-1*H*-pyrrole

11.16.2. Background

Among different types of cyclization reactions for preparing nitrogen heterocycles, one of the most popular methods for the synthesis of substituted pyrroles is known as the Paal-Knorr reaction. In this experiment, a 1,4-dicarbonyl compound is heated with a primary amine to produce a pyrrole derivative.

11.16.3. Procedure

In a round-bottom flask fitted with a reflux condenser, add 186 mg (2.0 mmol) of aniline, 228 mg (2 mmol) of hexane-2,5-dione, 0.5 ml of MeOH, and 1 drop of HCl (conc.). Heat the mixture at reflux for 15 min, and then add 5.0 ml of HCl 0.5 M (keep the mixture cool in an ice bath). Collect the crystals by means of vacuum filtration, and recrystallize from 1 ml of a mixture of MeOH/water (9:1). The yield is approximately 52% (178 mg; see ref. [7]).

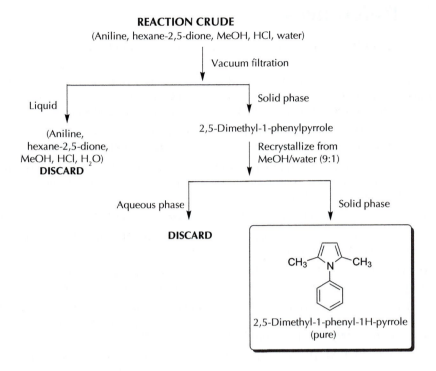

REACTION CRUDE
(Aniline, hexane-2,5-dione, MeOH, HCl, water)

Vacuum filtration

Liquid

(Aniline,
hexane-2,5-dione,
MeOH, HCl, H$_2$O)
DISCARD

Solid phase

2,5-Dimethyl-1-phenylpyrrole

Recrystallize from
MeOH/water (9:1)

Aqueous phase

DISCARD

Solid phase

2,5-Dimethyl-1-phenyl-1H-pyrrole
(pure)

Table 11.17: Physico-chemical properties of the reagents used.

Compound	M$_w$	M.p. (°C)	B.p. (°C)	Density (g·ml^{-1})	Danger[a] (GHS)
Aniline	93.13	−6	184	1.022	
Hexane-2,5-dione	114.14	6	191	-	
MeOH	32.04	−98	64.7	0.791	
HCl	36.46	−30	>100	1.200	
2,5-Dimethyl-1-phenyl-1H-pyrrole	171.24	48–52	156–160	-	Non-hazardous

[a] For brevity, only GHS icons are indicated. The information offered in the Material Safety Data Sheet (MSDS) should be consulted.

11.17. References

1. A. I. Vogel and B. S. Furniss, *Vogel's Textbook of Practical Organic Chemistry*, Longman, 1989.

2. R. A. W. Johnstone, A. H. Wilby, and I. D. Entwistle, *Heterogeneous catalytic transfer hydrogenation and its relation to other methods for reduction of organic compounds*, Chemical Reviews **85** (1985), no. 2, 129–70, DOI 10.1021/cr00066a003.

3. I. O. Kady, *Microscale electrophilic aromatic substitution of p-toluidine*, Journal of Chemical Education **72** (1995), no. 1, A9-A10, DOI 10.1021/ed072pA9.

4. J. A. Charonnat and C. S. Steinberg, *Microscale synthesis of 2,3:5,6-di-O-isopropylidene-α-D-mannofuranose ('diacetone mannose')*, Journal of Chemical Education **73** (1996), no. 8, A170-A171, DOI 10.1021/ed073pA170.

5. K. H. Meyer, *Anthrone*, Organic Syntheses **8** (1928), 8–9, DOI 10.15227/orgsyn.008.0008.

6. R. G. Fowler, *Microscale reactions of vanillin*, Journal of Chemical Education **69** (1992), no. 2, A43-A44, A46, DOI 10.1021/ed069pA43.

7. R. Al-Awar and G. H. Jr. Wahl, *Microscale syntheses of heterocyclic compounds*, Journal of Chemical Education **67** (1990), no. 3, 265–266, DOI 10.1021/ed067p265.

8. D. M. Birney and S. D. Starnes, *Parallel combinatorial esterification: a simple experiment for use in the second-semester organic chemistry laboratory*, Journal of Chemical Education **76** (1999), no. 11, 1560, DOI 10.1021/ed076p1560.

9. A. M. Hartel and J. M. Hanna, *Preparation of oil of wintergreen from commercial aspirin tablets. A microscale experiment highlighting acyl substitutions*, Journal of Chemical Education **86** (2009), no. 4, 475, DOI 10.1021/ed086p475.

10. G. W. V. Cave and C. L. Raston, *Efficient synthesis of pyridines via a sequential solventless aldol condensation and Michael addition*, Journal of the Chemical Society, Perkin Transactions 1 **0** (2001), 3258–3264, DOI 10.1039/B107302H.

11. D. R. Palleros, *Solvent-free synthesis of chalcones*, Journal of Chemical Education **81** (2004), no. 9, 1345, DOI 10.1021/ed081p1345.

12. A. M. Van Arendonk and M. E. Cupery, *The reaction of acetophenone derivatives with sodium hypochlorite*, Journal of the American Chemical Society **53** (1931), no. 8, 3184–3186, DOI 10.1021/ja01359a506.

13. J. Wilbrand and F. Beilstein, *Vorläufige notiz über nitrodracylsäure*, Justus Liebigs Annalen der Chemie **126** (1863), no. 2, 255–256, DOI 10.1002/jlac.18631260217.

14. N. M. Weinshenker, *9-Bromoanthracene*, Organic Preparations and Procedures **1** (1969), no. 1, 33–34, DOI 10.1080/00304946909458344.

15. C. F. Cullis and J. W. Ladbury, *Kinetic studies of the oxidation of aromatic compounds by potassium permanganate. Part IV. n- and iso-Propylbenzene*, Journal of the Chemical Society **0** (1955), 4186–4190, DOI 10.1039/JR9550004186.

16. D. Landini and F. Rolla, *Addition of hydrohalogenic acids to alkenes in aqueous-organic, two-phase systems in the presence of catalytic amounts of onium salts*, Journal of Organic Chemistry **45** (1980), no. 17, 3527–3529, DOI 10.1039/JR9550004186.

Chapter 12

Green Chemistry

Chemistry is a science that provides unquestionable contributions to improve the quality of human life and well-being, devising creative solutions in such diverse fields as drug preparation, providing cures for many diseases, developing pesticides that have enabled crops to feed the world's population, or producing new materials to help many sectors of the population.

However, Chemistry in society often looks bad in the media, where *chemical* is synonymous with something dangerous, harmful, or negative. This frequently leads to broadcasting only the more negative aspects of this branch of science, such as the toxicity of some substances or pollution resulting from misuse or improper handling of chemicals in general.

The term "Green Chemistry" deals with the design of chemical products and processes that reduce or eliminate the use and production of toxic or hazardous substances. It cannot be considered a branch of chemistry but rather a code of conduct to try to reduce the environmental impact of any chemical process.

12.1. Introduction

Given the challenges to develop better habits, and due to the social consciousness that emerged in the 1990s concerning problems arising from the use, storage, or transport of chemicals, the need has arisen to redirect some of the practices common in laboratories used in teaching, research, and industry. As a basis for these, new approaches to the development of chemical processes, the following documents and initiatives appeared:

- Pollution Prevention Act (1990) EPA.[1]

- Financing Program of projects and grants, the "Presidential Green Chemistry Challenge" from the US government.[2]

- Code of conduct for members of the American Chemical Society.

---

[1] http://www.epa.gov/opptintr/p2home/p2policy/act1990.htm (accessed July 14, 2015).
[2] http://www.epa.gov/greenchemistry/pubs/pgcc/presgcc.html (accessed July 14, 2015).

The term "Green Chemistry" deals with the design of chemical products and processes that reduce or eliminate the use and production of toxic or hazardous substances. Thus, Green Chemistry cannot be considered as a branch of Chemistry as occurs with the traditional classification of Chemistry in Analytical Chemistry, Physical Chemistry, Inorganic Chemistry, Organic Chemistry, and Chemical Engineering but rather as a code of conduct meant to reduce the environmental impact of any chemical process, whether at the laboratory scale or industrial scale. Synonyms of Green Chemistry are Sustainable Chemistry or low-environmental-impact Chemistry [1].

The main fields of action of Green Chemistry can be summarized by the following points:

- Use of alternatives to the current raw materials, less toxic and with manufacturing processes that present less environmental impact than current ones, based mainly on renewable raw materials.

- Development of safe reagents to replace the toxic or hazardous reagents currently used.

- Replacement of hazardous solvents by others involving less risk in use and handling.

- Development of alternative reaction conditions compared to the present ones, to consume less energy, shorten reaction times, and simplify isolation and purification of final products.

Since the birth of Green Chemistry, an exponential growth in scientific publications of all kinds has occurred, resulting in a multitude of specific journals, books, and monographs on the topic as well as many scientific networks and organizations interested in this subject. Furthermore, the principles of Green Chemistry have been introduced into the curriculum of chemists and chemical engineers as part of the subjects in undergraduate studies as well as specific courses, postgraduate studies, etc., and is today an important part of the training of future professionals.

12.2. The 12 principles of Green Chemistry

Paul Anastas is considered the founder of the Green Chemistry. In his book *Green Chemistry: Theory and Practice*, he develops the so-called 12 principles of Green Chemistry [6].

1. Prevention:

   It is better to prevent waste than to try to clean it up after it is formed.

2. Atom economy:

   Synthetic methods should be designed to incorporate, in the final product, a maximum of all the materials used in the process, minimizing the formation of byproducts.

3. Using methodologies that generate products with reduced toxicity:

   Whenever possible, synthetic methods should be designed to use and generate substances that possess little or no toxicity to human health and the environment.

4. Generate effective but non-toxic products:

   Chemical products should be designed to maintain efficacy while reducing toxicity.

5. Reduce the use of auxiliary substances:

   Whenever possible, substances that are not essential (solvents, reagents to perform separations, etc.) should be avoided, and those used should be as safe as possible.

6. Reduce energy consumption:

   Energy requirements should be catalogued for their environmental and economic impact. This impact should be minimized. Synthesis methods should be carried out at r.t. and atmospheric pressure.

7. Use of renewable raw materials:

   Raw materials should be renewable rather than exhaustible whenever technically and economically feasible.

8. Avoid unnecessary derivatization:

   Derivatization (blocking groups, protection / deprotection, temporary modification of physical / chemical processes) should be avoided where possible.

9. Enhancing catalysis:

   The use of catalysts should be as selective as possible and should be as reusable as possible, instead of stoichiometric reagents.

10. Generate biodegradable products:

    Chemicals are designed so that at the end of their function they do not persist in the environment and they are transformed into innocuous degradation products.

11. Develop analytical methodologies for real-time monitoring:

    The analytical methodologies will be further developed to allow monitoring and real-time control of the process, prior to the formation of hazardous substances.

12. Minimize the potential for chemical accidents:

    Substances used are selected based on the chemical processes so that the risk of chemical accidents is minimized, including releases, explosions, and fires.

12.3. Goals of Green Chemistry

The principles and guidelines of Green Chemistry are intended to fulfill the following goals for any chemical process, whether industrial or laboratory scale:

- Make better use of available resources for the development of a chemical process.

- Reduce waste generated in any preparation or handling of chemicals.

- Materials should be prepared by improved processes that reduce unwanted effects on the environment.

- Replace toxic reagents and products with others that have the same properties and applications but have less impact on the environment.

- Reduce the energy required to produce substances of interest, either by the use of much faster processes or by the use of renewable energies involving lower energy cost with equal efficiency.

- Reduce toxicity or general danger for a given compound substance and the compound itself.

- Reduce costs by eliminating any manipulation that is not strictly necessary and decreasing time invested in the preparation of a substance.

- Encourage all necessary actions to use chemicals compatible with sustainable development.

12.4. Parameters to evaluate chemical processes

In Green Chemistry the environmental impact of a reaction or the efficiency of a reaction or a chemical process can be quantified by a number of parameters [5]. Of all those that have been defined, here we will highlight the "E factor" and "atom economy."

12.4.1. E factor

One of the first parameters for evaluating the environmental impact of a chemical transformation is the term "E factor", which was introduced by R. Sheldon [4]. The concept is quite simple and easy to understand, and it has been applied mainly to the industrial sector. It is a measure of the waste generation in chemicals manufacture. The E factor is calculated by dividing the total mass of waste produced in the preparation of a compound by the total mass of product produced or synthesized. Waste is defined as everything but the desired product. Chemical yield is considered together with reagents, solvent losses, all process aids, and, in principle, even fuel used, although this item is often difficult to quantify. There is one exception: water is generally excluded from the calculation of the "E factor." For example, when considering an aqueous waste stream, only the inorganic salts and organic compounds contained in the water are counted, while the water itself is excluded. The lower the E value, the lower the waste produced and the lower the environmental impact of the process.

$$E \text{ Factor} = \frac{\text{Total mass of wastes (Kg)}}{\text{Mass of products (Kg)}} \qquad (12.1)$$

Table 12.1 lists the estimated values of E factors for different chemical industry branches.

Table 12.1: E factor.

Industry segment	Product tonnage	E factor (Kg waste/kg product)
Oil refining	10^6–10^8	<0.1
Bulk chemicals	10^4–10^6	<1–5
Fine chemicals	10^2–10^4	5–>50
Pharmaceuticals	10–10^3	25–>100

12.4.2. Atom economy

The atom economy concept is due to B. M Trost [2,3]. This is probably one of the most useful parameters for the analysis of reactions, allowing an assessment of the amount of waste generated in a reaction or a sequences of reactions. The atom economy has had a great influence in the further development of organic synthesis as a whole, because it focuses on the design of the synthetic methods according to the concept of sustainability.

The calculation of atom economy quantifies the use made of each of the atoms of a reactive, indicating which of them is actually incorporated into the final product. With the concept of atom economy, some approximations are assumed to simplify the calculations. Since it is a measure of how reagents are incorporated to the end products, when calculating, it neither counts the amount of solvent employed, the excess organic reagents, the catalysts, or the inorganic salts that may be added to the reaction nor intervenes in the reaction yield.

For a simple reaction A + B $\longrightarrow$ C atom economy calculation is as follows:

$$EA = \frac{\text{Molecular weight of C}}{\sum \text{Molecular weight of reactants (A + B)}} \times 100 \qquad (12.2)$$

As an example of calculating of atom economy, the preparation of ethylene oxide (oxirane) by the traditional method can be considered. Under the assumption that the reaction yield was 100%, the atom economy of the process would be of 25.46%:

$$C_2H_4 + Cl_2 + Ca(OH)_2 \longrightarrow C_2H_4O + CaCl_2 + H_2O$$

$$EA = \frac{\text{Molecular weight of } C_2H_4O}{\sum \text{Molecular weight of } C_2H_4 + Cl_2 + Ca(OH)_2} \times 100\% \qquad (12.3)$$

$$EA = \frac{44.05u}{28.05u + 70.91u + 74.09u} \times 100\% = 25.46\% \qquad (12.4)$$

12.5. Green Chemistry for undergraduate laboratories

To apply the principles of Green Chemistry to an undergraduate Organic Chemistry laboratory may present two issues. First, regarding the laboratory management, by an evaluation of the techniques to be used, by the improvement of safety process with the available resources, and by the calibration of the toxicity for reagents and solvents commonly used. Second, and as a result of this evaluation, the progressive replacement of the most aggressive chemicals by less harmful substances to the environment and people, which perform the same function, as well as the development of new techniques that maintain, or even increase the training level of students.

In this sense, two types of actions can be taken, following the trends in most universities and other educational organizations interested in spreading the principles of Green Chemistry in laboratories.

- Learning microscale techniques involving the acquisition of the same skills as in conventional laboratory techniques but with the consumption of smaller amounts of reagents and solvents and minimal waste generation for each experiment.

- Development of specific practices of Green Chemistry, with alternative methods and procedures to the traditional approach.

In this textbook, both points of view have been considered. Part of the text has been dedicated to describing microscale experiments. A total of 16 undergraduate experiments are included. The specific laboratory glassware and the lab basic operations are treated. On the other hand, a group of 30 experiments are described with the goal of treating more common organic reactions, a variety of functional groups and their transformations and some unconventional procedures under the perspective of Green Chemistry principles. In this field, some experimental methods such as microwave and ultrasound irradiation are very popular and have been used in many labs, but we have intentionally omitted them to be able to make all the experiments with conventional and affordable lab equipment.

12.6. References

1. F. G. Calvo-Flores and J. A. Dobado, *Química sostenible: una alternativa creíble*, Anales de la Real Sociedad Española de Química **3** (2008), 205–210.

2. B. M. Trost, *The atom economy: a search for synthetic efficiency*, Science **254** (1991), no. 5037, 1471–1477, DOI 10.1126/science.1962206.

3. B. M. Trost, *Atom economy - a challenge for organic synthesis: homogeneous catalysis leads the way*, Angew. Chem., Int. Ed. Engl. **34** (1995), no. 3, 259–281, DOI 10.1002/anie.199502591.

4. Roger A. Sheldon, *The E Factor: fifteen years on*, Green Chemistry **9** (2007), no. 12, 1273–1283, DOI 10.1039/b713736m.

5. Francisco G. Calvo-Flores, *Sustainable Chemistry Metrics*, ChemSusChem **2** (2009), no. 10, 905–919, DOI 10.1002/cssc.200900128.

6. P. T. Anastas and J. C. Warner, *Green Chemistry: Theory and Practice*, Oxford University Press, 1998.

Chapter 13

Green Chemistry Experiments

A total of 30 experiments ranging from alternative to conventional synthetic processes to be carried out under the principles of Green Chemistry are presented in this chapter.

The use of solvents with low environmental impact, such as water for reactions of C–C bond formation, or acetic acid in the derivatization of functional groups for use as protecting groups is emphasized. Also, biodegradable polymers are prepared and catalysts or reagents that are more environmentally friendly are used. The main goal is to show students that they can perform experiments similar to those usually done in a laboratory but replacing the more toxic or hazardous reagents and reaction conditions involving environmental risks with less dangerous components while yielding similar efficiency.

In the field of undergraduate experiments of green organic synthesis, some techniques, such as the use of microwave or ultrasound irradiations, are commonly used. In our case, these very useful and environmentally friendly methods have been omitted, in order to be able to perform all these experiments with conventional equipment available in any undergraduate laboratory, but many of these experiments could be adapted, especially for microwave irradiation.

13.1. Oxidative cleavage of alkenes: producing adipic acid with H_2O_2/Na_2WO_4

Estimated time	Difficulty	Basic lab operations
3 h		✔ Gravity filtration (see p. 100). ✔ Vacuum filtration (see p. 103). ✔ Recrystallization from water (see p. 105). ✔ Reflux (see p. 89).

13.1.1. Goal

To prepare adipic acid from cyclohexene by oxidation with H_2O_2 and sodium tungstate using a sustainable alternative to the traditional method using potassium permanganate.

Cyclohexene ... H_2O_2/Na_2WO_4 / Aliquat 336/KHSO$_4$ ... Adipic acid

13.1.2. Background

Adipic acid is a substance used for the preparation of nylon, polyurethanes, and other products of commercial interest. The industrial synthesis of adipic acid uses nitric acid oxidation of a mixture of cyclohexanone and cyclohexanol termed "KA oil" (ketone and alcohol-oil). This procedure generates nitrous oxide, a substance that increases greenhouse gases, damages the ozone layer, and causes acid rain. On a laboratory scale, it has been traditionally prepared from cyclohexene by oxidation with potassium permanganate. An alternative with less environmental impact than the traditional method involves the preparation of adipic acid from cyclohexene oxidation with H_2O_2 and sodium tungstate. This procedure is considered a green method and involves a series of events that make a low environmental impact:

- The H_2O_2 is a green oxidant easy to handle and produces only water as a byproduct of the reaction.

- The sodium tungstate is used in catalytic amounts, is of low toxicity, and can be used several times (recyclable).

- This method is industrially applicable in laboratory scale experiments both at miniscale and microscale.

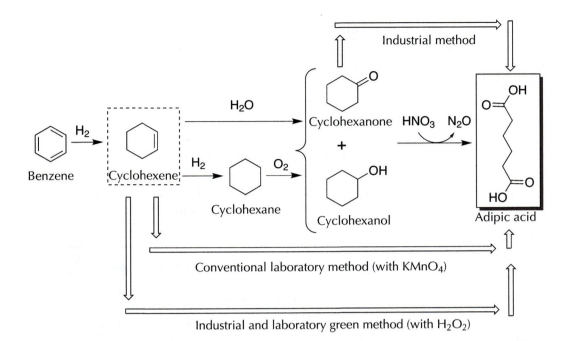

13.1.3. Procedure

To a 50 ml round-bottom flask, add 0.5 g of tricaprimethyl ammonium chloride (Aliquat-336®), 0.5 g of sodium tungstate dihydrate ($Na_2WO_4 \cdot 2\,H_2O$), 11.98 g of H_2O_2 30%, and 0.37 g of potassium hydrogen sulfate ($KHSO_4$). Stir the mixture at r.t. using magnetic stirring for 2 min. Then add 2 g of cyclohexene, attach the flask to a water condenser and a drying tube to minimize odors released by the reaction crude. Keep the reaction mixture at reflux for 2 h.

> Noticeboard 13.1.1
>
> **DANGER!**
>
> Gloves should be used due to both the toxicity of Aliquat 336 and because H_2O_2 can cause burns on contact with skin. In addition, potassium bisulfate is acidic and must be handled with caution.

The reaction progress can be monitored (visually) stopping the agitation. At first, the formation of two layers is observed. When transforming cyclohexene to adipic acid, which is soluble in water, the reaction crude will be transformed to present a majority aqueous phase and an oily rest for the phase-transfer catalyst.[1]

---

[1] The phase-transfer catalyst remaining in the round-bottom flask and the tungsten catalyst of the mother liquor from the crystallization of succinic acid can be recycled.

Once reflux has been completed, while still hot, transfer only the aqueous layer (with a Pasteur pipette) to an Erlenmeyer to avoid coprecipitation of Aliquat-336®. Adipic acid appears as a precipitate on cooling the aqueous layer. Filter the solid produced under vacuum, wash with 5 ml of cold water, and purify by recrystallization from water. Once dried, weigh and calculate the yield (approximately 70%; see refs. [1,2]).

Table 13.1: Physico-chemical properties of the reagents used.

Compound	M_w	M.p. (°C)	B.p. (°C)	Density (g·ml^{-1})	Danger[a] (GHS)
Aliquat®336	-	−6	-	-	
Cyclohexene	82.14	−104	83	0.779	
Adipic acid	146.14	151–154	337.5	1.36	
$KHSO_4$	136.17	214	-	2.320	
H_2O_2 (30%)	34.01	-	-	1.110	
$Na_2WO_4 \cdot 2\,H_2O$	329.85	698	-	4.180	

[a] For brevity, only GHS icons are indicated. The information offered in the Material Safety Data Sheet (MSDS) should be consulted.

13.2. Halogen addition to alkenes: addition of bromine to cyclohexene

Estimated time	Difficulty	Basic lab operations
2 h		✔ Liquid-liquid extraction with reaction (see p. 111). ✔ Washing in a separatory funnel (see p. 111). ✔ Rotary evaporator (see p. 117).

13.2.1. Goal

To perform the bromination of an alkene, avoiding using Br_2 reagent.

Cyclohexene   1,2-Dibromocyclohexane

13.2.2. Background

The addition of halogens to alkenes is a process of great interest in Synthetic Organic Chemistry. The classical procedure is performed from the alkene and halogen. For bromination, the reagent used is bromine (Br_2), which carries certain drawbacks and risks in handling. At r.t., bromine is a reddish-brown liquid that has a strong tendency to evaporate. It is often marketed in glass ampoules, which must be carefully opened, and contents quickly added to an organic solvent, usually CCl_4, in a fume hood. Its fumes are toxic and can cause skin burns.

$$2HBr + H_2O_2 \longrightarrow Br_2 + 2H_2O$$

The procedure described in this experiment involves bromination of an alkene, generating elemental bromine from HBr solution and H_2O_2. The halogen is generated in situ, making it easier to measure, transfer, and store reagents that generate bromine than bromine itself.

13.2.3. Procedure

Add to a 100 ml round-bottom flask 2.4 ml of H_2O_2 30% and a stir bar. Place the flask on a hot plate, fixing it to a support with a clamp. The flask is attached a water-jacketed condenser. There is no need to connect it to the cooling circuit.

When stirring begins, carefully add, with an eyedropper, 40 drops (approximately 1.6 ml) of HBr in acetic acid from the top of the condenser, so that the drops fall directly without touching the walls. The reaction mixture will take on a reddish color due to the formation of Br_2.

DANGER!

Work in a fume hood and wear gloves.

Once the addition of HBr is finished, add 40 drops (approximately 1.4 ml) of cyclohexene, again taking care that the drops fall directly into the flask without touching the walls of the condenser. When the reaction mixture changes color (becomes transparent first and then yellow), the reaction is finished.

Transfer the contents of the flask into a separatory funnel and extract with two portions of 15 ml of CH_2Cl_2. It is desirable for the first amount of CH_2Cl_2 to be used to wash the flask containing the crude reaction in order to dissolve the product that may remain inside. Combine the organic layers and wash once with a solution of $NaHSO_3$ in the funnel. After washing, transfer to a 100 ml Erlenmeyer. Dry the solution with anhydrous sodium sulfate, remove the desiccant by gravity filtration into a tared round-bottom flask, and remove the solvent under reduced pressure while keeping the bath of the rotary evaporator at 40 °C, as the reaction product has a b.p. of 145 °C. Weigh the flask and calculate the yield (see ref. [11]).

Table 13.2: Physico-chemical properties of the reagents used.

Compound	M_w	M.p. (°C)	B.p. (°C)	Density (g·ml^{-1})	Danger[a] (GHS)
H_2O_2 (30%)	34.01	-	-	1.110	
Hydrobromic acid (48%) (HBr)	80.91	-	100	1.490	
Cyclohexene	82.14	−104	83	0.779	
CH_2Cl_2	84.93	−97	40.0	1.33	
$NaHSO_3$ (40%)	104.06	-	-	1.48	
Na_2SO_4	142.04	884	-	2.630	Non-hazardous

[a] For brevity, only GHS icons are indicated. The information offered in the Material Safety Data Sheet (MSDS) should be consulted.

13.3. Green epoxidation: cyclohexene reaction with Oxone®

Estimated time	Difficulty	Basic lab operations
1.5 h		✔ Gravity filtration (see p. 100). ✔ Vacuum filtration (see p. 103). ✔ Recrystallization from organic solvent (see p. 105). ✔ Reflux (see p. 89).

13.3.1. Goal

Epoxidation of an alkene using an easy-to-handle oxidant such as Oxone®.

13.3.2. Background

The typical procedure of alkene epoxidation calls for peracids in an organic solvent. One of the most common experimental methods consists of the reaction of peracetic acid in acetic acid. The peracid is generated in situ by the addition of hydrogen peroxide. The proposed green oxidation of one alkene to give the corresponding epoxide is to use the handle itself and save oxidant Oxone® ($2KHSO_5 \cdot KHSO_4 \cdot K_2SO_4$). Epoxidation takes place in a tandem reaction. Oxone® reacts with acetone and sodium bicarbonate to produce dimethyldioxirane, and, in the presence of an alkene such as cyclohexene, it is converted to the corresponding epoxide.

13.3.3. Procedure

In a round-bottom flask equipped with a stir bar, prepare a solution of cyclohexene (10 mmol, 0.82 g) in acetone (30 ml). To this solution, add $NaHCO_3$ (4 g, 47.6 mmol) and cool the mixture solution in an ice bath to 0 °C. In another flask, add dropwise a solution of Oxone® (8.0 g, 13.0 mmol) in water (30 ml). When the addition is complete, remove the mixture from the ice bath, and allow it to reach r.t. while stirring. After 30 min, the reaction is complete. Transfer the reaction mixture to a separatory funnel, and extract with two aliquots of diethyl ether (2×25 ml). Combine the organic layers and place back into the separatory funnel and wash with 20 ml of water. Transfer the organic layer to an Erlenmeyer and then dry over anhydrous $MgSO_4$, gravity filter, and transfer to a round-bottom flask. The solvent is removed in a rotatory evaporator to give the crude epoxide in nearly quantitative yield (see ref. [16]).

Table 13.3: Physico-chemical properties of the reagents used.

Compound	M_w	M.p. (°C)	B.p. (°C)	Density (g·ml^{-1})	Danger[a] (GHS)
Oxone®	614.78	-	-	-	
Cyclohexene	82.14	−104	83	0.779	
Diethyl ether	74.12	−116	34.6	0.71	
NaHCO$_3$	84.01	300	-	2.160	Non-hazardous
MgSO$_4$	120.37	1124	-	1.070	Non-hazardous

[a] For brevity, only GHS icons are indicated. The information offered in the Material Safety Data Sheet (MSDS) should be consulted.

13.4. Acylation of aromatic amines: obtention of acetanilide with acetic acid and Zn

Estimated time	Difficulty	Basic lab operations
3 h	⚗	✔ Vacuum filtration (see p. 103). ✔ Recrystallization from water (see p. 105).

13.4.1. Goal

To synthesize aromatic amide without using acetic anhydride.

Aniline → (AcOH, Zn) → N-Phenylacetamide

13.4.2. Background

Acetamides are usually synthesized from an amine (aromatic or aliphatic) with acetic anhydride. Acetic anhydride, a reagent widely used in the Organic Chemistry laboratories, is irritating to mucous membranes, is corrosive, and has an unpleasant odor that requires work in a fume hood. In addition, it reacts violently with water. As an alternative to this procedure, the synthesis of an aromatic amide (acetanilide) is proposed by using acetic acid as the acylating agent, catalyzed with zinc.

13.4.3. Procedure

In a 100 ml round-bottom flask, prepare a mixture of 10 ml of aniline, 0.5 g of zinc dust, and 30 ml of acetic acid. Connect the flask to a water-jacketed condenser and heat in a water bath at 60 °C for 2 h with magnetic stirring.

Upon completion of the heating time, pour the crude reaction mixture carefully into a 250 ml Erlenmeyer containing 100 ml of cold water. Cool the Erlenmeyer in an ice-water bath, and stir the mixture with a glass rod. After a few minutes, acetanilide crystals will begin to form. Remove them by vacuum filtration, and wash the solid with cold water until the odor from acetic acid disappears. Recrystallize from water by adding activated carbon if the product contains colored impurities (see ref. [13]).

Table 13.4: Physico-chemical properties of the reagents used.

Compound	M_w	M.p. (°C)	B.p. (°C)	Density (g·ml^{-1})	Danger[a] (GHS)
Aniline	93.13	−6	184	1.022	
Zinc (dust)	65.39	420	907	7.133	
Glacial acetic acid	60.05	16.2	118	1.049	
Acetanilide	135.16	113–115	304	-	
Active carbon	12.01	3,550	-	0.25–0.60	Non-hazardous

[a] For brevity, only GHS icons are indicated. The information offered in the Material Safety Data Sheet (MSDS) should be consulted.

13.5. Solvent-free reductive amination: preparation of dibenzylamine hydrochloride

Estimated time	Difficulty	Basic lab operations
4 h		✔ Gravity filtration (see p. 100). ✔ Vacuum filtration (see p. 103). ✔ Recrystallization from organic solvent (see p. 105). ✔ Reflux (see p. 89).

13.5.1. Goal

To perform a reductive amination in the solid phase.

13.5.2. Background

Reductive amination is the conversion of an aldehyde or a ketone into an amine via an intermediate imine. This reaction is considered the most remarkable way to synthesize complex amines, and a majority of amines prepared in the pharmaceutical industry are made by this procedure. This experiment consists of the synthesis of *N*-benzylamine by the reaction of equimolar amounts of benzaldehyde and benzylamine to give the corresponding imine, which is converted into the amine, roughly mixing the reagents in a mortar in a solvent-free transformation. The first step is to mix the reagents, and the second one is to add *p*-toluensulfonic acid and NaBH$_4$. The transformation of reagents is easily detected by physical changes, which take place without solvent. The amine is isolated as *N*-benzylamine hydrochloride.

13.5.3. Procedure

In a mortar and pestle, place 0.51 ml (approximately 5 mmol) of the benzaldehyde and bezylamine (0.58 ml, 5 mmol). Use a separate 1 ml syringe for each. Take turns gently mixing the reaction for 15 min until a solid (slushy consistency) forms. This indicates that the imine has formed. Weigh 5 mmol of p-toluenesulfonic acid (PTSA) and 5 mmol of $NaBH_4$, and mix them together well in a watch glass with a glass rod. Put this solid mixture into the mortar, and take turns gently grinding for 25–30 min. Transfer the resulting mixture into a 50 or 100 ml beaker with a spatula, and wash the remaining material from the mortar and pestle into the beaker with 20 ml of 5% $NaHCO_3$. Transfer the solution to a separatory funnel and add 20 ml of ethyl acetate. Remove the aqueous layer and label the beaker with the aqueous layer for proper disposal later. Transfer the organic layer to a 50 ml Erlenmeyer, and dry the solvent by placing one or two scoops of potassium carbonate into the organic layer and swirling the container. Gravity filter by placing a filter paper and funnel into a previously weighed 50 ml round-bottom flask. Concentrate the product to dryness by placing the round-bottom flask on the rotary evaporator and removing the ethyl acetate. Measure the mass of the round-bottom flask containing the product. Add 5 ml of EtOH to the round-bottom flask, and transfer the product to a 25 ml Erlenmeyer flask using an automatic pipette. Dropwise add 0.5 ml of concentrated HCl. At this point, some crystals will form in the flask. When this occurs, collect the crystals by vacuum filtration using a Büchner or a Hirsch funnel, and keep the vacuum on for a few minutes until the crystalline salt is dry. Transfer the solid to a 25 ml Erlenmeyer flask, and recrystallize from EtOH (approximately 10 ml). Collect the purified crystals by vacuum filtration using a Hirsch funnel, washing them with 1–2 ml of EtOH, and allow them to air dry at least overnight. The next day, weigh the final salt and measure the m.p. (see ref. [31]).

Table 13.5: Physico-chemical properties of the reagents used.

Compound	M_w	M.p. (°C)	B.p. (°C)	Density (g·ml^{-1})	Danger[a] (GHS)
Benzaldehyde	106.12	−26	178–179	1.044	!
Benzylamine	107.15	10	184–185	0.981	!
p-Toluenesulfonic acid (<5% H_2SO_4)	172.2	106–107	-	1.240	!
$NaBH_4$	37.8	400	-	1.07	
HCl	36.46	−30	>100	1.200	!
EtOH	46.07	−114.1	78.5	0.790	

[a] For brevity, only GHS icons are indicated. The information offered in the Material Safety Data Sheet (MSDS) should be consulted.

13.6. Friedel-Crafts alkylation: xylene reaction with 2-bromopropane catalyzed with graphite

Estimated time	Difficulty	Basic lab operations
3 h		✔ Vacuum filtration (see p. 103). ✔ Reflux (see p. 89).

13.6.1. Goal

To conduct the Friedel-Crafts reaction under low environmental impact with a reusable catalyst.

p-Xylene   2-Bromopropane   2-Isopropyl-1,4-dimethylbenzene

13.6.2. Background

Alkylation of aromatic compounds is a remarkable reaction both at laboratory and at industrial scale. Such reactions are catalyzed by Lewis acids. One of the most widely used for its effectiveness is $AlCl_3$. When the reaction occurs between an alkyl halide and an aromatic derivative, Lewis acid weakens the carbon-halogen (C–X) bond, forming an electrophilic carbocation that reacts with the aromatic ring. Often, transpositions occur in the carbocation formation, leading to mixtures of replacement product. Moreover, once the reaction ends, treatment with an aqueous medium to remove $AlCl_3$ from the reaction crude is required.

In the search for new synthetic procedures respectful of environment, graphite has been found to be capable of catalyzing the alkylation reaction of aromatic compounds with secondary, tertiary, and benzyl halides, while primary halides do not react. The reaction mechanism is not fully understood.

This procedure is within the principles advocated by Green Chemistry, because, under these conditions, the formation of byproducts and waste is minimized, the toxicity of the compounds being handled is reduced, and especially catalyst can be re-used.

13.6.3. Procedure

Add 5 ml of *p*-xylene, 0.44 ml of 2-bromopropane, and 0.5 g of graphite to a 100 ml round-bottom flask equipped with a stir bar and then attach a water condenser and reflux for 90 min. At the end of the reaction time, cool the flask to r.t.

Filter the reaction crude in vacuo with a Hirsch funnel or a filter plate. Wash the graphite with 15 ml of hexane, which joins to the filtrate. Transfer the result of filtration and wash to an round-bottom flask, and evaporate under reduced pressure (rotary evaporator) to remove the solvents. Weigh the reaction product, which is a liquid, to calculate the yield. Once the catalyst is dried, it is ready for use again in another reaction (see ref. [8]).

Table 13.6: Physico-chemical properties of the reagents used.

Compound	M_w	M.p. (°C)	B.p. (°C)	Density (g·ml^{-1})	Danger[a] (GHS)
p-Xylene	106.17	13.0	138.4	0.860	🔥 !
2-Bromopropane	122.99	−89	59	1.310	🔥 ☣
Graphite	12.01	3,800	-	1.900	!
Hexane	86.18	−95	69	0.659	🔥 ☣ ! 🌿

[a] For brevity, only GHS icons are indicated. The information offered in the Material Safety Data Sheet (MSDS) should be consulted.

13.7. Regiospecific nitration of phenols: obtention of *ortho*-nitrophenol

Estimated time	Difficulty	Basic lab operations
1 h		✔ Gravity filtration (see p. 100). ✔ Rotary evaporator (see p. 117). ✔ Reflux (see p. 89). ✔ Chromatography (see p. 135).

13.7.1. Goal

Regiospecific preparation of o-nitrophenol using $Ni(NO_3)_2$ and p-toluenesulfonic acid.

Phenol   + p-Toluensulfonic acid (catalyst) $\xrightarrow[\text{H}_3\text{C}\quad\text{CH}_3]{Ni(NO_3)_2}$ o-Nitrophenol

13.7.2. Background

Conventional reactions of nitration of aromatic compounds involve the use of HNO_3 as the nitrating agent in the presence of H_2SO_4. In this experiment a nitration process that is fast and regiospecific and that has low environmental impact is proposed. The reaction is performed with nickel nitrate and is catalyzed with p-toluenesulfonic acid in acetone. Under these conditions, the nitration of phenol occurs only at the *orto*-position with respect to the hydroxyl group. The reaction can be performed both at r.t. and at reflux, although to reflux, the reaction time is relatively short.

13.7.3. Procedure

In a 100 ml round-bottom flask provided with a stir bar, dissolve 94 mg (1 mmol) of phenol in 15 ml of acetone. Add to the solution, 290.8 mg of $Ni(NO_3)_2 \cdot 6\,H_2O$ (1 mmol) and a catalytic amount of p-toluenesulfonic acid (5 mg). Reflux the reaction for 30 min. Remove the solvent under reduced pressure (rotary evaporator), treat the residue with 20 ml of CH_2Cl_2, and wash the solution with water (20 ml) in a separatory funnel. Dry the organic layer (lower) on anhydrous sodium sulfate for several minutes. Then remove the desiccant by gravity filtration, filtering in a tared round-bottom flask. Remove the solvent under reduced pressure (rotary evaporator), weigh the solid, and calculate the yield (estimated yield 85%, m.p. = 44–45 °C). Optionally, purify by column chromatography using hexane:ethyl acetate (98:2) as eluent. (See ref. [6]).

Table 13.7: Physico-chemical properties of the reagents used.

Compound	M_w	M.p. (°C)	B.p. (°C)	Density (g·ml^{-1})	Danger[a] (GHS)
CH_2Cl_2	84.93	−97	40.0	1.33	
Phenol	94.11	40–42	182	1.07	
p-Toluenesulfonic acid monohydrate	190.22	100–106	-	-	
Acetone	58.08	−94	56	0.791	
Na_2SO_4	142.04	884	-	2.630	Non-hazardous
$Ni(NO_3)_2 \cdot 6\,H_2O$	290.79	56	136.7	2.050	
o-Nitrophenol	139.11	45	214	-	
Hexane	86.18	−95	69	0.659	
Ethyl acetate	88.11	−84	77.1	0.902	

[a] For brevity, only GHS icons are indicated. The information offered in the Material Safety Data Sheet (MSDS) should be consulted.

13.8. Oxidation of aromatic aldehydes: reaction of benzaldehyde with Oxone®

Estimated time	Difficulty	Basic lab operations
2 h		✔ Vacuum filtration (see p. 103). ✔ Recrystallization from water (see p. 105). ✔ Reflux (see p. 89).

13.8.1. Goal

To perform the oxidation of an aldehyde to carboxylic acid, avoiding the use of transition metals (Cr, Mn, etc.), specifically the oxidation by Oxone®.

Aldehyde → Carboxylic acid

13.8.2. Background

Oxone® is the trade name of a triple salt with a molecular mass of $614\ \mathrm{g \cdot mol^{-1}}$, formed by potassium peroxymonosulfate ($KHSO_5$), potassium hydrogen sulfate ($KHSO_4$), and sulfate potassium (K_2SO_4) in a 2:1:1 ratio, respectively. The strongly oxidizing character of this product is given by the presence of $KHSO_5$, which can oxidize numerous compounds such as alkenes (transforming terminal alkenes into epoxides) or transforming aldehydes to yield carboxylic acid. Using Oxone®, oxidants such as transition metals (Cr and Mn) are avoided, and the procedure has a lower environmental impact.

13.8.3. Procedure

To a 50 ml round-bottom flask, add 1.0 g of benzaldehyde, 7.25 g of Oxone®, and 25 ml of deionized water. Attach a water condenser (reflux) to the flask, and heat in a water bath at 60 °C for 75 min with magnetic stirring.

Afterward, cool the reaction in an ice bath for 15 min until a precipitate appears. Filter the solid under vacuum, and if necessary, drag the remaining solid residues in the flask with the minimum amount of water. Finally, flush the solid in a Büchner with 10 ml of cold water. Dry the solid by pressing it between two pieces of filter paper, weigh it, and determine the yield. Recrystallize the benzoic acid from water and determine the m.p. (see ref. [16]).

Table 13.8: Physico-chemical properties of the reagents used.

Compound	M_w	M.p. (°C)	B.p. (°C)	Density (g·ml^{-1})	Danger[a] (GHS)
Oxone®	614.78	-	-	-	
Benzaldehyde	106.12	−26	178–179	1.044	

[a] For brevity, only GHS icons are indicated. The information offered in the Material Safety Data Sheet (MSDS) should be consulted.

13.9. Green synthesis of ethers: preparation of benzyl butyl ether

Estimated time	Difficulty	Basic lab operations
1.5 h	⚗️⚗️	✔ Gravity filtration (see p. 100). ✔ Liquid-liquid extraction (see p. 111). ✔ Recrystallization from organic solvent (see p. 105). ✔ Rotary evaporator (see p. 117). ✔ Vacuum distillation (see p. 125). ✔ Reflux (see p. 89). ✔ Chromatography (see p. 135).

13.9.1. Goal

To perform ether synthesis using phase-transfer catalyst in a water-basic solution.

Benzyl butyl ether

13.9.2. Background

In the usual Williamson synthesis of an ether, a primary alkyl halide reacts with an alkoside. The reaction conditions must be rigorous, paying special attention to the anhydrous solvent. Even a trace of water can spoil the reaction. However, the Williamson synthesis may be carried out under milder conditions by employing a phase-transfer catalyst. This procedure is highly efficient and can be used with conventional solvents (without being anhydrous). In this experiment, benzyl butyl ether is synthesized from benzylchloride and butanol with a phase-transfer catalyst such as tetrabutylammonium hydrogen sulfate.

13.9.3. Procedure

In a round-bottom flask, make a 50% solution of NaOH by dissolving 20 g of NaOH (0.5 mol) in water (20 ml). Cool to r.t. and add 7.4 g (9.2 ml, 0.1 mol) of 1-butanol. A thick, white slurry will form, followed by benzyl chloride (15.2 g, 13.8 ml, 0.12 mol) and tetrabutylammonium hydrogen sulfate (1.7 g, 0.005 mol). Fit with a reflux condenser, heat the mixture in a water bath to 75 °C for 15 min with stirring. The crude is cooled to r.t. and transferred to a separatory funnel. The final product is extracted with methylene chloride (2 × 30 ml). Organic extracts are dried over anhydrous $MgSO_4$. When dried, remove the desiccant by gravity filtration on a tared round-bottom flask. The solvent is eliminated under vacuum (rotary evaporator) to give a liquid. The process can provide good yield (50–90%, average 67%). Benzyl butyl ether can be purified by vacuum distillation (m.p. = 111–112 °C, 23 mm Hg; see ref. [30]).

Noticeboard 13.9.1

 DANGER!

Benzyl chloride has a strong, irritating odor that causes tearing of the eyes, can be harmful when inhaled, and may be absorbed through the skin.

Table 13.9: Physico-chemical properties of the reagents used.

Compound	M_w	M.p. (°C)	B.p. (°C)	Density (g·ml^{-1})	Danger[a] (GHS)
NaOH	40.00	318	1,390	2.130	
Butan-1-ol	74.12	−90	116–118	0.810	
Benzyl chloride	126.58	177–181	-	1.1	
Tetrabutylammonium hydrogensulfate	339.53	-	169–171	1.01	!
CH$_2$Cl$_2$	84.93	−97	40.0	1.33	
MgSO$_4$	120.37	1124	-	1.070	Non-hazardous

[a] For brevity, only GHS icons are indicated. The information offered in the Material Safety Data Sheet (MSDS) should be consulted.

13.10. Synthesis of an acetal in water: preparation of 5,5-bis(hydroxymethyl)-2-phenyl-1,3-dioxane

Estimated time	Difficulty	Basic lab operations
2 h	⚗	✔ Gravity filtration (see p. 100). ✔ Vacuum filtration (see p. 103). ✔ Recrystallization from organic solvent (see p. 105). ✔ Reflux (see p. 89).

13.10.1. Goal

To synthesize an acetal from benzaldehyde in aqueous media.

13.10.2. Background

Acetals are common carbonyl compound derivatives that are often used in Organic Synthesis as protecting groups for aldehydes and ketones, as well as in many other reactions. Cyclic acetals are formed by the acid-catalyzed reaction of an aldehyde or ketone with a diol —e.g., ethylene glycol. The formation of acetals and ketals entails a dehydration process, because a water molecule is formed from the starting materials. The reaction is an equilibrium process that is usually performed in an organic solvent, in which water is removed from the reaction to shift the equilibrium toward product formation. Despite this, in this experiment, acetal formation is performed in water with pentaerythritol and benzaldehyde in the presence of an acid. The reaction takes place with good yields, owing to the insolubility of the final product in the aqueous reaction medium. Two remaining hydroxyl groups remain unreacted in the pentaerythritol moiety.

13.10.3. Procedure

Pentaerythritol (1.8 g, 13 mmol) and water (26 ml) are placed in a 50 ml round-bottom flask or Erlenmeyer flask equipped with a magnetic stir bar and a thermometer. The solid is dissolved by gently heating the mixture to 35 °C in a water bath on a hot plate/stirrer. Upon dissolution of pentaerythritol, concentrated HCl (2 drops, 0.1 ml) is added, followed by benzaldehyde (1.4 ml,

14 mmol). The mixture is heated at 35 °C for 1 h, during which a solid precipitates. The solid is filtered under vacuum and washed with 1 ml of cold water and air-dried in the filter. Recrystallization of the solid from 12 ml of toluene gives 5,5-bis(hydroxymethyl)-2-phenyl-1,3-dioxane, as a colorless solid, which is air-dried on the filter. The recrystallized final product gives an m.p. of 135–137 °C (134–135 °C; see ref. [17]).

Table 13.10: Physico-chemical properties of the reagents used.

Compound	M_w	M.p. (°C)	B.p. (°C)	Density (g·ml^{-1})	Danger[a] (GHS)
Pentaerythritol	136.15	276	-	-	Non-hazardous
Benzaldehyde	106.12	−26	178–179	1.044	❗
Toluene	92.14	−93	110.6	0.867	🔥☣❗
HCl	36.46	−30	>100	1.200	🧪❗

[a] For brevity, only GHS icons are indicated. The information offered in the Material Safety Data Sheet (MSDS) should be consulted.

13.11. Reduction of a ketone alkaline solution: cyclohexanone reaction with NaBH$_4$

Estimated time	Difficulty	Basic lab operations
2 h		✔ Gravity filtration (see p. 100). ✔ Vacuum filtration (see p. 103). ✔ Recrystallization from organic solvent (see p. 105). ✔ Reflux (see p. 89).

13.11.1. Goal

To reduce a cyclic ketone with sodium borohydride in an alkaline water solution.

13.11.2. Background

Sodium borohydride (NaBH$_4$) is a reagent that transforms aldehydes and ketones to the corresponding alcohol, primary or secondary, respectively. The conventional procedure is usually performed in MeOH or even with EtOH, but frequently the reaction with cyclohexanone as the starting product takes place with very poor yields. MeOH and EtOH react partially with NaBH$_4$, producing hydrogen. NaBH$_4$ is very soluble and relatively stable in aqueous alkali. For example, in a 0.2 M aqueous solution of NaOH, less than 5% decomposition results in 2 weeks. The main limitation of water as a solvent is aldehydes and ketones is that most are insoluble in water. Cyclohexanone can be reduced easily in an aqueous medium, using a stirred solution of aqueous NaBH$_4$ stabilized with sodium hydroxide. This procedure is not necessarily a general one, but it can be applied to carbonylic compounds, partially water soluble.

13.11.3. Procedure

Prepare a solution of 30 ml of water and 5 drops of 10% NaOH with a Pasteur pipette, in a 100 ml Erlenmeyer flask equipped with a stir bar. Add NaBH$_4$ (1.5 g, 0.04 mol) to this solution with stirring. Place the flask in an ice-water bath, keeping the temperature below 40 °C, throughout the process, and add dropwise cyclohexanone (9.8 g, 0.1 mol) while stirring, from a Pasteur pipette. After the addition is complete, stir the mixture for another 10 min at r.t. Finally, dropwise slowly add 10 ml of HCl 6 M with stirring. Check the pH value to be sure that the solution is acidic, and add 10 g of NaCl and stir until the solution is saturated. The mixture is transferred to a funnel and extracted with diethylether (3 × 15 ml). Combine the organic extracts and dry in an appropriate flask with anhydrous sodium sulfate. Separate the drying agent by gravity filtration in a tared round-bottom flask. Fit a distillation apparatus to the flask and collect the fraction boiling at 15–65 °C, such as cyclohexanol (estimated yield 70%; see ref. [21,27]).

Table 13.11: Physico-chemical properties of the reagents used.

Compound	M_w	M.p. (°C)	B.p. (°C)	Density (g·ml^{-1})	Danger[a] (GHS)
NaBH$_4$	37.8	400	-	1.07	
Cyclohexanone	98.14	−47	155	0.947	
Diethyl ether	74.12	−116	34.6	0.71	
HCl	36.46	−30	>100	1.200	
NaOH	40.00	318	1,390	2.130	
NaCl	58.44	801	1,413	2.165	Non-hazardous

[a] For brevity, only GHS icons are indicated. The information offered in the Material Safety Data Sheet (MSDS) should be consulted.

13.12. Solvent-free Baeyer–Villiger reaction: oxidation of 4-*tert*-butylcyclohexanone

Estimated time	Difficulty	Basic lab operations
3 h	⚗⚗	✔ Gravity filtration (see p. 100). ✔ Vacuum filtration (see p. 103). ✔ Recrystallization from organic solvent (see p. 105). ✔ Reflux (see p. 89).

13.12.1. Goal

To synthesize a bulk chemical from a renewable feedstock.

4-*tert*-Butylcyclohexanone   *m*-CPBA   γ-*tert*-Butyl-ε-caprolactone

13.12.2. Background

The Baeyer-Villiger oxidation is a rearrangement transformation of ketones to give esters. This is a suitable and efficient method to synthesize lactones from cyclic ketones in a one-step reaction. In this experiment, 4-*tert*-butylcyclohexanone is transformed into γ-*tert*-butyl-ε-caprolactone using *m*-chloroperoxybenzoic acid (*m*-CPBA) as the oxidizing reagent under solvent-free conditions.

13.12.3. Procedure

To a 50 ml round-bottom flask, add 0.5 g (3.2 mmol) of 4-*tert*-butylcyclohexanone and 1.12 g (6.5 mmol) of *m*-chloroperoxybenzoic acid. Immediately attach a reflux condenser to the flask. After approximately 2 min of heating, the crude reaction will begin to become warm and will bubble. Once the reaction has begun, stir the mixture every 5 min for 30 min. After this period, treat the reaction mixture with an aqueous solution of sodium bisulfite (20 ml of 20% $NaHSO_3$). Transfer the mixture to a separatory funnel and rinse the reaction flask with ether (2 × 20 ml). Transfer the ether extracts again to the separatory funnel and wash with portions of an aqueous sodium bicarbonate solution (1.1 g of $NaHCO_3$ in 200 ml of H_2O) in a separatory funnel (4 × 50 ml). Dry the ether layer on anhydrous sodium sulfate. The desiccant is eliminated by gravity filtration on a tared round-bottom flask. Remove the ether by vacuum distillation (rotary evaporator) to give a solid residue. Wash the off-white residue with a small amount of cold pentane and collect the solid by vacuum filtration (m.p. = 57.5–58.5 °C). Expected yields are in the 50–80% range (see ref. [19]).

Table 13.12: Physico-chemical properties of the reagents used.

Compound	M_w	M.p. (°C)	B.p. (°C)	Density (g·ml^{-1})	Danger[a] (GHS)
4-*tert*-Butylcyclohexanone	154.25	113–117	225.1	-	❗
m-Chloroperbenzoic acid	172.57	69–71	-	-	🔥 ❗
NaHSO$_3$ (40%)	104.06	-	-	1.48	🧪 ❗
NaHSO$_4$	104.06	300	-	-	🧪 ❗
MgSO$_4$	120.37	1124	-	1.070	Non-hazardous
Diethyl ether	74.12	−116	34.6	0.71	🔥 ❗
Pentane	72.15	−130	36.1	0.626	🔥 ☠ ❗ 🌿
Hexane	86.18	−95	69	0.659	🔥 ☠ ❗ 🌿
γ-*tert*-Butyl-ϵ-caprolactone	170.25	57.5–58.5	-	-	❗

[a] For brevity, only GHS icons are indicated. The information offered in the Material Safety Data Sheet (MSDS) should be consulted.

13.13. Enzymatic reduction of β-ketoesters: synthesis of chiral ethyl 3-hydroxybutanoate

Estimated time	Difficulty	Basic lab operations
5 h		✔ Gravity filtration (see p. 100). ✔ Vacuum filtration (see p. 103). ✔ Recrystallization from organic solvent (see p. 105). ✔ Reflux (see p. 89).

13.13.1. Goal

To performe a stereoselective reduction of a carbonyl compound with a biological reagent.

Ethyl acetoacetate → Baker's yeast / H₂O/sucrose → Ethyl (*S*)-3-hydroxybutyrate

13.13.2. Background

The reduction of ethyl acetoacetate using common baker's yeast is an example of enzymatic reduction of carbonyl compounds to alcohols in the presence of ester groups. The procedure performed in water shows the stereoselective power of biochemical systems. In this case, the reduction of ethyl 3-acetoacetate with baker's yeast gives the corresponding (*S*)-hydroxy ester with a yield from 30 to 70% and a 58–97% e.e.

13.13.3. Procedure

A) Yeast reduction of a ketone (first session):
To a 500 ml Erlenmeyer flask equipped with a stir bar, add 150 ml of water, and warm to 35 °C using a hot plate set on low (monitor temperature with a thermometer). When the temperature is stabilized at 35 °C, add sucrose (7 g) and baker's yeast (7 g). Let this solution sit for 15 min at 35 °C. In a test tube, dissolve ethyl acetoacetate (3 g) in hexane (8 ml). Add this solution to the yeast mixture, and stir for 3 h while maintaining the temperature at 35 °C. At the end of the period, store the reaction mixture at r.t. until the next lab meeting (see ref. [23]).

B) Isolation of the alcohol product (second session):
To the yeast solution, add Celite® (5 g) and stir for 1 min. Let the solid settle

as much as possible (wait about 5 min). While the solution is settling, set up a vacuum filtration apparatus with a trap using the large Büchner funnel with the corresponding filter paper. First, decant and filter as much of the clear supernatant liquid as possible before adding the Celite® slurry. Wash the Celite® residue with 20 ml of water, and filter the solution once more using the plastic steri-cup filtration apparatus. To the filtered solution, add 20 g of NaCl and swirl the solution until it dissolves. Extract the aqueous solution with diethyl ether (2×50 ml) using a 250 ml separatory funnel. An emulsion occasionally forms. If this happens drain off the lower aqueous layer up to the emulsion. By gently stirring the emulsion with a stirring rod, help break it up. If necessary, transfer the emulsified portions to glass centrifuge tubes, and centrifuge the mixture in order to separate it. Dry the combined ether extracts in an Erlenmeyer flask over anhydrous magnesium sulfate (approximately 1 g) for 5 min. Eliminate the desiccant by gravity filtration to a tared beaker, and evaporate under vacuum (rotary evaporator) until the volume of liquid remains constant (approximately 1–2 ml). This is the final product, ethyl 3-hydroxybutanoate (an alcohol). Determine and record the weight of the product and calculate the yield (see ref. [23]).

Table 13.13: Physico-chemical properties of the reagents used.

Compound	M_w	M.p. (°C)	B.p. (°C)	Density (g·ml^{-1})	Danger[a] (GHS)
Hexane	86.18	−95	69	0.659	🔥☠!🌿
Ethyl acetoacetate	130.14	−43	181	1.029	!
Celite® S	-	-	-	-	!
NaCl	58.44	801	1,413	2.165	Non-hazardous
Diethyl ether	74.12	−116	34.6	0.71	🔥!
MgSO$_4$	120.37	1124	-	1.070	Non-hazardous

[a] For brevity, only GHS icons are indicated. The information offered in the Material Safety Data Sheet (MSDS) should be consulted.

13.14. C−C Bond forming: suzuki reaction

Estimated time	Difficulty	Basic lab operations
2 h		✔ Gravity filtration (see p. 100). ✔ Vacuum filtration (see p. 103). ✔ Recrystallization from organic solvent (see p. 105). ✔ Reflux (see p. 89).

13.14.1. Goal

To perform a C−C coupling reaction in aqueous medium to obtain a bisphenyl derivative.

Phenylboronic acid	4-Iodophenol		4-Phenylphenol

K_2CO_3 / $Pd(C)$

13.14.2. Background

The Suzuki reaction allows the formation of C−C bonds of type σ under mild conditions and with good yields from a boronic acid or ester and vinyl or aryl halide in basic medium, in the presence of Pd(0). This reaction has been widely used to synthesize various polyolefins, styrene derivatives, and substituted biphenyls.

13.14.3. Procedure

Add to a 50 ml round-bottom flask, in this order, 122 mg of phenylboronic acid, 414 mg of potassium carbonate, 220 mg of 4-iodophenol, and 10 ml of deionized water. Weigh in a suitably sized container 3 mg Pd on C 10%, add 1 ml of deionized water, and stir gently by hand to form a slurry that is then transferred to the reaction flask.

> **Noticeboard 13.14.1**
>
> ⚡ **DANGER!**
>
> Perform this experiment in a fume hood.

Couple the flask to a water-jacketed condenser, and reflux the mixture on a hot plate with a magnetic stirrer vigorously for 30 min (until a precipitate appears). After this time, switch off the plate and allow to cool to r.t. Add HCl 2 M to an acidic pH (check with indicator paper). Separate the resulting solid, still containing the catalyst, by filtering with a Hirsch funnel. Wash the solid with 10 ml of water. Then, in a Hirsch funnel, add 10 ml of MeOH, and collect the filtrate in a clean container. Add to the resulting MeOH solution 10 ml of deionized water to obtain the precipitate of the product. Purify by recrystallization, heating in a water bath container with the precipitate and the MeOH/H$_2$O mixture. If necessary, add 1 to 2 ml more of hot MeOH, to finish dissolving the solid. Filter under vacuum with a Hirsch funnel, air dry the solid (can recover the next day). Weigh and calculate the yield (see ref. [5]).

Table 13.14: Physico-chemical properties of the reagents used.

Compound	M_w	M.p. (°C)	B.p. (°C)	Density (g·ml^{-1})	Danger[a] (GHS)
Phenylboronic acid	121.93	216–222	-	-	!
4-Iodophenol	220.01	92–94	138	-	!
4-Phenylphenol	1170.21	164–166	321	-	
K$_2$CO$_3$	138.21	891	-	-	!
Pd(C)	106.42	-	-	-	Non-hazardous
HCl	36.46	−30	>100	1.200	!
MeOH	32.04	−98	64.7	0.791	

[a] For brevity, only GHS icons are indicated. The information offered in the Material Safety Data Sheet (MSDS) should be consulted.

13.15. C=C Bond formation in the absence of solvent: the Wittig reaction

Estimated time	Difficulty	Basic lab operations
1.5 h		✔ Vacuum filtration (see p. 103). ✔ Chromatography (see p. 135).

13.15.1. Goal

To conduct a reaction of carbon-carbon double bond C=C forming in the absence of solvent.

4-Bromobenzaldehyde   Benzyltriphenyl-   4-Bromostilbene
                             phosphonium chloride

13.15.2. Background

The Wittig reaction converts carbonyl compounds into alkenes. In such reactions, phosphorus ylides are produced from a phosphonium salt and a base. This ylide reacts with a carbonyl compound to give an alkene. Although, the Wittig reaction has a low atom economy, because triphenylphosphine oxide is formed as byproduct; however, this reaction can be improved in the absence of solvent.

13.15.3. Procedure

In a porcelain mortar having an outer diameter of 80 mm, mix 200 mg of benzyltriphenylphosphonium chloride, 95 mg of 4-bromobenzaldehyde, and 425 mg of potassium phosphate. Grind the mixture in the mortar, doing circular movements with the pestle for 20 min. Stop the grinding periodically to gently scrape the inner surface of the mortar and the solid that remains stuck to the pestle with a spatula, to improve the mixing process. Resume grinding, so that the actual mixing time is at least 15 min, to be sure the reaction is complete. Check by TLC that the reaction has completed, using as eluent a mixture of ethyl acetate:hexane (1:4). To do this, take a few mg of the solid and dissolve in a few drops of EtOH, click the TLC plate with the capillary. Also, taking 1 mg of 4-bromobenzaldehyde dissolved in a few drops of EtOH, to use as a reference.

Add to the mortar 10 ml of deionized water and with a spatula scrape both sides of the mortar and pestle until it clears all solid and emulsion. The solid obtained is a mixture of E and Z isomers that are isolated by vacuum filtration. Drag the solid if necessary that has been glued to the wall of the mortar with 5 ml of extra water. The E isomer can be isolated as follows:

Transfer the solid to the isomer mixture to a test tube, add 2 ml of EtOH and heat in a water bath at 70–80 °C until the solid is dissolved. Allow the solution to cool to r.t. Finally, cool the test tube in an ice bath until a solid corresponding to 4-bromoestilbene or (E)-(4-bromophenyl)-2-phenylethene appears (m.p. = 134–135 °C; see ref. [18]).

Table 13.15: Physico-chemical properties of the reagents used.

Compound	M_w	M.p. (°C)	B.p. (°C)	Density (g·ml^{-1})	Danger^a (GHS)
EtOH	46.07	−114.1	78.5	0.790	🔥
4-Bromobenzaldehyde	185.02	55–58	-	-	☠ !
Benzyltriphenyl-phosphonium chloride	388.87	300	-	-	☠
K$_3$PO$_4$	212.27	-	-	2.564	🧪
Hexane	86.18	−95	69	0.659	🔥 ☠ ! 🌿
Ethyl acetate	88.11	−84	77.1	0.902	🔥 !
4-Bromostilbene	259.14	138–142	341.99	1.372	See MSDS

^{a} For brevity, only GHS icons are indicated. The information offered in the Material Safety Data Sheet (MSDS) should be consulted.

13.16. Diels-Alder in water: reaction of 9-anthracenemethanol with *N*-methylmaleimide

Estimated time	Difficulty	Basic lab operations
2 h		✔ Vacuum filtration (see p. 103). ✔ Reflux (see p. 89). ✔ Chromatography (see p. 135).

13.16.1. Goal

To perform a Diels-Alder (DA) reaction without using high aromatic b.p. solvents as commonly used in organic reactions, but instead using water as the solvent.

9-(Hydroxymethyl) anthracene   *N*-Methylmaleimide

13.16.2. Background

Diels-Alder (DA) reactions are [4+2] cycloadditions that allow the formation of six-membered ring systems from a diene and a dienophile.

Diene + Dienophile

They are considered some of the most notable synthetic processes for the formation of C–C bonds, and because of their effectiveness, relatively little energy is required for its implementation, enabling high atom economy. Many DA reactions can be performed in water, not only because of the environmental benefits presented by the use of this universal solvent but because a significant acceleration often occurs in the reaction rate of such transformations due to the hydrophobic effects that occur in the water. In the following experiment, the DA reaction is carried out between 9-anthracenemethanol and *N*-methylmaleimide.

13.16.3. Procedure

Place 100 mg of 9-anthracenemethanol and 50 ml of water in a 100 ml round-bottom flask with magnetic stirring. Add 160 mg of *N*-methylmaleimide, and then attach a water condenser and heat the reaction to reflux for approximately 1 h. The reaction course is monitored by TLC, using a mixture of ethyl acetate/hexane (1:1) as an eluent. Ensure proper water evaporation when an aliquot of the reaction crude is poked with a capillary, drying the TLC plate well with an air dryer.

When the reaction is complete, cool the flask in an ice bath until the appearance of a white solid, and vacuum filter using a Hirsch funnel. Dry the solid, weigh, determine the m.p., and calculate the yield (see ref. [2]).

Table 13.16: Physico-chemical properties of the reagents used.

Compound	M_w	M.p. (°C)	B.p. (°C)	Density (g·ml^{-1})	Danger[a] (GHS)
N-Methylmaleimide	111.1	94–96	-	-	☠ !
9-(Hydroxymethyl)-anthracene	208.26	162–164	-	-	Non-hazardous
Ethyl acetate	88.11	−84	77.1	0.902	🔥 !
Hexane	86.18	−95	69	0.659	🔥 ☠ ! ☠

[a] For brevity, only GHS icons are indicated. The information offered in the Material Safety Data Sheet (MSDS) should be consulted.

13.17. Green synthesis of a *p*-cymene: preparation of *p*-cymene from limonene

Estimated time	Difficulty	Basic lab operations
4 h	⚗️⚗️	✔ Gravity filtration (see p. 100). ✔ Vacuum filtration (see p. 103). ✔ Recrystallization from organic solvent (see p. 105). ✔ Reflux (see p. 89).

13.17.1. Goal

Synthesis of a bulk chemical from a renewable feedstock.

Limonene

Dehydrogenation

Palladium catalyst on charcoal

p-Cymene   +   H$_2$

13.17.2. Background

D-Limonene is the major constituent of citrus peel oils (90–95%). Some 50,000 tons of limonene are produced annually by the citrus industry, this being an available renewable feedstock that can be used as the starting material for preparing several products and intermediates. Perhaps the most notable is *p*-cymene, which has applications in the fragrance and pharmaceutical industries as a solvent, and as a starting material for poly(ethylene terephthalate) (PET), which is a widely used polymer that can be prepared by the direct esterification of terephthalic acid and ethylene glycol or by transesterification of dimethyl terephthalate with ethylene glycol. In both cases, the starting materials are currently petroleum derivatives synthesized from *p*-xylene by oxidation. First, *p*-xylene is oxidized to produce terephthalic acid (TA), which is then esterified to dimethylterephthalate (DMT). This transformation can be performed by a two-step sequence or in a one-step reaction in which the oxidation is conducted using a cobalt catalyst in the presence of MeOH. However, it is also possible to prepare terephthalic acid by the oxidation of *p*-cymene, which in turn can be produced from the dehydrogenation of limonene. In this experiment, the intermediate *p*-cymene is prepared from limonene, which is isolated from orange peels.

13.17.3. Procedure

A) Isolation of limonene (Part 1):

Grate the citrus fruit rind using the finest texture of a common cheese grater. Care must be taken during the grating of the fruit. It is essential to grate only the flavedo, the colored portion of the peel, avoiding abrading the albedo, or white portion of the inner peel. It is also essential to avoid abrading the pulp to avoid excessive water contamination. Place approximately, 2.5 g of the finely grated peel into a suitable separatory funnel. The rind should then be extracted three times with 7 ml portions of pentane for 10 min intervals, being certain to frequently vent the funnel (pentane is used for the extraction as opposed to the higher boiling hexane because pentane could be more readily removed by evaporation without risking oxidation of the terpene extracts). The combined extracts are then dried over anhydrous sodium sulfate (approximately 1 g) for 15 min. Filter the resulting solution to ensure complete removal of the sodium sulfate. Transfer the extract to a tared 50 ml beaker, and remove the solvent over a low heat (approximately 35 °C), using a very gentle stream of nitrogen to avoid evaporation of the volatile components of the citrus essential oils. After evaporation of the pentane, weigh the essential oils and calculate the yield (see refs. [24,28]).

B) Transformation of limonene onto *p*-cymene (Part 2):

Place limonene (20 ml, 0.125 mol) in a 3-necked 100 ml round-bottom flask fitted with a condenser with an oil trap. One neck is fitted with a Suba-Seal through which nitrogen gas was passed (through a syringe needle). The final neck is fitted with a glass stopper, through which samples are taken. The limonene is heated, with magnetic stirring to 100 °C, and 2 ml of limonene should be withdrawn by glass pipette. Pd on charcoal (0.1 g of 5% wt) is carefully added to the flask through the side arm via a glass funnel; the residue remaining in the glass funnel is washed into the flask using the 2 ml of limonene previously withdrawn. The reaction is carried out under a nitrogen atmosphere at 100 °C for 3 h. At the end of reaction, allow the reaction mixture to cool before being filtered and weighed (see refs. [24,28]).

Table 13.17: Physico-chemical properties of the reagents used.

Compound	M_w	M.p. (°C)	B.p. (°C)	Density (g·ml^{-1})	Danger^a (GHS)
Pentane	72.15	−130	36.1	0.626	
Na$_2$SO$_4$	142.04	884	-	2.630	Non-hazardous
(R)-(+)-Limonene	136.23	-	176-177	-	
Pd(C)	106.42	-	-	-	Non-hazardous
Nitrogen	28.01	−210	−196	0.970	

^{a} For brevity, only GHS icons are indicated. The information offered in the Material Safety Data Sheet (MSDS) should be consulted.

13.18.   Oxidative coupling reaction in water: synthesis of racemic 1,1′-bi-2-naphthol

Estimated time	Difficulty	Basic lab operations
1.5 h	⚗⚗	✔ Gravity filtration (see p. 100). ✔ Vacuum filtration (see p. 103). ✔ Recrystallization from organic solvent (see p. 105). ✔ Reflux (see p. 89).

13.18.1.   Goal

Synthesis of racemic 1,1′-bi-2-naphthol by an oxidative coupling reaction in aqueous media.

racemic
1,1′-bi-2-Naphthol

13.18.2.   Background

1,1′-Bi-2-naphthol is a compound that can be used as a ligand for transition-metal. 1,1′-Bi-2-naphthol is an example of molecule that presents atropisomers. Enantiomeric pure complex of such derivatives can be used as a catalyst in some asymmetric syntheses but may undergo racemization upon heating. Racemic 1,1′-bi-2-naphthol can be prepared using iron(III) salts as an oxidant. The mechanism involves a process of complexation of iron(III) with the hydroxyl group of 2-naphthol, followed by a radical coupling reaction of the naphthol rings started by iron(III) that is reduced to iron(II). In this case, the reaction is performed in water.

13.18.3.   Procedure

To a 100 ml Erlenmeyer flask, add a solution of $FeCl_3 \cdot 6H_2O$ (11.4 g, 42 mmol) in 40 ml of water and 2-naphthol (3 g, 21 mmol) finely powdered. Heat the Erlenmeyer in a water bath at 50 °C for 2 h while stirring the mixture with a magnetic stirrer. Cool the resulting green suspension to r.t. and separate the solids by vacuum filtration. Dissolve the solid residue with 50 ml of CH_2Cl_2 and transfer the solution to a separatory funnel.[2] Extract the organic solution with HCl 1 M (2×40 ml) and with water (2×40 ml) to remove the inorganic salts. Transfer the organic solution to an appropriate flask, and dry it with anhydrous magnesium sulfate. Remove the desiccant by gravity filtration, in a tared round-bottom flask. Eliminate the solvent under vacuum by rotary evaporator to recover a light-brown solid (1,1'-bi-2-naphthol), which can be purified by recrystallization from a minimum amount of hot toluene (estimated yield 70%; see ref. [22]).

Table 13.18: Physico-chemical properties of the reagents used.

Compound	M_w	M.p. (°C)	B.p. (°C)	Density ($g \cdot ml^{-1}$)	Danger[a] (GHS)
$FeCl_3 \cdot 6\,H_2O$	270.3	280-285	37	1.820	
β-Naphthol	144.17	120–122	285–286	1.280	
CH_2Cl_2	84.93	−97	40.0	1.33	
HCl	36.46	−30	>100	1.200	
$MgSO_4$	120.37	1124	-	1.070	Non-hazardous
Toluene	92.14	−93	110.6	0.867	

[a] For brevity, only GHS icons are indicated. The information offered in the Material Safety Data Sheet (MSDS) should be consulted.

---

[2]Care is needed because the mixture may not dissolve very well in CH_2Cl_2 because of the water and inorganic compounds that are present.

13.19. Organic compounds from recycled polymer: phthalate derived from PET

Estimated time	Difficulty	Basic lab operations
4 h	🔺	✔ Gravity filtration (see p. 100). ✔ Vacuum filtration (see p. 103). ✔ Washing in a separatory funnel (see p. 111). ✔ Rotary evaporator (see p. 117). ✔ Reflux (see p. 89).

13.19.1. Goal

Recycled polyethylene terephthalate (PET) by a depolymerization reaction and use of the product for the synthesis of a diester of high added value.

13.19.2. Background

A large number of synthetic polymers are materials that are usually obtained from monomers originating from the petrochemical industry.

Once the lifetimecycle of objects made from polymers has expired, they pose a serious environmental problem, because many of them have a strong resistance to biodegradation. Thus they can remain in landfills without significant alterations for years. Today, the technologies developed for recycling different types of plastic polymers are of great interest for yielding other products or materials as alternatives to burning or incineration. PET is a thermoplastic polyester that is widely used for the manufacture of packaging for beverages or food products, textiles, etc. It is produced from dimethyl terephthalate and ethylene glycol. Depolymerizing PET yields the starting monomer (dimethyl terephthalate), which can be again used to synthesize new molecules of PET.

13.19.3. Procedure

A) Preparation of PET samples:

From bottles of soda or mineral water previously cleaned without labels, using scissors, cut plastic into pieces under 3×3 cm to obtain a mass of approximately 5 g.

B) Hydrolysis of PET samples:

To a 100 ml round-bottom flask, add 35 ml of pentan-1-ol (mixture of isomers or pentan-1-ol), 5.0 g of PET (0.052 mol equivalent ester), and 4.4 g of KOH (0.079 mol). Reflux the mixture with magnetic stirring. While at first the polymer does not dissolve, shortly after dissolution a dense white emulsion forms. If magnetic stirring is stopped by the viscosity of the reaction crude, add more solvent in portions of 5 ml. Once the stirring resumes, maintain the reflux for 1.5 h. Afterward, cool the reaction to r.t., and add 25 ml of water so that the white solid (terephthalate potassium salt) is dissolved. Remove any remaining suspended solid by vacuum filtration as the water/pentan-1-ol mixture filters very slowly by gravity. Transfer the filtrate to a separatory funnel. Decant the aqueous (lower) layer and wash the organic (top) with another 25 ml of water. Collect in a beaker both aqueous layers, and add diluted HCl very slowly with magnetic stirring to neutrality.[3] Once the terephthalic acid has precipitated, filter the crystals under vacuum and wash with portions of 5 ml of acetone. Weigh, calculate yield, and store for the next step (see ref. [14]).

C) Producing dimethyl terephthalate:

In a round-bottom flask, add 2.7 g of terephthalic acid and 20 ml MeOH. Cool the mixture in an ice bath and slowly add with a dropper 3 ml of concentrated H_2SO_4. Add a stir bar into the flask and reflux (using a drying tube) for 90 min. Cool the flask contents to r.t. Filter the resulting solid under vacuum and wash with two portions of 10 ml of CH_2Cl_2 to dissolve

---

[3]If the acid is added too quickly, a solid crystal of very small size will form and will be difficult to filter.

all the dimethyl terephthalate that may have crystallized. Transfer the filtrate together with the CH_2Cl_2 portions to a separatory funnel, and wash successively with 25 ml of water, twice with 25 ml of 10% NaOH, and finally with 25 ml of brine. Dry the organic layer (bottom) over anhydrous sodium sulfate, and remove the drying agent by gravity filtration. Transfer the dried solution to a round-bottom flask, remove the solvent under reduced pressure (rotary evaporator), weigh, and calculate the yield. Finally, transfer the solid to a Petri dish for storage (see ref. [14]).

Table 13.19: Physico-chemical properties of the reagents used.

Compound	M_w	M.p. (°C)	B.p. (°C)	Density (g·ml^{-1})	Danger[a] (GHS)
Acetone	58.08	−94	56	0.791	🔥 ❗
Pentan-1-ol	88.15	−78	136–138	0.811	🔥 ❗
CH_2Cl_2	84.93	−97	40.0	1.33	☣
NaOH	40.00	318	1,390	2.130	🧪
KOH	56.11	361	1,320	2.044	🧪 ❗
MeOH	32.04	−98	64.7	0.791	🔥 ☠ ☣
PET $(C_{10}H_{12}O_6)_n$	-	250–255	-	1.680	Non-hazardous
H_2SO_4 96–98%	98.08	3	-	1.80–1.84	🧪
HCl	36.46	−30	>100	1.200	🧪 ❗
Na_2SO_4	142.04	884	-	2.630	Non-hazardous
Terephthalic acid	166.13	>300	-	1.500	❗
Dimethyl terephthalate	194.18	139–141	282	1.350	Non-hazardous

[a] For brevity, only GHS icons are indicated. The information offered in the Material Safety Data Sheet (MSDS) should be consulted.

13.20. Crosslinked biodegradable polymers: producing slime

Estimated time	Difficulty	Basic lab operations
1 h		-

13.20.1. Goal

To produce a biodegradable polymer from polyvinyl alcohol (PVA) and borax.

13.20.2. Background

This name is applied to a non-toxic and biodegradable polymer prepared from polyvinyl alcohol (PVA), and borax, food coloring (tartrazine), or fluorescein can be added to the polymer. PVA is a polymer that in turn is synthesized from polyvinyl acetate by treating this with NaOH in MeOH. This is not a polymerization process, but rather the transformation of one polymer into another. Depending on the type of PVA used, it is a more or less dense material that, according to its characteristics, is used for the manufacture of toys, fluid puncture repair kits, or special effects in the entertainment industry for viscous blood and masks to characterize zombies and corpses, simulate wounds, etc. Borax is the sodium tetraborate decahydrate ($Na_2B_4O_7 \cdot 10H_2O$) that, when dissolved in water, is hydrolyzed to boric acid and OH^- anions, yielding a pH of about 9.13. Subsequently, the boric acid reacts with water to form the borate anion. When an aqueous solution of borax and PVA is mixed, a borate ion is sandwiched between the polymer chains, forming covalent bonds, and leading to a change in its properties to a viscous material.

13.20.3. Procedure

Dissolve 1 g of borax in 25 ml of water. Simultaneously, prepare another solution of 4 g of PVA in 100 ml of water. To achieve complete dissolution, it is convenient to gently warm the mixture at a temperature not exceeding 50 °C.

At this point, add a few drops of food coloring (tartrazine) or fluorescein to the dissolution of alcohol, and mix both solutions. A slime forms quickly and can be stored in a plastic bag with self-closing mechanism to prevent drying. In the case of the addition of fluorescein, the fluorescence of the material can be checked with a UV lamp, illuminating a prepared sample (see ref. [7]).

Table 13.20: Physico-chemical properties of the reagents used.

Compound	M_w	M.p. (°C)	B.p. (°C)	Density (g·ml^{-1})	Danger^a (GHS)
Tartrazine	534.36	-	-	-	☠
Borax	201.22	741	-	2.367	☠
Polyvinyl alcohol	-	200	-	1.269	Non-hazardous
Fluorescein	332.31	320	-	-	☠
MeOH	32.04	−98	64.7	0.791	🔥☠☠
NaOH	40.00	318	1,390	2.130	⚠

^{a} For brevity, only GHS icons are indicated. The information offered in the Material Safety Data Sheet (MSDS) should be consulted.

13.21. Polymers from renewable raw materials: starch

Estimated time	Difficulty	Basic lab operations
2 h		✔ Gravity filtration (see p. 100).

13.21.1. Goal

To produce biodegradable polymers from renewable raw materials with different mechanical properties by the method of preparation.

Starch

13.21.2. Background

Starch is a natural polymer present in cereals or potatoes, consisting of glucose units. It is a fundamental component of the human diet, but in turn can be used as a basis for the development of semisynthetic polymers with multiple applications, by adding molecules such as urea or glycerol that are inserted in the chains of amylose and/or amylopectin, modifying mechanical properties of the natural polymer. Based on these properties, different polymers will be prepared in this experiment.

13.21.3. Procedure

Potato starch can be extracted by grating 100 g of potatoes using a kitchen grater. Place the grated potato in a mortar, add 100 ml of water, and crush using the pestle until the grated potato forms a paste as homogeneous as possible. Filter the mixture through a colander over a beaker. The operation is performed twice on each potato sample to increase efficiency in the production of starch. The mixture is allowed to stand until it forms a paste deposit in the beaker. Decant to remove most of the supernatant with the aid of a Pasteur pipette, avoiding stirring. Remove the water that could not be removed previously by decantation, and, on this material, perform the following operation twice to provide two samples of starch.[4]

A) Preparation of a brittle polymer:
    To a starch sample obtained according to the above procedure, add 20 ml of water and 3 ml of HCl (0.1 M). Gently heat the mixture for 15 min, trying not to bring it to a boil. Dropwise, add a solution of NaOH (0.1 M) until neutral, checking the result with a piece of indicator paper. The result is a viscous liquid, to which 2 drops of food coloring (tartrazine), are added. Mix using a glass rod to form a smooth paste. Spread on a watch glass to a film and place in an oven at 60 °C for approximately 1.5 h, or allow to air dry overnight.

B) Preparation of a flexible polymer:
    A similar procedure to the above case is followed but adding 2 ml of glycerol as a plasticizer.

Compare the properties of the materials produced by the two different methods.

Table 13.21: Physico-chemical properties of the reagents used.

Compound	M_w	M.p. (°C)	B.p. (°C)	Density (g·ml^{-1})	Danger[a] (GHS)
Potato starch	-	-	-	1.500	Non-hazardous
Tartrazine	534.36	-	-	-	☠
Glycerol	92.09	20	182	-	Non-hazardous
HCl	36.46	−30	>100	1.200	⬟ !
NaOH	40.00	318	1,390	2.130	⬟

[a] For brevity, only GHS icons are indicated. The information offered in the Material Safety Data Sheet (MSDS) should be consulted.

---

[4]Also, this experiment can be done directly from 1.25 g of commercial potato starch or corn starch.

13.22. Organocatalysis: reaction of 4-nitrobenzaldehyde and dimedone

Estimated time	Difficulty	Basic lab operations
2.5 h	⚗⚗	✔ Gravity filtration (see p. 100). ✔ Vacuum filtration (see p. 103). ✔ Recrystallization from organic solvent (see p. 105). ✔ Reflux (see p. 89).

13.22.1. Goal

To perform a tandem aldol condensation Michael addition reaction using an organocatalyst and to learn how to recover the organocatalyst.

13.22.2. Background

Organocatalysts are small organic molecules that can catalyze reactions in the absence of metals or metal ions. Proline is a chiral organic compound that can be used as a catalyst in many reactions such as aldol condensation, Mannich reactions, or Michael additions, and it should attach itself at more than one point in a symmetrical compound. Proline is nontoxic, inexpensive, and readily available in both enantiomeric forms. Reactions with proline do not require inert conditions, as do so many organometal catalysts, and most reactions carried out with this compound can be at r.t. Proline is an example of organocatalyst as a green alternative to classic organocatalysts.

On the other hand, one of the fields of interest of Green Chemistry is the use of low-environmental-impact solvents. In this sense, Polyethylene glycol (PEG-

400) is a non-volatile and non-flammable solvent that possesses remarkable properties, such as recyclability, ease of work-up, thermal stability, and low cost. In addition, the solvating capability of PEG-400 often makes a reaction system homogeneous, allowing molecular interactions to be more efficient. Under the proposed reaction conditions, 4-nitrobenzaldehyde (1 mol) reacts with dimedone (2 mol). In the first step, a Knoevenagel reaction occurs, when proline reacts with the aldehyde to form an enamine to give the condensation product. In a next step, a second molecule of dimedone is added to the double bond by a Michael reaction.

13.22.3. Procedure

In a 25 ml round-bottom flask, equipped with a stir bar, place 4-nitrobenzaldehyde (151 mg, 1 mmol), 5,5-dimethylcyclohexane-1,3-dione (dimedone, 280 mg, 2 mmol), ($\pm$)-proline (57 mg, 0.5 mmol), and PEG-400 (2 ml); an automatic-delivery pipette is recommended. Stir the reaction mixture with a stirring plate for 30 min at r.t., checking the reaction by TLC (hexane/ethyl acetate, 2:1, spot visualization by UV), and continue stirring for an additional 1 h. A new TLC plate will indicate whether the reaction is complete. Once the reaction is finished, remove the flask from the magnetic stirring plate and put it in an ice-cold water until a white precipitate is formed (about 10 min). Return the flask to the stirrer/hot plate and stir for 10 min (a white precipitate will form). Collect the solid by vacuum filtration using a Hirsch funnel, rinsing any residual product from the reaction flask with no more than 5 ml of ice-water bath to avoid re-solubilization. The white solid is recrystallized from absolute EtOH to give 2'-((4-nitrophenyl)methylene)bis(3-hydroxy-5,5-dimethylcyclohex-2-enone) (typical yields 30–80%; m.p. = 188–190 °C; see ref. [25]).

Noticeboard 13.22.1

 Catalyst-solvent recovery:

PEG-400/($\pm$)-proline mixture can be recycled after the reaction. Pour the filtrate from the vacuum filtration flask of the former step into a 50 ml round-bottom flask and evaporate the water using a rotary evaporator. The residue can be placed into a vial for future use.

Table 13.22: Physico-chemical properties of the reagents used.

Compound	M_w	M.p. (°C)	B.p. (°C)	Density ($g \cdot ml^{-1}$)	Danger[a] (GHS)
4-Nitrobenzalde-hyde	151.12	-	103–106	-	❗
Dimedone	140.18	164–166	233.7	-	Non-hazardous
(±)-Proline	115.13	-	208	-	Non-hazardous
PEG-400	400	4–6	-	1.1254	❗
EtOH	46.07	−114.1	78.5	0.790	🔥
Hexane	86.18	−95	69	0.659	🔥💣❗🌿
Ethyl acetate	88.11	−84	77.1	0.902	🔥❗

[a] For brevity, only GHS icons are indicated. The information offered in the Material Safety Data Sheet (MSDS) should be consulted.

13.23. Knoevenagel-Pinner reactions in water: synthesis of 7-hydroxy-3-carboxycoumarin

Estimated time	Difficulty	Basic lab operations
4 h		✔ Gravity filtration (see p. 100). ✔ Vacuum filtration (see p. 103). ✔ Recrystallization from organic solvent (see p. 105). ✔ Reflux (see p. 89).

13.23.1. Goal

To synthesis of 7-hydroxy-3-carboxycoumarin in water by a tandem reaction.

13.23.2. Background

Knoevenagel reaction is a condensation between an aldehyde or a ketone with an active hydrogen compound in the presence of a basic catalyst to yield α,β-unsaturated compounds. Usually the catalyst is a weakly basic amine, and the active hydrogen compound bears electron-withdrawing groups such as CO_2R, COR, CHO, CN, or NO_2.

The reaction is usually performed in aprotic organic solvents such as DMF, MeCN, or pyridine, which play the double role of solvent and catalyst. Research on this carbon-carbon formation method continues to find new catalysts and reaction conditions according to Green Chemistry strategies such as microwave and ultrasound irradiation, solvent-free conditions, solid-phase synthesis, or use of water as an inexpensive and environmentally benign solvent.

In this experiment, 7-hydroxy-3-carboxycoumarin is prepared by a one-pot consecutive process in water using starting materials 2,4-dihydroxybenzaldehyde and malononitrile. In the first step, a Knoevenagel reaction takes place, to produce an intermediate imine derivative that is hydrolyzed in situ to give the final product (Pinner reaction).

The procedure highlights the advantages of using water, with the possibility of controlling the pH and isolating the reaction product without using any organic solvent. The coumarin ring system is present in many natural products with useful pharmacological properties —for example, as an antitumoural, as anti-HIV agents, or as central nervous system effects.

13.23.3. Procedure

Under a well-ventilated fume hood, vigorously stir finely powered 2,4-dihydroxy-benzaldehyde (1.38 g, 10 mmol), malononitrile (0.80 g, 12.5 mmol), and 0.05 M aqueous $NaHCO_3$ solution (50 ml) for 1.5 h at r.t. in a 100 ml round-bottom flask fitted with a magnetic stir bar and reflux condenser. Then add concentrated HCl (1.25 ml), heat the heterogeneous mixture, and stir for 1 h at 90 °C. After cooling the mixture, add 1 M aqueous $NaHCO_3$ solution (20 ml) and heat the mixture again for 2 h at 90 °C while stirring. The sodium salt of 7-hydroxy-3-carboxycoumarin is water soluble. Acidify the final solution, cooled to r.t., stir with concentrated HCl to pH = 2.0, and refrigerate at 0–5 °C. Separate the precipitate 7-hydroxy-3 carboxycoumarin from the aqueous medium by vacuum filtration using a Büchner funnel under reduced pressure and then dry (1.75 g, yield 85%). The crude coumarin has a purity higher than 98% and can be further purified by recrystallization from H_2O/AcOH 8:2; m.p. = 248–250 °C.

The process can be stopped after the Knoevenagel condensation and the Pinner reaction (1.5 h) or after the hydrolysis of the imino derivative (2.5 h) to produce 7-hydroxy-3-cyano-2-iminocoumarin (m.p. = 250 °C from DMF/H_2O 9:1) and 7-hydroxy-3-cyanocoumarin (m.p. = 273–275 °C from H_2O/AcOH 9:1), respectively, with high yield and purity (>95%) by vacuum filtration of the aqueous phase. These can be suitable stopping points for two or three lab periods. In this case the aqueous reaction mixture is stored at r.t. and the subsequent reactions carried out the next day (see ref. [20]).

Table 13.23: Physico-chemical properties of the reagents used.

Compound	M_w	M.p. (°C)	B.p. (°C)	Density (g·ml^{-1})	Danger[a] (GHS)
2,4-Dihydroxyben-zaldehyde	138.12	135–137	220–228	-	!
Malononitrile	66.06	30–32	220	1.049	☠
$NaHCO_3$	84.01	300	-	2.160	Non-hazardous
HCl	36.46	−30	>100	1.200	!
Glacial acetic acid	60.05	16.2	118	1.049	🔥

[a] For brevity, only GHS icons are indicated. The information offered in the Material Safety Data Sheet (MSDS) should be consulted.

13.24. Biginelli reaction: synthesis of tetrahydro-pyrimidinone

Estimated time	Difficulty	Basic lab operations
1.5 h	⚗	✔ Gravity filtration (see p. 100). ✔ Vacuum filtration (see p. 103). ✔ Recrystallization from organic solvent (see p. 105). ✔ Reflux (see p. 89).

13.24.1. Goal

To perform one-pot synthesis of a tetrahydropyrimidinone.

13.24.2. Background

The Biginelli reaction, first described in 1893, is a three-component reaction between an aldehyde, a β-ketoester, and urea to yield in a one-pot procedure dihydropyrimidones. This reaction is a quick and easy method to synthesize heterocycles, compounds with a great potential for pharmaceutical application.

13.24.3. Procedure

In a mortar, grind together equivalent amounts of benzaldehyde (0.5 mol), ethyl acetoacetate (0.5 mol), and urea (0.5 mol) with a small amount of p-toluenesulfonic acid (as an acid catalyst), for about 3–5 min to give a light-yellow solid mass that will be mostly the target tetrahydropyrimidinone. Wash the crude product with cold water to remove the color, and purify by recrystallization from acetone/EtOH (m.p. = 208–210 °C) (estimated yield 94%; see ref. [15]).

Table 13.24: Physico-chemical properties of the reagents used.

Compound	M_w	M.p. (°C)	B.p. (°C)	Density ($g \cdot ml^{-1}$)	Danger[a] (GHS)
Ethyl acetoacetate	130.14	−43	181	1.029	❗
Benzaldehyde	106.12	−26	178–179	1.044	❗
Urea	60.06	132–135	-	1.335	❗
p-Toluenesulfonic acid (<5% H_2SO_4)	172.2	106–107	-	1.240	❗ 🔧
Acetone	58.08	−94	56	0.791	🔥 ❗
EtOH	46.07	−114.1	78.5	0.790	🔥

[a] For brevity, only GHS icons are indicated. The information offered in the Material Safety Data Sheet (MSDS) should be consulted.

13.25. Multi-component synthesis in water: Passerini reaction

Estimated time	Difficulty	Basic lab operations
1 h		✔ Gravity filtration (see p. 100). ✔ Vacuum filtration (see p. 103). ✔ Recrystallization from organic solvent (see p. 105). ✔ Reflux (see p. 89).

13.25.1. Goal

To develop a multi-component reaction (MCR), as an example of a high-atom-economy reaction. Also, in this special case, the reaction is performed in aqueous medium, which means dramatically reducing the environmental impact of the process.

13.25.2. Background

MCRs are reactions in which three or more molecules interact with each other to give a single compound that incorporates most of the atoms from each of the starting materials, so that these reactions have high atom economy.

The three types of compounds involved in the Passerini reaction are an isocyanide, an aldehyde (or ketone), and a carboxylic acid. It is usually performed in an organic solvent such as CH_2Cl_2 or MeOH. Reaction times are typically on the order of a day, so it is unusual to conduct such reactions in a practice session, because this takes far longer than a typical laboratory session. However, this reaction can be conducted in an aqueous medium and not only is a green alternative but also represents a reduced reaction time while offering excellent yields at r.t. This method is more environmentally friendly than traditional, since water is non-toxic and safe and it is not necessary to heat using volatile and flammable organic solvents.

13.25.3. Procedure

In a 50 ml Erlenmeyer flask, add 20 ml of water, benzoic acid (0.61 g, 0.50 mmol) and benzaldehyde (0.51 ml, 0.50 mmol). Stir the mixture for a few minutes, and then add *tert*-butyl isocyanide (0.57 ml, 0.50 mmol). After adding all reagents, stir the reaction vigorously at r.t. until the appearance of a white precipitate (approximately 25 min).

Collect the white solid from the reaction crude by vacuum filtration. Transfer the solid to a round-bottom flask and dissolve in hot EtOH at reflux. The addition of water leads to the formation of a white solid compound that is analytically pure and is isolated by vacuum filtration. Dry, weigh, and calculate the yield. The typical yield of this reaction in water ranges from 85 to 95%; see ref. [9]).

Table 13.25: Physico-chemical properties of the reagents used.

Compound	M_w	M.p. (°C)	B.p. (°C)	Density (g·ml^{-1})	Danger[a] (GHS)
tert-Butyl iso-cyanide	83.13	-	91		🔥☠
EtOH	46.07	−114.1	78.5	0.790	🔥
Benzoic acid	122.12	125	249	1.08	🗯 !
Benzaldehyde	106.12	−26	178–179	1.044	!
tert-Butylisocyanide	83.13	-	91	-	🔥☠

[a] For brevity, only GHS icons are indicated. The information offered in the Material Safety Data Sheet (MSDS) should be consulted.

13.26. Preparation of a dipyrrol derivative in water: *meso*-diethyl-2,2′-dipyrromethane

Estimated time	Difficulty	Basic lab operations
4 h		✔ Gravity filtration (see p. 100). ✔ Vacuum filtration (see p. 103). ✔ Recrystallization from organic solvent (see p. 105). ✔ Reflux (see p. 89).

13.26.1. Goal

To synthesize *meso*-diethyl-2,2′-dipyrromethane in water with negligible subsequent purification.

1-*H*-Pyrrole   Pentan-3-one   *meso*-Diethyl-2,2′-dipyrromethane

13.26.2. Background

Dipyrromethanes are used as intermediates for the synthesis of porphyrins and other value-added compounds such as fluorescent markers or coordination compounds. The syntheses of such compounds are generally based on the acid-catalyzed condensation of pyrrole with aldehydes or ketones in an organic solvent. Careful control over the reaction is necessary to stop it when the dipyrromethane concentration is at its maximum, and for the correct purification of the final product, distillation of the excess pyrrole is required. In particular, *meso*-diethyl-2,2′-dipyrromethane from pyrrole and 3-pentanone can be synthesized in water with very high efficiency. This green synthesis of the dipyrromethane derivative involves negligible subsequent purification, since the final product separates from the aqueous reaction media spontaneously in a one-pot reaction with good yields.

13.26.3. Procedure

In a 250 ml round-bottom flask, place a solution of 3-pentanone (3.8 ml, 0.036 mol) in 100 ml of hot water (90–100 °C) and 0.5 ml of 37% aqueous HCl, followed by the dropwise addition of pyrrole (5 ml, 0.072 mol). Fit a reflux condenser and heat the reaction under reflux for 45–50 min: a suspension will form during the reaction time. After it is heated, let the crude cool to 40–50 °C, and then transfer the upper layer to another flask using a separatory funnel and allow to cool to r.t. Over several hours (one day is preferable), let the product crystallize into large pale-white crystals, giving the desired *meso*-diethyl-2,2′-dipyrromethane (m.p. = 109 °C; expected yield near 80%; see ref. [29]).

Table 13.26: Physico-chemical properties of the reagents used.

Compound	M_w	M.p. (°C)	B.p. (°C)	Density (g·ml^{-1})	Danger[a] (GHS)
Pyrrole	67.09	−23	131	-	! ⚗🔥
3-Pentanone	86.13	−42	101.5	0.813	! 🔥
HCl	36.46	−30	>100	1.200	⚗ !

[a] For brevity, only GHS icons are indicated. The information offered in the Material Safety Data Sheet (MSDS) should be consulted.

13.27. Ionic liquid in water: one-pot 1-butyl-3-methylimidazolium derivative preparation

Estimated time	Difficulty	Basic lab operations
2.5 h		✔ Gravity filtration (see p. 100). ✔ Vacuum filtration (see p. 103). ✔ Recrystallization from organic solvent (see p. 105). ✔ Reflux (see p. 89).

13.27.1. Goal

To synthesize imidazolium ionic liquid in water.

13.27.2. Background

The so-called ionic liquids are salts in which at least one of their components is an organic compound that has a delocalized charge. This characteristic prevents the formation of stable crystal lattices because both the cation and the anion are poorly coordinated. Most of these salts, called r.t. ionic liquids (RTILs), are liquids below 100 °C, or even at r.t. The nature and the properties of ionic liquids are diverse, so they can be used as solvents, separation media, electrolytes, or lubricants, and, depending on their structure, they can be miscible with water or organic solvents. In addition, some transition metal catalysts are soluble in ionic liquids, and both may be recycled after extraction with water. The potential of these new solvents is enormous in many chemical processes, in work-up procedures, and in many cases, they are considered green solvents because of their origin (it is possible to prepare from renewable sources) and their properties (non-volatile compounds, reusable many times, and biodegradable). The only limitation for their massive use in many conventional organic reactions at a large scale is their high price.

In this experiment, the synthesis of an ionic liquid is described starting from 1-methylimidazole, 1-bromobutane, and potassium hexafluorophosphate (KPF_6) in an organic solvent-free reaction.

13.27.3. Procedure

In a 25 ml round-bottom flask magnetically stirred on a hot plate, place water (1 ml), 1-methylimidazole (1 ml), and 1-bromobutane (1.35 ml) (use separately marked pipettes to measure each reagent). Adjust a condenser on the flask and reflux the reaction mixture for 1.5 h. By that time, a homogeneous solution will result (no oily appearance). Allow the reaction mixture to cool to r.t., and add distilled water (10 ml) followed by KPF_6 (2.3 g). If necessary, gently break the chunks of KPF_6, if any, with a glass rod or a spatula, and swirl the resulting mixture for 10–15 min. Two clear phases should appear. Transfer the whole mixture into a 125 ml separatory funnel, wash the 25 ml round-bottom flask with CH_2Cl_2 (25 ml), and add it to the separatory funnel. Perform the liquid-liquid extraction.

> **Noticeboard 13.27.1**
>
> ⚠ **DANGER!**
>
> Do not forget to vent!

Drain the CH_2Cl_2 layer into a 100 ml Erlenmeyer flask, and wash to extract the aqueous layer (upper) with another portion of fresh CH_2Cl_2 (10 ml). Combine the organic layers of CH_2Cl_2, and dry them with anhydrous Na_2SO_4; 1–2 scoops should be enough. Eliminate the desiccant by gravity filtration, and pour the filtrate into a dry 25 ml and tared round-bottom flask.[5] Eliminate the solvent under vacuum (rotary evaporator) and weigh the flask again to calculate the yield (estimated 76%; see ref. [18]).

Table 13.27: Physico-chemical properties of the reagents used.

Compound	M_w	M.p. (°C)	B.p. (°C)	Density (g·ml^{-1})	Danger[a] (GHS)
1-Methylimidazole	82.10	−6	198	1.03	
1-Bromobutane	137.02	−112	100–104	1.276	🔥 ! 🌳
CH_2Cl_2	84.93	−97	40.0	1.33	⬥
Na_2SO_4	142.04	884	-	2.630	Non-hazardous

[a] For brevity, only GHS icons are indicated. The information offered in the Material Safety Data Sheet (MSDS) should be consulted.

---

[5]Wash and dry the round-bottom flask beforehand.

13.28. Beckmann rearrangement: synthesis of laurolactam from cyclododecanone

Estimated time	Difficulty	Basic lab operations
3 h		✔ Gravity filtration (see p. 100). ✔ Vacuum filtration (see p. 103). ✔ Liquid-liquid extraction with reaction (see p. 111). ✔ Washing in a separatory funnel (see p. 111). ✔ Recrystallization from organic solvent (see p. 105). ✔ Rotary evaporator (see p. 117). ✔ Reflux (see p. 89). ✔ Chromatography (see p. 135).

13.28.1. Goal

To produce a cyclic amide (lactam), key in nylon synthesis from a cyclic ketone, via a Beckmann rearrangement.

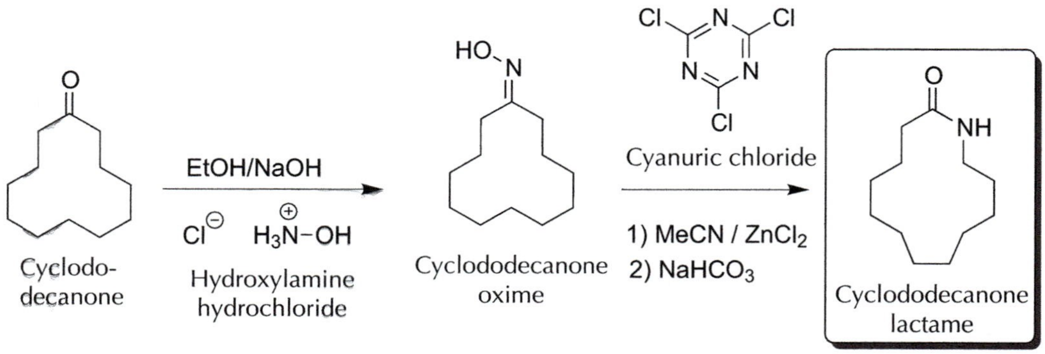

13.28.2. Background

Laurolactam (12-aminododecalactam) is used to synthesize polyamide-12, also known as Nylon-12, which is a semi-crystalline polyamide with very high tenacity and good chemical resistance. In this experiment, laurolactam is synthesized by a process of low environmental impact, in two steps from cyclodecanone, the key step of the synthesis being a Beckmann rearrangement.

13.28.3. Procedure

A) Producing cyclododecanone oxime:

In a round-bottom flask, dissolve 1.5 g of cyclododecanone in approximately 8 ml of 95% EtOH. Add 0.6 g of hydroxylamine hydrochloride, 25 ml of deionized water, and 15 ml of aqueous NaOH (10% by weight), with magnetic stirring. Couple a water condenser and heat the reaction to reflux while stirring. Completion of the reaction is detected by the formation of crystals floating on the surface of the reaction mixture, approximately 30 min after beginning the reflux. After 30 min of refluxing, remove the condenser and cool the mixture in an ice-water bath, until crystallization is complete. Filter the resulting crystals, and dry under vacuum. To purify the product, recrystallize from 95% EtOH (10 ml EtOH per gram of recovered crystals). Add deionized water (15 ml of water per gram of crystals recovered of crude oxime), and cool the mixture in an ice-water bath until complete crystallization. Filter the solid and, under vacuum, dry the crystals produced in the same way as before (see ref. [4]).

B) Conversion to cyclododecanone lactam (laurolactam):
To a round-bottom flask, add cyclododecanone oxime produced in the previous step. For this, add 12 ml of a solution of acetonitrile containing 8.0 mg of cyanuric chloride (2,4,6-trichloro-1,3,5-triazine) and 11.5 mg of anhydrous zinc chloride. Using the same reflux equipment as in the previous step, heat the mixture to reflux (82 °C) for 1 h. Check the completion of the reaction by TLC using as the eluent a mixture of diethyl ether/hexane (1:1) (see Section 4.14.1, p. 135). The reaction is complete when the spot of the oxime disappears and a more polar product appears. After the reaction ends, remove the flask from the hot plate, and add 20 ml of an aqueous solution saturated with $NaHCO_3$. Transfer the reaction mixture to a 100 ml separatory funnel, and extract the product with two portions of 15 ml of ethyl acetate. Combine the extracts of ethyl acetate in a 100 ml Erlenmeyer, and dry the ethyl acetate by adding Na_2SO_4 anhydrous. Filter the drying agent by gravity, collecting it in a tared round-bottom flask, and remove the solvent under reduced pressure (rotary evaporator). Recrystallize from 95% EtOH (7 ml EtOH per gram of product). Calculate the yield and determine the m.p. (lit. 148–149 °C). Weigh and calculate the yield (see ref. [4]).

Table 13.28: Physico-chemical properties of the reagents used.

Compound	M_w	M.p. (°C)	B.p. (°C)	Density (g·ml^{-1})	Danger[a] (GHS)
EtOH	46.07	−114.1	78.5	0.790	
Acetonitrile	41.05	−48	81–82	-	
Ethyl acetate	88.11	−84	77.1	0.902	
Hexane	86.18	−95	69	0.659	
Diethyl ether	74.12	−116	34.6	0.71	
NaOH	40.00	318	1,390	2.130	
Na$_2$SO$_4$	142.04	884	-	2.630	Non-hazardous
Cyclododecanone	182.3	59–61	85	-	
NaHCO$_3$	84.01	300	-	2.160	Non-hazardous
Cyanuric chloride	184.41	145–147	190	-	
Hydroxylamine hydrochloride	69.49	155–157	-	1.670	
ZnCl$_2$	136.30	293	732	2.907	
Cyclododecanone oxime	197.32	-	288.6	1.014	Non-hazardous
Laurolactam	197.32	150–153	348	0.973	Non-hazardous

[a] For brevity, only GHS icons are indicated. The information offered in the Material Safety Data Sheet (MSDS) should be consulted.

13.29. Fluorescent natural product: preparation of 7-hydroxy-4-methyl-2*H*-chromen-2-one

Estimated time	Difficulty	Basic lab operations
2 h		✔ Gravity filtration (see p. 100). ✔ Vacuum filtration (see p. 103). ✔ Recrystallization from organic solvent (see p. 105). ✔ Reflux (see p. 89).

13.29.1. Goal

To synthesise fluorescent 7-hydroxy-4-methyl-2*H*-chromen-2-one in a solvent-free reaction.

13.29.2. Background

Coumarins are natural products occurring in a variety of plants, including those used as traditional herbal medicines dating to as early as 1,000 B.C. Coumarins have been used for many practical applications, such as cosmetics, sunscreens, flavorings, laser dyes, pharmaceuticals, and well-known anticoagulants. The coumarin 4-methylumbelliferone is used as a choleretic and antispasmodic drug and as a standard for the fluorometric determination of enzyme activity, such as β-galactosidase. The target product is non-toxic, is easy to prepare and isolate, and can be rapidly synthesized by Pechmann condensation heating resorcinol, and ethyl acetoacetate, in the presence of a strong-acid ion exchange resin (Dowex 50WX4). Also it displays pH-dependent fluorescence that is easily viewed using a standard UV lamp.

13.29.3. Procedure

In a 50 ml Erlenmeyer flask, place ethyl acetoacetate (1.0 ml, 1.02 g, 7.8 mmol), resorcinol (800 mg, 7.3 mmol), and Dowex 50WX4 beads (1.0 g). Place the flask on a hot plate set to the lowest setting. The reaction proceeds at a reasonable rate at temperatures as low as 80 °C, and the start of the reaction is indicated by gentle bubbling. Swirl the mixture occasionally or stir with a glass rod until bubbling ceases and the mixture solidifies to a tan solid (typically 20–30 min). Add a small volume (2–3 ml) of hot 95% EtOH to dissolve the solid and then use a Pasteur pipette to transfer the hot solution from the Dowex beads to a clean Erlenmeyer flask. If desired, the beads may be rinsed with an additional portion of hot EtOH (95%). While the mixture is heating, add hot water until the solution becomes slightly cloudy, remove the flask from the hot plate, and allow to cool slowly to r.t. Isolate the white to off-white crystalline precipitate by vacuum filtration and wash with water. The product is allowed to dry before determining the m.p. (179–183 °C; estimated yields 50 to 65%; see ref. [26]).

Table 13.29: Physico-chemical properties of the reagents used.

Compound	M_w	M.p. (°C)	B.p. (°C)	Density (g·ml^{-1})	Danger[a] (GHS)
Resorcinol	110.11	178	109–112	1.272	⚠ ! ☠
Dowex 50WX4	-	-	-	-	Non-hazardous
Ethyl acetoacetate	130.14	−43	181	1.029	!
EtOH	46.07	−114.1	78.5	0.790	🔥

[a] For brevity, only GHS icons are indicated. The information offered in the Material Safety Data Sheet (MSDS) should be consulted.

13.30. Photochemical solid phase: [2+2] cycloaddition of cinnamic acid

Estimated time	Difficulty	Basic lab operations
15 days	⚗	✔ Vacuum filtration (see p. 103). ✔ Rotary evaporator (see p. 117).

13.30.1. Goal

To perform a photochemical [2+2] cycloaddition reaction with the use of sunlight.

trans-Cinnamic acid   *trans*-Cinnamic acid   Sunlight   Truxillic acid

13.30.2. Background

Cycloaddition reactions are processes having high atom economy, since all atoms of the starting materials are incorporated in the final molecule. The [2+2] cycloaddition is a reaction type that fails to react conventional alkenes under thermal conditions. However, a synthetic tool may be useful under photochemical conditions. In the case of *trans*-cinnamic acid, a total of eleven stereoisomers can form: six dimers of the head-head type (truxinic acids) and five dimers of the head-tail type (truxillic acids). Both families of compounds are natural products found, for example, in coca leaves. Dimerization of *trans*-cinnamic acid in solution does not occur but, under photochemical conditions described in this experiment does take place and surprisingly produces a single stereoisomer.

13.30.3. Procedure

Weigh 1.5 g of *trans*-cinnamic acid, and place in a 100 ml Erlenmeyer. Add approximately 2 ml of THF. Heat (water bath, see p. 83), mixing until the cinnamic acid is dissolved. When it has dissolved, remove the Erlenmeyer flask from the heat and gently stir by hand to form on the flask wall a film with the solid as homogeneous as possible. Otherwise, repeat the operation to achieve the goal. Allow the contents of the flask to dry completely, leaving the flask at r.t. for 30 min. It is then covered with a septum and is labeled with the name and date. Then the flask is left exposed to sunlight for 15 days on a window sill (facing south) or under the lamplight, rotating the flask at 7 days.

After this time, add 15 ml of toluene and gently heat (water bath at 40 °C) to dissolve the remaining cinnamic acid. Separate the solid by vacuum filtration, and wash with 10 ml of additional toluene. Pool the toluene fractions, and transfer to a tared round-bottom flask. Remove the solvent under reduced pressure (rotary evaporator), weigh the product, and the calculate the yield. Determine the m.p. (286 °C; see ref. [24], p. 206).

Table 13.30: Physico-chemical properties of the reagents used.

Compound	M_w	M.p. (°C)	B.p. (°C)	Density (g·ml^{-1})	Danger[a] (GHS)
Toluene	92.14	−93	110.6	0.867	🔥☠ !
Cinnamic acid	148.16	132–135	300	1.248	!
THF	72.11	−108.0	65–67	0.89	🔥 !
Truxillic acid	296.32	-	470.46	1.324	See MSDS

[a] For brevity, only GHS icons are indicated. The information offered in the Material Safety Data Sheet (MSDS) should be consulted.

13.31. References

1. S. M. Reed and J. E. Hutchison, *Green chemistry in the organic teaching laboratory: an environmentally benign synthesis of adipic acid*, Journal of Chemical Education **77** (2000), no. 12, 1627, DOI 10.1021/ed077p1627.

2. K. Sato, M. Aoki, and R. Noyori, *A green route to adipic acid: direct oxidation of cyclohexenes with 30 percent hydrogen peroxide*, Science **281** (1998), no. 5383, 1646–1647, DOI 10.1126/science.281.5383.1646.

3. K. M. Doxsee and J. E. Hutchison, *Green Organic Chemistry: Strategies, Tools, and Laboratory Experiments*, Thomson-Brooks/Cole, 2004.

4. D. F. Taber and P. J. Straney, *The synthesis of laurolactam from cyclododecanone via a Beckmann rearrangement*, Journal of Chemical Education **87** (2010), no. 12, 1392–1392, DOI 10.1021/ed100599q.

5. E. Aktoudianakis, E. Chan, A. R. Edward, I. Jarosz, V. Lee, L. Mui, S. S. Thatipamala, and A. P. Dicks, *"Greening up" the Suzuki reaction*, Journal of Chemical Education **85** (2008), no. 4, 555, DOI 10.1021/ed085p555.

6. V. Anuradha, P. V. Srinivas, P. Aparna, and J. Madhusudana Rao, *p-Toluenesulfonic acid catalyzed regiospecific nitration of phenols with metal nitrates*, Tetrahedron Letters **47** (2006), no. 28, 4933–4935, DOI 10.1016/j.tetlet.2006.05.017.

7. F. G. Clavo-Flores and J. Isac, *Introducción a la química de los polímeros biodegradables: una experiencia para alumnos de segundo ciclo de la ESO y Bachillerato*, Anales de la Real Sociedad Española de Química **109** (2013), no. 1, 38–44.

8. G. A. Sereda and V. B. Rajpara, *A green alternative to aluminum chloride alkylation of xylene*, Journal of Chemical Education **84** (2007), no. 4, 692, DOI 10.1021/ed084p692.

9. M. M. Hooper and B. DeBoef, *A green multicomponent reaction for the organic chemistry laboratory. The aqueous Passerini reaction*, Journal of Chemical Education **86** (2009), no. 9, 1077, DOI 10.1021/ed086p1077.

10. S. H. Leung and S. A. Angel, *Solvent-free Wittig reaction: a green organic chemistry laboratory experiment*, Journal of Chemical Education **81** (2004), no. 10, 1492, DOI 10.1021/ed081p1492.

11. L. C. McKenzie, L. M. Huffman, and J. E. Hutchison, *The evolution of a green chemistry laboratory experiment: greener brominations of stilbene*, Journal of Chemical Education **82** (2005), no. 2, 306, DOI 10.1021/ed082p306.

12. L. C. McKenzie, L. M. Huffman, J. E. Hutchison, C. E. Rogers, T. E. Goodwin, and G. O. Spessard, *Greener solutions for the organic chemistry teaching lab: exploring the advantages of alternative reaction media*, Journal of Chemical Education **86** (2009), no. 4, 488, DOI 10.1021/ed086p488.

13. V. K. Redasani, V. S. Kumawat, R. P. Kabra, P. Kansagara, and S. J. Surana, *Applications of green chemistry in organic synthesis*, International Journal of ChemTech Research **2** (2010), no. 3, 1856-1859.

14. D. Kaufman, G. Wright, R. Kroemer, and J. Engel, *New compounds from old plastics: recycling PET plastics via depolymerization. An activity for the undergraduate organic lab*, Journal of Chemical Education **76** (1999), no. 11, 1525, DOI 10.1021/ed076p1525.

15. A. K. Bose, S. Pednekar, S. N. Ganguly, G. Chakraborty, and M. S. Manhas, *A simplified green chemistry approach to the Biginelli reaction using Grindstone Chemistry*, Tetrahedron Lett. **45** (2004), no. 45, 8351–8353, DOI 10.1016/j.tetlet.2004.09.064.

16. W. C. Broshears, J. J. Esteb, J. Richter, and A. M. Wilson, *Simple epoxide formation for the organic laboratory using oxone*, Journal of Chemical Education **81** (2004), no. 7, 1018–1019, DOI 10.1021/ed081p1018.

17. D. M. Collard, A. G. Jones, and R. M. Kriegel, *Synthesis and spectroscopic analysis of a cyclic acetal: a dehydration performed in aqueous solution*, Journal of Chemical Education **78** (2001), no. 1, 70–72, DOI 10.1021/ed078p70.

18. S. V. Dzyuba, K. D. Kollar, and S. S. Sabnis, *Synthesis of imidazolium room-temperature ionic liquids: exploring green chemistry and click chemistry paradigms in undergraduate organic chemistry laboratory*, Journal of Chemical Education **86** (2009), no. 7, 856–858, DOI 10.1021/ed086p856.

19. J. J. Esteb, J. N. Hohman, D. E. Schlamadinger, and A. M. Wilson, *A Solvent-Free Baeyer–Villiger Lactonization for the Undergraduate Organic Laboratory: Synthesis of γ-t-Butyl-ϵ-caprolactone*, Journal of Chemical Education **82** (2005), no. 12, 1837.

20. F. Fringuelli, O. Piermatti, and F. Pizzo, *One-pot synthesis of 7-hydroxy-3-carboxycoumarin in water*, Journal of Chemical Education **81** (2004), no. 6, 874–876, DOI 10.1021/ed081p874.

21. N. J. Hudak and A. H. Sholes, *Reduction of cyclohexanone with sodium borohydride in aqueous alkaline solution-a beginning organic chemistry experiment*, Journal of Chemical Education **63** (1986), no. 2, 161, DOI 10.1021/ed063p161.

22. K. K. W. Mak, *Synthesis and resolution of the atropisomeric 1,1'-bi-2-naphthol: An experiment in organic synthesis and 2-D NMR spectroscopy*, Journal of Chemical Education **81** (2004), no. 11, 1636–1640, DOI 10.1021/ed081p1636.

23. J. Patterson and S. Th. Sigurdsson, *Use of enzymes in organic synthesis: reduction of ketones by Baker's yeast revisited*, Journal of Chemical Education **82** (2005), no. 7, 1049–1050, DOI 10.1021/ed082p1049.

24. D. C. Smith, S. Forland, E. Bachanos, M. Matejka, and V. Barrett, *Qualitative Analysis of Citrus Fruit Extracts by GC/MS: An Undergraduate Experiment*, The Chemical Educator **6** (2001), no. 1, 28–31, DOI 10.1007/s00897000450a.

25. J. M. Stacey, A. P. Dicks, A. A. Goodwin, B. M. Rush, and M. Nigam, *Green Carbonyl Condensation Reactions Demonstrating Solvent and Organocatalyst Recyclability*, Journal of Chemical Education **90** (2013), no. 8, 1067–1070, DOI 10.1021/ed300819r.

26. D. M. Young, J. J. C. Welker, and K. M. Doxsee, *Green Synthesis of a Fluorescent Natural Product*, Journal of Chemical Education **88** (2011), no. 3, 319–321, DOI 10.1021/ed1004883.

27. N. M. Zaezek, *Another look at reduction of cyclohexanone with sodium borohydride in aqueous alkaline solution*, Journal of Chemical Education **63** (1986), no. 10, 909, DOI 10.1021/ed063p909.

28. D. M. Roberge, D. Buhl, J. P. M. Niederer, and W. F. Holderich, *Catalytic aspects in the transformation of pinenes to p-cymene.*, Appl. Catal., A **215** (2001), no. 1-2, 111–124, DOI 10.1016/S0926-860X(01)00514-2.

29. A. J. F. N. Sobral, *Synthesis of meso-diethyl-2,2'-dipyrromethane in water. An experiment in green organic chemistry.*, Journal of Chemical Education **83** (2006), no. 11, 1665–1666, DOI 10.1021/ed083p1665.

30. J. W. Hill and J. Corredor, *An Ether Synthesis Using Phase Transfer Catalysis*, Journal of Chemical Education **57** (1980), no. 11, 822, DOI 10.1021/ed057p822.

31. S. W. Goldstein and A. V. Cross, *Solvent-Free Reductive Amination: An Organic Chemistry Experiment*, Journal of Chemical Education **07** (2015), no. 92, 1214—1216, DOI 10.1021/ed5006618.

Index